Geoenvironmental Engineering

Other Titles of Interest

Field Guide to Environmental Engineering for Development Workers: Water, Sanitation, and Indoor Air, by James R. Mihelcic, Lauren M. Fry, Elizabeth A. Myre, Linda D. Phillips, and Brian Barkdoll (ASCE Press 2009). In this complete handbook for international engineering service projects, James Mihelcic and his coauthors provide the tools necessary to implement the right technology in developing regions around the world. (ISBN 978-0-7844-0985-5)

Sustainable Solid Waste Management, edited by Jonathan W. C. Wong, Rao Y. Surampalli, Tian C. Zhang, Rajeshwar D. Tyagi, and Ammaiyappan Selvam (ASCE/EWRI 2016). This book describes basic principles and recent advances for handling solid waste in an environmentally sustainable way. (ISBN 978-0-7844-1410-1)

Water Balance Covers for Waste Containment: Principles and Practice, by William H. Albright, Craig H. Benson, and W. Joseph Waugh (ASCE Press 2010). Albright, Benson, and Waugh present the latest research regarding water balance covers for solid waste sites, along with case studies drawn from current field testing. (ISBN 978-0-7844-1070-7)

Groundwater Contamination by Organic Pollutants: Analysis and Remediation, edited by Jagath J. Kaluarachchi (ASCE/EWRI 2001). MOP 100 surveys important technologies used in the remediation of contaminated aquifers. (ISBN 978-0-7844-0527-7)

Carbon Capture and Storage: Physical, Chemical, and Biological Methods, edited by Rao Y. Surampalli, Tian C. Zhang, R. D. Tyagi, Ravi Naidu, B. R. Gurjar, C. S. P. Ojha, Song Yan, Satinder K. Brar, Anushuya Ramakrishnan, and C. M. Kao (ASCE/EWRI 2015). This book contains 18 invited chapters that provide in-depth information on the principles of carbon capture and storage technology and recent advances and future research and development in the CCS technology field. (ISBN 978-0-7844-1367-8)

ASCE Manuals and Reports on Engineering Practice No. 159

Geoenvironmental Engineering

Site and Contaminant Characterization, Containment Facilities, Solid Waste Materials, and Contaminated Ground Interventions

Edited by
Dimitrios Zekkos, Ph.D., P.E.

Sponsored by the
Geoenvironmental Engineering Technical Committee of the
Technical Coordination Council of the
Geo-Institute of the
American Society of Civil Engineers

Published by the American Society of Civil Engineers

Library of Congress Cataloging-in-Publication Data

Names: Zekkos, Dimitrios, editor. | American Society of Civil Engineers. Geo-Institute. Geoenvironmental Engineering Committee, sponsoring body.

Title: Geoenvironmental engineering : site and contaminant characterization, containment facilities, solid waste materials, and contaminated ground interventions / edited by Dimitrios Zekkos, Ph.D., P.E. ; sponsored by the Geoenvironmental Engineering Technical Committee of the Technical Coordination Council of the Geo-Institute of the American Society of Civil Engineers.

Description: Reston, Virginia : American Society of Civil Engineers, 2024. | Series: ASCE manuals and reports on engineering practice ; no. 159 | Includes bibliographical references and index. | Summary: "MOP 159 provides an overview of a variety of aspects of geoenvironmental engineering, outlining design procedures and recommendations in engineering practice where appropriate"-- Provided by publisher.

Identifiers: LCCN 2024009442 | ISBN 9780784416235 (soft cover) | ISBN 9780784485491 (PDF)

Subjects: LCSH: Environmental geotechnology. | Soil remediation. | Soil pollution. | Sanitary landfills. | Refuse and refuse disposal.

Classification: LCC TD171.9 .G4555 2024 | DDC 628--dc23/eng/20240404

LC record available at https://lccn.loc.gov/2024009442

Published by American Society of Civil Engineers
1801 Alexander Bell Drive
Reston, Virginia 20191-4382
www.asce.org/bookstore | ascelibrary.org

***Errata:* Errata, if any, can be found at https://dx.doi.org/10.1061/9780784485491**

ISBN 978-0-7844-1623-5 (soft cover)
ISBN 978-0-7844-8549-1 (PDF)
ISBN 978-0-7844-8550-7 (ePub)

Manufactured in the United States of America.
28 27 26 25 24 1 2 3 4 5

MANUALS AND REPORTS ON ENGINEERING PRACTICE

(As developed by the ASCE Technical Procedures Committee, July 1930, and revised March 1935, February 1962, and April 1982)

A manual or report in this series consists of an orderly presentation of facts on a particular subject, supplemented by an analysis of limitations and applications of these facts. It contains information useful to the average engineer in his or her everyday work, rather than findings that may be useful only occasionally or rarely. It is not in any sense a "standard," however, nor is it so elementary or so conclusive as to provide a "rule of thumb" for nonengineers.

Furthermore, material in this series, in distinction from a paper (which expresses only one person's observations or opinions), is the work of a committee or group selected to assemble and express information on a specific topic. As often as practicable, the committee is under the direction of one or more of the Technical Divisions and Councils, and the product evolved has been subjected to review by the Executive Committee of the Division or Council. As a step in the process of this review, proposed manuscripts are often brought before the members of the Technical Divisions and Councils for comment, which may serve as the basis for improvement. When published, each manual shows the names of the committees by which it was compiled and indicates clearly the several processes through which it has passed in review, so that its merit may be definitely understood.

In February 1962 (and revised in April 1982), the Board of Direction voted to establish a series titled "Manuals and Reports on Engineering Practice," to include the manuals published and authorized to date, future Manuals of Professional Practice, and Reports on Engineering Practice. All such manual or report material of the Society would have been refereed in a manner approved by the Board Committee on Publications and would be bound, with applicable discussion, in books similar to past manuals. Numbering would be consecutive and would be a continuation of present manual numbers. In some cases of joint committee reports, bypassing of journal publications may be authorized.

A list of available Manuals of Practice can be found at https://ascelibrary.org/page/books/s-mop.

CONTENTS

LIST OF CONTRIBUTORS xvii

BLUE-RIBBON PANEL REVIEWERS xxi

PREFACE xxiii

ACKNOWLEDGMENTS xxv

PART 1: INTRODUCTION 1

1. CONTAMINANT TRANSPORT AND FATE 3
Joe Scalia and Charles Shackelford
   1.1 Introduction 3
   1.2 Mass Transport 3
   1.3 Fate 15
   References 19

PART 2: WASTE AND SITE CHARACTERIZATION 25

2. FIELD TESTING FOR GEOTECHNICAL CHARACTERIZATION OF MUNICIPAL SOLID WASTE 27
Neven Matasovic
   2.1 Introduction 27
   2.2 Nonintrusive Field Measurement Techniques 28
      2.2.1 General 28
      2.2.2 Wave Propagation Velocity Measurements 28
      2.2.3 Electrical Resistivity and Conductivity Surveys 31
      2.2.4 Surface Deformation Measurements 32
      2.2.5 Strong Ground Motion Measurements 33

2.2.6 Shear Strength Estimation Based on Stability Observations 33
2.3 Intrusive Field Measurement Techniques 35
2.3.1 General 35
2.3.2 Borehole Sampling and Testing 36
2.4 Hybrid Field Measurement Techniques 49
2.4.1 General 49
2.4.2 Cone Penetration Test Soundings 49
2.4.3 Dilatometer and Seismic Dilatometer Testing 51
2.4.4 In Situ Measurement of Shear Modulus Reduction 51
2.5 Recommendations 53
Acknowledgments 56
References 56

3. LABORATORY TESTING FOR CHEMICAL CHARACTERIZATION OF SOLIDS, LIQUIDS, AND GASES 63
Marisa Chrysochoou and Nefeli Bompoti
3.1 Introduction 63
3.2 Solid Analysis Methods 64
3.2.1 Bulk Chemical Analysis 64
3.2.2 Mineralogical Analysis 66
3.3 Liquid Analysis Methods 67
3.3.1 General Parameters 68
3.3.2 Analysis of Metals and Metalloids 69
3.3.3 Analysis of Organics 71
3.3.4 Analysis of Other Compounds and Parameters 78
3.4 Gas Analysis 80
3.4.1 Questionnaire 81
References 83

PART 3: WASTE MATERIAL PROPERTIES 85

4. HAZARDOUS WASTE 87
Prabir K. Kolay and Rudolph Bonaparte
4.1 Introduction 87
4.2 Definition and Identification Process for Hazardous Waste 88
4.2.1 Listed Hazardous Waste 89
4.2.2 Characteristic Hazardous Waste 90
4.2.3 Hazardous Wastes, Constituents, Chemicals, and Substances 91
4.3 Cradle-to-Grave Regulatory Framework 92
4.3.1 Hazardous Waste Generation and Transportation 92
4.3.2 Hazardous Waste Recycling, Treatment, Storage, and Disposal 92
4.4 Requirements for Hazardous Waste Landfills 93
4.4.1 Double-Liner System 93

4.4.2 Action Leakage Rate and Response Action Plan 95
4.4.3 Closure and Postclosure Care 97
4.5 Resource Conservation and Recovery Act Land Disposal Restrictions 100
4.5.1 Basic Requirements and Treatment Standards 100
4.5.2 Alternative Land Disposal Restrictions for Contaminated Soils 101
4.6 Remediation of Contamination at Resource Conservation and Recovery Act Hazardous Waste Treatment, Storage, and Disposal Facilities 102
4.7 Remediation of Contamination at Non-Resource Conservation and Recovery Act Sites under Comprehensive Environmental Response Compensation and Liability Act 103
4.7.1 Comprehensive Environmental Response Compensation and Liability Act Basics 103
4.7.2 Comprehensive Environmental Response Compensation and Liability Act Removal Actions and Remedial Actions 104
4.7.3 Comprehensive Environmental Response Compensation and Liability Act Applicable or Relevant and Appropriate Requirements 108
4.7.4 Redevelopment of Remediated Comprehensive Environmental Response Compensation and Liability Act Sites 109
4.8 Special Requirements for Managing and Disposing Polychlorinated Biphenyl Wastes 110
4.9 Special Requirements for Mixed Wastes 111
References 112

5. MUNICIPAL SOLID WASTE: CHARACTERIZATION AND ENGINEERING PROPERTIES 117
Dimitrios Zekkos, Xuede Qian, Jeffrey Fassett and Christopher Bareither
5.1 Introduction 117
5.2 Municipal Solid Waste Composition 119
5.3 Solid Waste Structure 119
5.4 Biogas Generation 120
5.5 Unit Weight 120
5.6 Moisture Content 120
5.7 Field Capacity 121
5.8 Hydraulic Conductivity 121
5.9 Shear Strength 122
5.10 Compressibility and Settlement 124
5.11 Dynamic Properties 125
References 125

6. COAL COMBUSTION RESIDUALS 129
Prabir K. Kolay, Bora Cetin and Lin Li
6.1 Background 129
6.2 Environmental Regulation 129
6.3 Characteristics and Properties of Coal Combustion Residuals 130
6.3.1 Size and Shape 130
6.3.2 Chemical Composition 131
6.3.3 Classification of Fly Ashes 131
6.3.4 Particle Size Distribution 133
6.3.5 Specific Gravity 133
6.3.6 Loss on Ignition 134
6.3.7 Compaction 134
6.3.8 Shear Strength 135
6.4 Beneficial Use of Coal Combustion Residuals 135
6.4.1 Reuse of Coal Combustion Residual (Particularly Fly Ash) 136
6.4.2 Environmental Concerns 139
References 140

7. WOODY BIOMASS FLY ASH: PROPERTIES AND ENGINEERING APPLICATIONS 145
Susan E. Burns, N. N. Nortey Yeboah, Xenia Wirth and Shaivan Shivaprakash
7.1 Introduction 145
7.2 Biomass Ash 146
7.3 Characterization of Physical and Chemical Properties 148
7.3.1 Particle Morphology and Particle Size 148
7.3.2 Specific Gravity 149
7.3.3 Chemical Composition and Classification of Biomass Fly Ash 149
7.3.4 Organic Content: Loss on Ignition versus Residual Organic Carbon 150
7.4 Engineering Properties and Applications 151
7.4.1 Biomass Fly Ash in Concrete 151
7.4.2 Biomass Fly Ash for Soil Improvement 152
7.4.3 Biomass Fly Ash as Mine Backfill 152
7.4.4 Biomass Fly Ash for Waste Solidification/Stabilization 152
7.4.5 Biomass Fly Ash as Sorptive Agents 153
References 153

PART 4: CONTAINMENT SYSTEMS 157

8. DESIGN, ANALYSIS, AND CONSTRUCTION OF LINER SYSTEMS FOR CONTAINMENT FACILITIES 159
John S. McCartney and Keaton Botelho
8.1 Introduction 159
8.2 Regulatory Requirements for Liners 160
8.2.1 Background 160
8.2.2 Prescriptive Designs 161
8.3 Geomembrane Liners 163
8.3.1 Overview 163
8.3.2 Hydraulic Performance 165
8.3.3 Mechanical Performance 166
8.4 Compacted Clay Liners 167
8.4.1 Overview 167
8.4.2 Hydraulic Performance 169
8.4.3 Mechanical Performance 170
8.4.4 Construction Issues and Specifications 170
8.5 Geosynthetic Clay Liners 171
8.5.1 Overview 171
8.5.2 Mechanical Performance 172
8.5.3 Construction Issues and Specifications 172
8.6 Performance of Composite Liners 173
8.6.1 Hydraulic Performance 173
8.6.2 Temperature and Desiccation Effects 174
8.7 Drainage Systems 175
8.7.1 Overview 175
8.7.2 Hydraulic Performance 175
8.8 Final Comments 176
References 176

9. FINAL COVERS FOR LANDFILLS AND WASTE PILES 183
Richard Thiel, Joe Scalia and N. Yesiller
9.1 Introduction 183
9.2 General Requirements for Final Covers 184
9.2.1 Definition of a Final Cover 184
9.2.2 Physical Isolation of Waste 184
9.2.3 Fluid Barrier 184
9.2.4 Regulatory Requirements 185
9.3 Types of Landfill Final Covers 186
9.3.1 Soil Covers 186
9.3.2 Water Balance Covers 187
9.3.3 Resistive Covers 191
9.4 Resistive Layers 196
9.4.1 Single Compacted Clay Liners 196
9.4.2 Single Geosynthetic Clay Liners 197

- 9.4.3 Single Geomembranes ... 198
- 9.4.4 Composite Covers ... 199
- 9.4.5 Exposed Geomembrane Covers ... 201
- 9.5 Additional Technical Issues for the Design of Final Covers ... 202
  - 9.5.1 Erosion Control and Surface Water Management ... 202
  - 9.5.2 Control of Fluid Pressures below and above Barrier Layers to Preserve Veneer Slope Stability ... 203
  - 9.5.3 Settlement ... 207
  - 9.5.4 Landfill Cover Thermal Regime ... 208
  - 9.5.5 Landfill Gas Control in Final Cover Systems ... 209
- 9.6 Postclosure Maintenance ... 211
  - 9.6.1 Documentation ... 211
  - 9.6.2 Inspections ... 212
- References ... 214

10. STATIC SLOPE STABILITY OF CONTAINMENT SYSTEMS ... 221

Kevin Foye

- 10.1 Introduction ... 221
- 10.2 Typical Applications ... 222
- 10.3 Limitations ... 223
- 10.4 Roles ... 223
- 10.5 Workflow ... 224
- 10.6 Methods of Analysis ... 225
  - 10.6.1 Limit Equilibrium ... 225
  - 10.6.2 Limit Analysis ... 230
  - 10.6.3 Finite-Element Analysis ... 230
  - 10.6.4 Other Computational Methods ... 231
  - 10.6.5 Three-Dimensional Analyses ... 231
  - 10.6.6 Probabilistic Analyses ... 232
- 10.7 Modes of Failure in Containment Systems ... 233
- 10.8 Inputs ... 236
  - 10.8.1 Slope Geometry ... 237
  - 10.8.2 Soil Shear Strength/Interface Shear Strength ... 238
  - 10.8.3 Pore Pressure ... 243
- 10.9 Forensic Evaluations ... 246
- 10.10 Soil Reinforcement ... 248
- 10.11 Review Guidelines ... 250
- 10.12 Documentation ... 254
- References ... 254

11. SEISMIC STABILITY OF WASTE CONTAINMENT SYSTEMS ... 259

Edward Kavazanjian Jr., Dimitrios Zekkos and George Zalachoris

- 11.1 Introduction ... 259
- 11.2 Hierarchy of Seismic Stability Analyses ... 259
- 11.3 Seismic Performance Standard ... 261
- 11.4 Design Criteria ... 263

11.5 Design Ground Motion ... 266
11.6 Pseudo-Static Limit Equilibrium ... 270
11.7 Newmark-Type Displacement Analysis ... 275
11.7.1 Simplified Newmark-Type Analysis Chart Procedures ... 275
11.7.2 Site-Specific Newmark-Type Analysis ... 277
11.8 Stress–Deformation Dynamic Analyses ... 279
References ... 282

PART 5: MINE WASTE GEOTECHNICS ... 287

12. MINE WASTE GEOTECHNICS: CHARACTERISTICS AND MANAGEMENT ... 289
Christopher A. Bareither
12.1 Introduction ... 289
12.2 Mine Waste Rock and Tailings ... 290
12.2.1 Waste Rock ... 292
12.2.2 Mine Tailings ... 293
12.3 Mine Waste Management ... 296
12.3.1 Waste Rock Piles ... 296
12.3.2 Tailing Storage Facilities ... 296
References ... 300

PART 6: CONTAMINATED GROUND INTERVENTIONS ... 303

13. EXCAVATION AND REPLACEMENT ... 305
Carsten H. Floess
13.1 Introduction ... 305
13.2 Excavation Support ... 306
13.3 Sheet Pile Installation ... 309
13.4 Construction Vibrations and Protection of Adjacent Structures ... 310
13.5 Groundwater Control ... 313
13.6 Subsurface Investigation and Design ... 313
13.7 Open Water Excavation (Dredging) ... 314
References ... 315

14. ELECTROKINETIC REMEDIATION ... 317
Arvin Farid
14.1 Introduction ... 317
14.2 How Electrokinetics Works ... 319
14.3 Electrokinetics Application in the Field ... 321
14.4 Electrokinetics Governing Factors ... 323
14.4.1 Electric Voltage and Current ... 323
14.4.2 Electrode Placement ... 324
14.4.3 Soil Type ... 324

14.4.4 Contamination Level ... 325
14.4.5 Cost Estimation ... 326
14.5 Examples of Electrokinetic Remediation ... 326
14.6 Future Prospect ... 329
References ... 330

15. GROUND IMPROVEMENT FOR GEOENVIRONMENTAL ENGINEERING PRACTICE ... 339
Anirban De
15.1 Introduction ... 339
15.2 Common Ground Improvement Methods ... 340
15.2.1 Introduction ... 340
15.2.2 Preloading ... 340
15.3 Conclusions ... 348
References ... 349

PART 7: BARRIERS ... 353

16. VERTICAL BARRIER WALLS FOR GEOENVIRONMENTAL APPLICATIONS ... 355
Jeffrey C. Evans and Kristin M. Sample-Lord
16.1 Introduction ... 355
16.2 Vertical Barrier Deployment Options ... 355
16.3 Vertical Barrier Types ... 357
16.3.1 Soil–Bentonite Slurry Trench Cutoff Walls ... 357
16.3.2 Cement–Bentonite Slurry Trench Cutoff Walls ... 361
16.3.3 Mix-in-Place Methods (Panel Mixing) ... 362
16.3.4 Mix-in-Place Methods (Continuous Trench Mixing) ... 364
16.3.5 Other Vertical Barrier Methods ... 366
16.4 Longevity Considerations ... 367
16.5 Summary ... 369
References ... 369

17. ZERO-VALENT IRON PERMEABLE REACTIVE BARRIERS FOR IN SITU TREATMENT OF ORGANICS AND METALS IN GROUNDWATER ... 373
Andrzej Przepiora, Claire F. Wildman and Megan Hart
17.1 Introduction ... 373
17.2 Zero-Valent Iron Permeable Reactive Barrier Applicability ... 374
17.2.1 Feasibility Evaluation ... 374
17.2.2 Zero-Valent Iron Materials ... 376
17.2.3 Contaminants Treated ... 377
17.2.4 Treatment of Chlorinated Aliphatic Hydrocarbons ... 378
17.2.5 Treatment of Dissolved Metals ... 382
17.3 Zero-Valent Iron Permeable Reactive Design ... 384
17.3.1 Predesign Considerations ... 385

17.3.2 Site Characterization for Permeable Reactive Barrier Design ... 388
17.3.3 Determination of Permeable Reactive Barrier Thickness for Chlorinated Aliphatic Hydrocarbons ... 390
17.3.4 Long-Term Performance of Zero-Valent Iron Permeable Reactive Barriers for Chlorinated Aliphatic Hydrocarbons ... 392
17.3.5 Determination of Zero-Valent Iron Width for Metals ... 396
17.4 Zero-Valent Iron Construction Methods ... 398
17.4.1 Trenched Zero-Valent Iron Permeable Reactive Barriers ... 399
17.4.2 Injected Zero-Valent Iron Permeable Reactive Barriers ... 400
17.5 Permeable Reactive Barrier Monitoring ... 401
17.5.1 Monitoring Well Network ... 401
17.5.2 Groundwater Sample Collection and Analysis Plan ... 402
17.6 Summary ... 403
References ... 404

INDEX ... 409

LIST OF CONTRIBUTORS

Editor

Dimitrios Zekkos, Ph.D., P.E., M.ASCE, *Professor, University of California at Berkeley; 2015–2021 Chair of the Geoenvironmental Engineering Technical Committee*

Authors

Chapter 1 Contaminant Transport and Fate

Joe Scalia, Ph.D., *Associate Professor, Department of Civil and Environmental Engineering, Colorado State University*

Charles Shackelford, Ph.D., P.E., F.ASCE *Professor, Department of Civil and Environmental Engineering, Colorado State University*

Chapter 2 Field Testing for Geotechnical Characterization of Waste

Neven Matasovic, *Principal/Director of Geotechnical Engineering, Geo-Logic Associates, Inc.*

Chapter 3 Geochemical Characterization of Solids, Gases, and Liquids

Marisa Chrysochoou, M.ASCE, *Professor, Department of Civil and Environmental Engineering, University of Connecticut*

Nefeli Bompoti, *Assistant Research Professor, Department of Civil and Environmental Engineering, University of Connecticut*

Chapter 4 Hazardous Waste

Prabir K. Kolay, Ph.D., P.E., F.ASCE, *Professor and Director, School of Civil, Environmental and Infrastructure Engineering, Southern Illinois University Carbondale*

Rudolph Bonaparte, Ph.D., P.E., NAE, *Chairman of the Board, Geosyntec Consultants*

Chapter 5 Municipal Solid Waste

Dimitrios Zekkos, Ph.D., P.E., M.ASCE, *Professor, Department of Civil and Environmental Engineering, University of California at Berkeley*

Xuede Qian, *Principal Geotechnical and Geoenvironmental Engineering Specialist, Michigan Department of Environment, Great Lakes, and Energy, Material Management Division*

Jeffrey Fassett, *Senior Consultant, Geosyntec Consultants*

Christopher Bareither, *Associate Professor, Department of Civil and Environmental Engineering, Colorado State University*

Chapter 6 Coal Ash

Prabir K. Kolay, Ph.D., P.E., F.ASCE, *Professor and Director, School of Civil, Environmental and Infrastructure Engineering, Southern Illinois University Carbondale*

Bora Cetin, Ph.D., M.ASCE, *Associate Professor, Department of Civil and Environmental Engineering, Michigan State University*

Lin Li, Ph.D., M.ASCE, *Professor and Head, Department of Civil and Architectural Engineering, Tennessee State University*

Chapter 7 Biomass Fly Ash

Susan E. Burns, F.ASCE, *Professor, School of Civil and Environmental Engineering, Georgia Institute of Technology*

N. N. Nortey Yeboah, M.ASCE, *Principal Engineer, R&D, Southern Company*

Xenia Wirth, M.ASCE, *Assistant Professor, Department of Civil and Environmental Engineering, California State University, Fullerton*

Shaivan Shivaprakash, *School of Civil and Environmental Engineering, Georgia Institute of Technology*

Chapter 8 Design of Base Containment Systems

John S. McCartney, Ph.D., P.E., F.ASCE, *Professor, Department of Structural Engineering, University of California San Diego*

Keaton Botelho, P.E., *Senior Engineer, Geosyntec Consultants*

Chapter 9 Review and Design of Covers

Richard Thiel, P.E., M.ASCE, *President, Thiel Engineering*

Joe Scalia, Ph.D., A.M.ASCE, *Associate Professor, Civil and Environmental Engineering, Colorado State University*

Nazli Yesiller, Ph.D., A.M.ASCE, *Director, Global Waste Research Institute, California Polytechnic State University*

Chapter 10 Static Stability of Containment Systems

Kevin Foye, Ph.D., P.E., M.ASCE, *CTI and Associates, Inc.*

Chapter 11 Seismic Stability of Containment Systems
Edward Kavazanjian, Jr., *Professor, School of Sustainable Engineering and the Built Environment, Arizona State University*
Dimitrios Zekkos, Ph.D., P.E., M. ASCE, *Professor, University of California at Berkeley*
George Zalachoris, *ElxisGroup S.A., Athens, Greece*

Chapter 12 Mine Waste Geotechnics
Christopher A. Bareither, Ph.D., P.E., *Associate Professor, Civil and Environmental Engineering, Colorado State University*

Chapter 13 Excavation and Replacement
Carsten H. Floess, Ph.D., P.E., D.GE, F.ASCE, *AECOM*

Chapter 14 Electrokinetics
Arvin Farid, *Professor, Civil Engineering, Boise State University*

Chapter 15 Ground Improvement for Geoenvironmental Engineering Practice
Anirban De, Ph.D., P.E., F.ASCE, *Professor and Chairman, Department of Civil and Environmental Engineering, Manhattan College*

Chapter 16 Cutoff (Impermeable) Barriers
Jeffrey C. Evans, Ph.D., P.E., D.GE, F.ASCE, *Professor, Emeritus*
Kristin M. Sample-Lord, Ph.D., P.E., M.ASCE, *Associate Professor, Villanova University*

Chapter 17 Permeable Barriers
Andrzej Przepiora, P.Geo., *Geosyntec Consultants, Inc.*
Claire F. Wildman, Ph.D., P.E., *Geosyntec Consultants, Inc.*
Megan Hart, Ph.D., R.G., *University of Missouri*

BLUE-RIBBON PANEL REVIEWERS

Craig Benson, Ph.D., P.E., NAE
Distinguished Professor Emeritus
Department of Civil and Environmental Engineering
University of Wisconsin–Madison

Rudolph Bonaparte, Ph.D., P.E., NAE
Chairman of the Board
Geosyntec Consultants
Atlanta, Georgia

James K. Mitchell, Ph.D., NAE
Distinguished Professor Emeritus
Department of Civil and Environmental Engineering
Virginia Tech

Krishna Reddy, Ph.D., P.E.
Professor
Department of Civil and Environmental Engineering
University of Illinois, Chicago

Note: Per ASCE policy, the Blue-Ribbon Panel Reviewers were not involved in any manner in the preparation of the manual. During the review process, Dr Rudoph Bonaparte contributed extensive text in the Chapter 4 on Hazardous Waste and thus an additional ad hoc Blue-Ribbon Panel member, Professor Krishna Reddy, was invited to review that specific Chapter.

PREFACE

The Geoenvironmental Engineering Manual of Practice represents a major effort undertaken by the Geoenvironmental Engineering Technical Committee of the Geo-Institute of the American Society of Civil Engineers (ASCE). The effort started in 2016 and involved 50 authors and reviewers. The majority of the membership of the Geoenvironmental Engineering Technical Committee was engaged in this effort as well as additional leading geoenvironmental engineering professionals who were not members of the Geoenvironmental Engineering Technical Committee with expertise on the topics covered.

The manual is aimed to be a consensus document that provides concise, straight-to-the-point documentation of aspects of geoenvironmental engineering with an emphasis on engineering practice. It also outlines, when appropriate, design procedures and recommendations for use in engineering practice with the goal of developing a technical and scientific resource for geoenvironmental engineering professionals. Although the manual is a valuable technical source of information, and when necessary, appropriate references to the scientific literature are provided, it is not intended to provide an exhaustive state-of-the-art treatment of each topic or to include a comprehensive synthesis of the literature. Instead, the goal is to reflect the state of practice in the United States of America.

The manual preparation process was the following: Each chapter was led by an invited lead author who prepared the main content of a chapter. Subsequently, a minimum of two additional reviewers, or contributors, critically reviewed and provided edits and amendments to the document. Because the goal of the manual is to provide a consensus on each treated topic, the reviewers/contributors were not anonymous, but worked synergistically with the lead author to make amendments. For each chapter, there were at least two rounds of review, although commonly, several more rounds of review were conducted, until reviewers and contributors concurred on the content of each chapter. For some chapters, depending on their contributions and the lead chapter author's decision, reviewers became coauthors. A minimum of one practitioner served as a reviewer/contributor to each chapter to ensure that the topic is presented in a manner that serves the professional community. During and following the completion of the reviews by all reviewers, the Editor also reviewed and provided edits for each chapter to ensure consistency throughout the manual. The entire manual was then submitted to a Blue-Ribbon Review Panel that consisted of leading geoenvironmental engineers in the field, who provided feedback and comments on the entire manual. These comments were addressed by the lead authors, and a revised document, along with responses to the review comments, was prepared and resubmitted to the Blue-Ribbon Panel Review that consisted of Prof. James K. Mitchell, Prof. Craig Benson, and Dr. Rudolph Bonaparte. Because Dr. Rudoph Bonaparte contributed extensive text in the chapter on Hazardous Waste, an additional ad hoc Blue-Ribbon Review Panel member, Prof. Krishna Reddy, re-reviewed that specific chapter.

The Geoenvironmental Engineering Manual of Practice covers a wide range of geoenvironmental engineering practice, including contaminant transport and fate, geomechanical and geochemical characterization, waste material characterization and properties, design of base and cover containment systems, static and seismic slope stability of waste containment systems, and permeable and impermeable (cutoff) barriers, as well as topics related to excavation, ground improvement, remediation, and mine waste geotechnics. Despite the breadth of content included in the manual, this manual does not exhaust the entire geoenvironmental engineering practice. In some cases, such as geoenvironmental remediation, we felt that extensive literature and manuals by the Environmental Protection Agency (USEPA) and others existed that treated these topics, whereas other topics were not completed during the manual preparation process. In the end, the Editor decided that additional chapters can be included in potential future revisions of the manual, and that the professional community would be better served by the publication of this first edition of the manual of geoenvironmental engineering practice rather than delay further publication of the important content that has been generated already.

ACKNOWLEDGMENTS

Chapter 1 was authored under the leadership of J. Scalia and C. Shackelford. Subsequently, K. Sample-Lord and E. Ghazanfari reviewed the manuscript independently and contributed to the topic.

Chapter 2 was authored under the leadership of N. Matasovic. Subsequently, A. Witthoeft, R. Thiel, A. De, R. Jeffrey, X. Fei, and T. Hadj-Hamou reviewed the manuscript independently and contributed to the topic.

Chapter 3 was authored under the leadership of M. Chrysochoou and N. Bompoti. Subsequently, J. Eun and A. Palomino reviewed the manuscript independently and contributed to the topic.

Chapter 4 was authored under the leadership of P. K. Kolay and R. Bonaparte. Subsequently, A. Rahmani and M. Azizul reviewed the manuscript independently and contributed to the topic.

Chapter 5 was authored under the leadership of D. Zekkos. Subsequently, X. Qian, J. Fassett, and C. Bareither reviewed the manuscript independently, contributed to the topic, and are listed as coauthors.

Chapter 6 was authored under the leadership of P. K. Kolay. Subsequently, B. Cetin and L. Li reviewed the manuscript independently, contributed to the topic, and are listed as coauthors.

Chapter 7 was authored under the leadership of S. E. Burns, N. N. N. Yeboah, and X. Wirth. Subsequently, J. Daniels and P. Kolay reviewed the manuscript independently and contributed to the topic.

Chapter 8 was authored under the leadership of J. S. McCartney. Subsequently, M. Malusis and S. Iravani reviewed the manuscript independently, contributed to the topic, and are listed as coauthors.

Chapter 9 was authored under the leadership of R. Thiel. Subsequently, J. Scalia and N. Yesiller reviewed the manuscript independently, contributed to the topic, and are listed as coauthors.

Chapter 10 was authored under the leadership of K. Foye. Subsequently, J. Evans and F. Settepani reviewed the manuscript independently and contributed to the topic.

Chapter 11 was authored under the leadership of E. Kavazanjian, Jr. Subsequently, D. Zekkos and G. Zalachoris reviewed the manuscript independently, contributed to the topic, and are listed as coauthors.

Chapter 12 was authored under the leadership of C. A. Bareither. Subsequently, L. de Mello and R. Gupta reviewed the manuscript independently and contributed to the topic.

Chapter 13 was authored under the leadership of C. H. Floess. Subsequently, B. Bate and R. Gupta reviewed the manuscript independently and contributed to the topic.

Chapter 14 was authored under the leadership of A. Farid. Subsequently, J. Wang and S. Pamucku reviewed the manuscript independently and contributed to the topic.

Chapter 15 was authored under the leadership of A. De. Subsequently, C. Woods and M. Flanagan reviewed the manuscript independently and contributed to the topic.

Chapter 16 was authored under the leadership of J. C. Evans and. K. M. Sample-Lord. Subsequently, A. Idil and G. Bonhoff reviewed the manuscript independently and contributed to the topic.

Chapter 17 was authored under the leadership of A. Przepiora and C. F. Wildman. Subsequently, M. Hart reviewed the manuscript independently, contributed to the topic, and is listed as coauthor. L. Li reviewed the manuscript independently and contributed to the topic.

PART 1
INTRODUCTION

CHAPTER 1
CONTAMINANT TRANSPORT AND FATE

Joe Scalia and Charles Shackelford

1.1 INTRODUCTION

The purpose of this chapter is to provide a brief overview of the concepts of contaminant transport and fate through porous media (i.e., soil). The presentation is limited to one-dimensional (1D) transport of aqueous miscible chemical species or solutes (e.g., dissolved contaminants) through water-saturated porous media. The chapter is intended only to introduce basic concepts and serve as a guide to the relevant literature. References are provided to furnish the interested reader with additional relevant, and substantially more comprehensive, literature on the topic.

1.2 MASS TRANSPORT

Transport of aqueous miscible chemical species or solutes typically is governed by three primary processes: advection, diffusion, and dispersion.

Advection is the migration of solutes together with the flow of water (solvent) in response to a hydraulic gradient, that is, the solute is carried along with the flowing water. Nonreactive solutes are transported at an average rate (assuming all pores are equally effective at conducting flow) equal to the seepage velocity of water, v_s [LT^{-1}, where L = units of length and T = units of time], defined as follows:

$$v_s = \frac{v}{n} \tag{1-1}$$

where

v [LT^{-1}] = Water flux, defined as the volume of water flow per unit total cross-sectional area of porous medium per unit time; and
n [L^3L^{-3}] = Total porosity of the porous medium (volume of voids per unit total volume of porous medium).

Note that the porosity is required in Equation (1-1) to account for the fact that the water flows only through the pores. The water flux is also known as the Darcian velocity (although v is not a true velocity) that is based on Darcy's law, which for 1D flow, can be written as follows:

$$q = vA = kiA = -k\frac{dh}{dx}A \tag{1-2}$$

where

q [L^3T^{-1}] = Volumetric flow rate of water,
A [L^2] = Total cross-sectional area (soil solids plus voids) perpendicular to the direction of flow,
k [LT^{-1}] = Hydraulic conductivity of the soil,
i [LL^{-1}] = Hydraulic gradient,
h [L] = Total hydraulic head, and
x [L] = Spatial coordinate in the direction of flow or seepage.

The negative sign in Equation (1-2) accounts for the fact that flow occurs from high h to low h. Fundamentally, v_s represents the average flow rate only through the void space of the porous media (i.e., pores), whereas v reflects the volumetric flow relative to the total cross-sectional area perpendicular to the direction of flow.

The hydraulic conductivity, k, of soils can be measured in the laboratory or the field. For engineered barrier soils, such as compacted clay liners and geosynthetic clay liners (GCLs), Daniel (1994) describes the laboratory apparatus and methods commonly used to measure k, and Daniel (1989) describes the field-testing procedures to measure k. An important consideration in the measurement of the k of barrier soils used to contain chemical solutions or waste leachates is the potential for adverse interactions between the containment liquids and the clays, referred to as chemical incompatibility, which can lead to significant increases (orders of magnitude) in k (e.g., Shackelford 1994). The macroscale k of coarse-grained (granular) soils in the field is generally determined via wells and pumping tests (e.g., Bear 1979, Freeze and Cherry 1979, Kruseman et al. 1994). In the laboratory, the hydraulic gradient, i, is generally controlled or measured, whereas in the field, i is typically determined via proximate wells or piezometers (Kruseman et al. 1994).

The transit or travel time, t [T], for a nonreactive solute to migrate a straight-line (macroscopic) distance between two points a distance L (i.e., Δx) [L] through saturated porous media due only to advection can be estimated as shown in the following equation:

$$t = \frac{L}{v_s} = \frac{nL}{ki} \tag{1-3}$$

The advective solute mass flux, J_A [$ML^{-2}T^{-1}$, where M is the units of mass], represents the mass of chemical or solute migrating per unit total cross-sectional area of the porous medium per unit time only by advection, or

$$J_A = nv_s c = vc = kic \tag{1-4}$$

where c [ML^{-3}] is the solute concentration defined as the mass of solute per unit volume of the soil pore water.

For porous media in which advection occurs through only a fraction of the void space, the total porosity (n) in the previous equations should be replaced with the effective porosity, n_e [L^3L^{-3}], defined as the volume of voids that conduct liquid flow relative to the total volume of the porous medium. In general, when an effective porosity exists, $n_e < n$. Additional discussion of n_e is provided by Freeze and Cherry (1979), Horton et al. (1985), and Peyton et al. (1986). Stephens et al. (1998) describe methods to calculate or estimate n_e.

Diffusion is the migration of solutes caused by the random motions of individual molecules of chemicals in the absence of advection (i.e., without a hydraulic gradient). Diffusive transport is conceptualized macroscopically as the migration of solutes in response to a gradient in concentration or, more precisely, a gradient in chemical potential (Robinson and Stokes 1959).

The diffusive mass flux, J_D [$ML^{-2}T^{-1}$], in a free solution (i.e., a bulk solution without the presence of a porous media) is described by Fick's first law as follows:

$$J_D = D_o i_c = -D_o \frac{dc}{dx} \tag{1-5}$$

where D_o [L^2T^{-1}] is the free-solution (aqueous-phase) diffusion coefficient (i.e., the diffusion coefficient of the solute in a bulk solution), and i_c is the concentration gradient given by $-dc/dx$ [ML^{-4}], where the negative sign reflects that transport occurs in the direction of decreasing concentration. Thus, the diffusive solute mass flux in aqueous solutions is proportional to the product of the magnitudes of D_o and i_c. Shackelford and Daniel (1991) tabulate D_o values for a variety of inorganic solutes, and Schwartz and Zhang (2003) and Yaws (2009) tabulate D_o values for organic solutes. Shackelford and Rowe (1998) summarize D_o values for a variety of both inorganic and organic solutes.

For diffusion in porous media, the effect of the existence of the solid phase is taken into account via a modified form of Fick's first law as follows:

$$J_D = n\tau D_o i_c = -n\tau D_o \frac{dc}{dx} \tag{1-6}$$

where the free-solution diffusive mass flux is reduced by the product of n (diffusion occurs only through pores) and a tortuosity factor, τ [L^2L^{-2}], which represents the increased distance of transport by solutes through the tortuous pathways or network of pores that exist in a saturated porous media. The tortuosity factor can be defined as follows:

$$\tau = \left(\frac{L}{L_e}\right)^2 \tag{1-7}$$

where L_e [L] is the effective or actual distance of transport between the same two macroscopic points defining the straight-line distance, L, taking into account obstructions to migration caused by the existence of the solid phase. Because of the presence of the solid phase, $L_e > L$, such that τ is < 1. Tortuosity factors reported in the literature typically range from 0.01 to 0.67 (Perkins and Johnson 1963, Daniel and Shackelford 1988, Shackelford 1989, Shackelford and Daniel 1991), although values of τ lower than 0.01 are possible. Because both n and τ are less than unity, J_D through porous media is lower than J_D in free solution, with all other factors being equal. Note that the quotient, L/L_e, is referred to as the tortuosity such that the tortuosity factor is the square of the tortuosity, and that in some literature, the tortuosity factor is defined as the inverse of τ [i.e., $(L_e/L)^2 > 1$] (Shackelford and Moore 2013).

Because τ is generally not known *a priori*, an effective diffusion coefficient, D^* [L^2T^{-1}], which includes the tortuosity factor, is defined as follows:

$$D^* = \tau D_o \tag{1-8}$$

such that J_D for saturated porous media is written as follows:

$$J_D = nD^* i_c = -nD^* \frac{dc}{dx} \tag{1-9}$$

Although n can generally be determined independently, n is also sometimes incorporated into an alternate definition of the effective diffusion coefficient (i.e., $D_e = n\tau D_o = nD^*$) (e.g., Shackelford 1988, Shackelford and Daniel 1991, Shackelford and Moore 2013). Some common methods for measuring D^* in the laboratory are described by Shackelford (1991). High-resolution characterization may be used to determine D^* for natural clays under field conditions, as described in Parker et al. (2022).

Mechanical dispersion is the physically driven mixing of fluids arising from heterogeneities in the flow systems of porous media that occurs across both microscopic and macroscopic scales. Processes that yield mechanical dispersion at the microscopic scale are illustrated schematically in Figure 1-1 (Fried 1975, Bear 1979, Freeze and Cherry 1979, Shackelford 1993). Variations of velocities within pore channels (Figure 1-1a), and through pores with varying areas (Figure 1-1b), which, by continuity, yield spatial variations in velocity, result in pore-scale mixing. Further mixing occurs because of the tortuous nature of flow paths (Figure 1-1c) through porous media.

At the macroscale, mechanical dispersion is attributed to variations in flow resulting from heterogeneity (i.e., macroscopic tortuosity) and anisotropy in k (Sale et al. 2013), such as the existence of low-permeability clay lenses within a conductive (granular) aquifer (Figure 1-2).

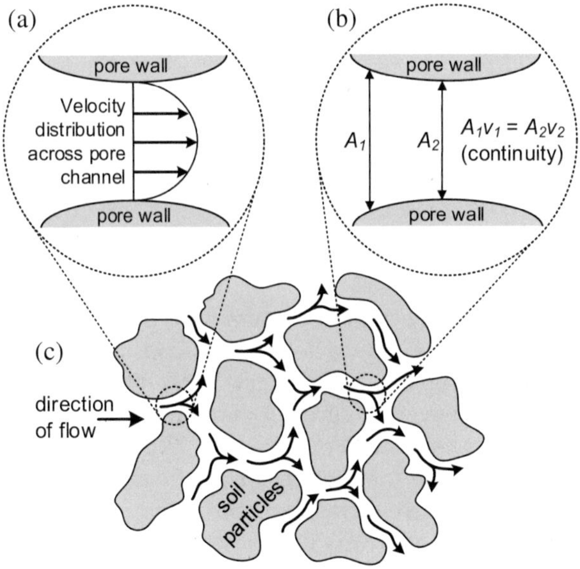

Figure 1-1. Microscopic dispersion in porous media caused by (a) velocity variation within a single pore, (b) variation in pore sizes, (c) tortuous flow paths. Source: Bear (1979), Freeze and Cheery (1979), Shackelford (1993).

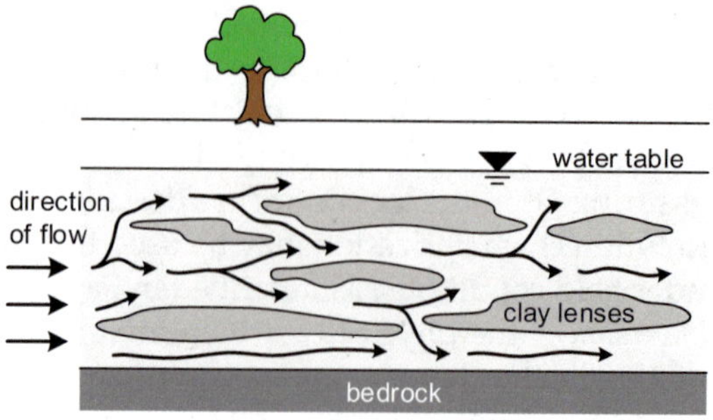

Figure 1-2. Macroscopic dispersion in porous media (at field or regional scale) caused by variations in flow resulting from heterogeneity.
Source: Shackelford (1993).

Mechanical dispersion yields reduced solute concentration in the core of a plume by simultaneously spreading the contaminant plume transversely and longitudinally. For 1D transport, longitudinal mechanical dispersive flux, J_M [$ML^{-2}T^{-1}$], is typically assumed to be a Fickian process that is expressed as follows (Shackelford 1993):

$$J_M = nD_m i_c = -nD_m \frac{dc}{dx} \tag{1-10}$$

where D_m [L^2T^{-1}] is the coefficient of mechanical dispersion, and the negative sign reflects that transport occurs in the direction of decreasing concentration (owing to solute spreading). Because dispersion results from variations in the magnitude of v_s, D_m is often defined as follows:

$$D_m = \alpha_L v_s^{\beta} \tag{1-11}$$

where α_L [L] is the longitudinal dispersity of the porous media, and β [dimensionless] is an empirically determined constant that commonly is assumed to be unity (Freeze and Cherry 1979), although the results of some studies have shown β to be greater than unity (Anderson and Cherry 1979, Anderson 1984, Bear and Verruijt 1987). Typically, α_L is assumed to equal $0.1L$, where L is the total length of the flow domain. However, in situations with low v_s, such as transport through low-*k* barrier materials, mechanical dispersion is relatively insignificant and, therefore, is typically ignored (Rowe 1987).

Total mass flux, J [$ML^{-2}T^{-1}$], of nonreactive solutes in the absence of coupled flow processes is the sum of advective, diffusive, and dispersive fluxes as follows:

$$J = J_A + J_D + J_M = nv_s c - nD^* \frac{dc}{dx} - nD_m \frac{dc}{dx} \tag{1-12}$$

The coefficients D^* and D_m typically are combined into a lumped coefficient referred to as the hydrodynamic dispersion coefficient, D_h [L^2T^{-1}], and J is rewritten more conveniently as follows:

$$J = nv_s c - nD_h \frac{dc}{dx} \tag{1-13}$$

The relative importance of advection and diffusion on total solute mass flux is dependent on both physical factors (k, i, n) and chemical factors (i_c, D_o). Under conditions typical of low permeability, engineered barriers, and for solutions with D_o in the range of about 6.0×10^{-10} to 3.0×10^{-9} m²/s (a common range for many inorganic and organic solutes), advection tends to be the dominant transport mechanism for barriers with k greater than approximately 1.0×10^{-9} m/s and diffusion tends to be the dominant transport mechanism for barriers with k less than approximately 1.0×10^{-10} m/s (Shackelford 1988). For k between 1.0×10^{-10} and 1.0×10^{-9} m/s, both advection and diffusion play prominent roles in governing contaminant transport. Given that many engineered barriers require k less than 1.0×10^{-9} m/s, ignoring diffusion can result in significant error that is unconservative (e.g., J is underpredicted, t is overpredicted) when evaluating the effectiveness of the barrier with respect to contaminant transport (Shackelford 1988, 1990).

Values of D_h can be determined in the laboratory by column testing, which essentially is the same as laboratory hydraulic conductivity testing, except that the influent and effluent chemistry must be controlled and measured, respectively. However, because values of α_L are scale-dependent (i.e., α_L increases with increasing distance of solute transport), laboratory determined values of D_h are generally considered unrepresentative, except in the case where the seepage rate is low and the distance of transport is short, which generally is applicable only for engineered containment barriers such as compacted clay liners, GCLs, and soil–bentonite backfills for vertical cutoff walls. In these cases, values of D_h typically are close to or the same as values of D^* because of the low seepage rates (e.g., Shackelford and Redmond 1995). Considerations for column testing of engineered clay-based barriers are described by Shackelford (2021), and examples of the use of laboratory column testing to determine D_h for engineered containment barriers include Visudmedanukul et al. (2007), Hong and Shackelford (2017), and Shackelford and Hong (2020).

Hydrodynamic dispersion can also be measured in the field typically using tracer experiments with pumping/injection wells and monitoring wells, and a nonreactive chemical tracer (Domenico and Schwartz 1990). Common types of field tracer tests include the natural gradient test (no pumping or injection), single-well pulse test, two-well tracer test, and single-well injection or withdrawal with multiple-observation wells. Nonreactive or "ideal" chemical tracers are chemical compounds or species that do not undergo chemical or biological transformations such that transport of the tracers reflects the physical migration pathways. Examples of tracers that may be considered to determine field hydrodynamic dispersion values include common ionic species such as chloride (Cl^-) and bromide (Br^-), radioisotopes of these species (e.g., ^{82}Br), and organic compounds such as rhodamine WT. However, field tracer experiments typically are limited in scale to less than a fraction of a kilometer. For contaminated field sites, hydrodynamic dispersion is often estimated by fitting a contaminant transport model to the observed spatial and temporal distribution of a given contaminant or tracer. However, the accuracy of this approach is dependent on the accuracy of the initial and boundary conditions and various hydraulic, transport, and fate parameters used in the model. Further information on the field measurement of hydrodynamic dispersion can be found in Domenico and Schwartz (1990).

The relative effects of advection and dispersion/diffusion can be illustrated by the schematic temporal profiles of solute concentration at a distance, L, within a porous medium from a constant source of solute concentration, c_o, resulting from 1D solute transport, as shown in Figure 1-3. Concentration Profile 1 represents the case of pure advection, or solute transport in the absence of any dispersion. Concentration Profile 2 illustrates the effect of dispersion, or spreading of the concentration profile, for the situation when mechanical dispersion dominates the hydrodynamic dispersion process and diffusion is negligible ($D_h \approx D_m$, $D^* \approx 0$), which typically is the situation for solute transport at relatively high seepage velocities associated with transport through granular soils such as sands

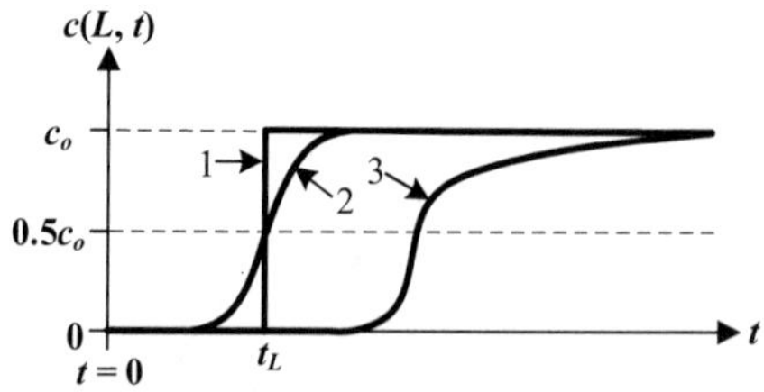

Figure 1-3. Schematic illustrations of temporal solute concentration profiles caused by 1D solute transport within a porous medium resulting from pure advection (Curve 1), advective-dominated hydrodynamic dispersion (Curve 2), and diffusive-dominated hydrodynamic dispersion (Curve 3).
Source: Shackelford (2021).

and gravels. In this case, the concentration profile is sigmoidal symmetric about the concentration $c(L,t) = 0.5c_o$ with respect to Profile 1 for pure advection. Finally, the concentration profile labeled 3 illustrates the effect of dispersion for the situation when mechanical dispersion is negligible and the hydrodynamic dispersion process is dominated by diffusion ($D_h \approx D^*$, $D_m \approx 0$), which typically is the situation for solute transport at low seepage velocities associated with transport through fine-grained soils such as silts and clays. In this case, the concentration profile is not sigmoidal symmetric, and the time required for the solute to reach the distance is longer on account of the slow seepage velocity.

Schematic scenarios illustrating the directions of advection and diffusion through an engineered soil barrier used for waste containment (e.g., landfill liner) under different hydraulic gradient conditions are shown in Figure 1-4. All scenarios assume that the concentration of solutes is greater within (above) the soil barrier than outside (below) the soil barrier and, therefore, diffusion occurs downward in all scenarios (i.e., independent of the direction of the hydraulic gradient).

Coupled flow processes are flows of one type that are driven by a gradient of another type (Mitchell and Soga 2005). For example, in addition to the flow of liquid in response to a hydraulic gradient in accordance with Darcy's law, liquid flow can also occur because of gradients in chemical concentration (chemico-osmosis), electrical potential (electro-osmosis), and temperature (thermo-osmosis). In addition, chemical flows caused by gradients in electrical potential (electro-phoresis) and temperature (Dufour effect) are possible. Mitchell and Soga (2005) provide an excellent review of coupled flow processes, including examples of applications of these processes in geotechnical engineering. A more recent, detailed review of the state-of-the-art in coupled flows as related to geotechnical engineering is provided by Shackelford et al. (2019).

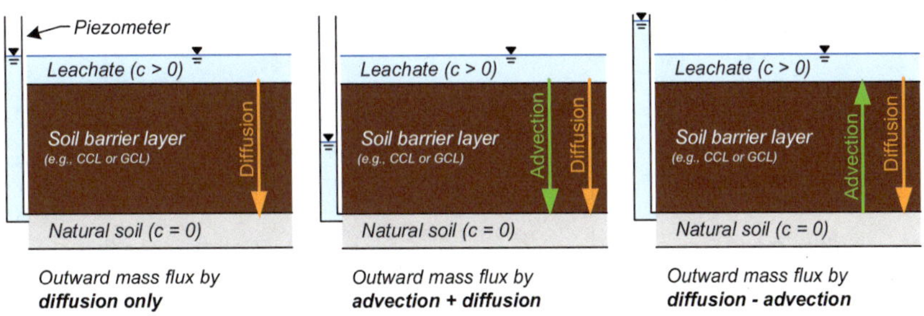

Figure 1-4. Schematic illustrations of the direction of advective and diffusive mass flux through a soil barrier layer (e.g., compacted clay liner, CCL, or geosynthetic clay liner, GCL) under different hydraulic gradient scenarios. Source: Shackelford (1989, 1993).

Coupled flow processes are often ignored in contaminant transport calculations (Shackelford 1993), largely because of their relative insignificance in high-permeability (granular) aquifers and the difficulty of accurately determining phenomenological coefficients (particularly under field conditions). However, coupled flow processes may be significant at low flow rates associated with clays including low-permeability-engineered barriers such as geosynthetic clay liners and compacted clay liners (Malusis et al. 2003, 2017; Manassero and Dominijanni 2003; Neuzil and Provost 2009; Bohnhoff and Shackelford 2013; Shackelford and Scalia 2016). Formulations for coupled solute transport are presented in Yeung (1990), Malusis and Shackelford (2002), Manassero and Dominijanni (2003), Malusis et al. (2012), Dominijanni et al. (2013), and Shackelford et al. (2019).

The transient (time-dependent) advective–dispersive–reactive transport of a solute through a porous medium can be considered on the basis of conservation of mass for a representative elementary volume (REV), as illustrated in Figure 1-5 (e.g., Freeze and Cherry 1979). An REV is defined as the smallest volume of a porous medium that allows application of the continuum approach to flow and transport through a porous media (Bear 1972, 1979; Bear and Verruijt 1987).

Conservation of mass can be written mathematically using the continuity equation:

$$\frac{dm}{dt} = -\nabla \cdot J \pm R \tag{1-14}$$

where m [ML^{-3}] is the total mass (adsorbed plus liquid phases) of solute per unit total volume (solids plus voids), R [ML^{-3}T^{-1}] represents a general source (> 0) or sink (< 0) term emanating from chemical, biological, and/or radioactive reaction(s), and ∇ is the mathematical symbol representing the divergence or vector gradient operator (i.e., $\nabla() = (\partial()/\partial x)\hat{i} + (\partial()/\partial y)\hat{j} + (\partial()/\partial x)\hat{k}$, where $\hat{i}$, $\hat{j}$, and $\hat{k}$ are the unit

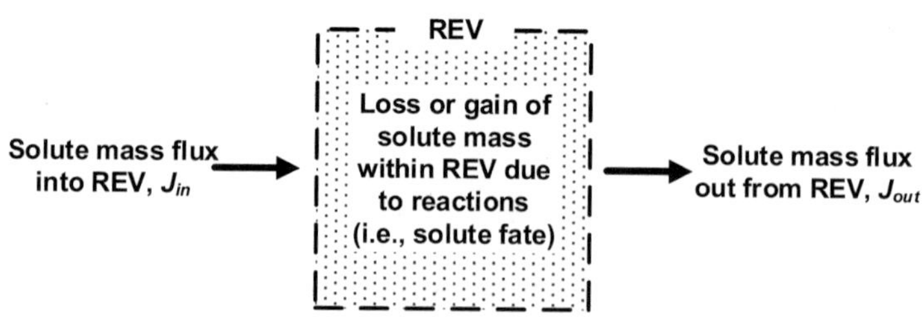

Figure 1-5. Conceptualization of the net rate of solute mass change within a representative elementary volume (REV) of a porous medium.

vectors in the x, y, and z directions, respectively). For saturated porous media considered in this section, the total mass of contaminant per unit total volume can be defined as follows:

$$m = nc + \rho_d S \quad (1\text{-}15)$$

where ρ_d [ML^{-3}] is the dry density of the soil defined as the dry mass of solids per unit total volume of the soil, and S is the solid-phase concentration of the contaminant [MM^{-1}], defined as the mass of contaminant on the solid phase per unit mass of the dry solid phase. Note that, in the groundwater literature, ρ_d is often referred to as the dry-bulk density or simply the bulk density.

In practice, the following assumptions are often made to simplify the governing (continuity) equation for transient contaminant transport:

- Coupled flow processes are ignored;
- Transport occurs in only one direction;
- Liquid flow is at a steady state;
- Porous media is homogeneous, isotropic, and nondeformable;
- Liquid is incompressible and contains only trace concentrations of solutes, such that changes in the liquid density caused by changes in solute concentrations are negligible; and
- Only linearly reversible, equilibrium reactions are included.

Based on these assumptions and incorporating J as defined previously, the continuity equation can be simplified to the following form (Freeze and Cherry 1979):

$$\frac{\partial c}{\partial t} = \frac{D_h \partial^2 c}{R_d \partial x^2} - \frac{v_s \partial c}{R_d \partial x} - \Lambda c \quad (1\text{-}16)$$

where R_d [dimensionless] is the retardation factor, and Λ [T^{-1}] is a lumped decay constant that potentially incorporates multiple decay processes such as radioactive decay and biological degradation. The preceding equation is commonly referred to as the 1D advection–dispersion–reaction equation (ADRE) or, when v_s is low and D_m is negligible (i.e., $D_h \approx D^*$), the one-dimensional advection–diffusion–reaction equation (Shackelford 1993). For solutes that sorb to the solid phase, the retardation factor is greater than unity ($R_d > 1$), whereas for nonsorbing solutes, $R_d = 1$. As a result, sorbing solutes travel at a rate, v_r ($= v_s/R_d$) [LT^{-1}] that is slower relative to that for nonsorbing solutes (i.e., $v_r < v_s$).

Relatively simple analytical solutions to the 1D ADRE with different initial and boundary conditions as well as different reaction processes are provided in van Genuchten (1981) and van Genuchten and Alves (1982). These analytical solutions can be used to determine contaminant transit times (e.g., Shackelford 1990) and contaminant transport through engineered containment barriers (e.g., Visudmedanukul et al. 2007).

In most cases, 1D analytical solutions to the ADRE are insufficient because of factors such as complicated and irregular geometry preventing simplification to a 1D problem, nonuniform soil properties, and so on. As a result, numerical (e.g., finite-difference, finite-element) models are frequently used in practice to assess advective–dispersive–reactive contaminant transport, and numerous such models are nowadays readily available (Anderson and Woessner 2002). Commonly used models are as follows:

- MT3D (Zheng 1990, Zheng and Wang 1999): 3D transport model using MODFLOW (McDonald and Harbaugh 1988) that simulates advection, dispersion, and chemical reactions (for more information: https://hydro.geo.ua.edu/mt3d/index.htm).
- RT3D: Reactive transport model building on MT3D for multispecies reactive transport (Clement 1997). RT3D is particularly useful for modeling biodegradation processes (for more information: https://bioprocess.pnnl.gov/).
- HYDRUS: 1D/2D/3D model for analysis of water and solute transport and fate in saturated and unsaturated porous media (Šimůnek et al. 1998, 2013; Šimůnek and van Genuchten 2008). HYDRUS is particularly useful for modeling reactive transport in unsaturated soils (for more information: https://www.pc-progress.com).
- COMSOL: Fully coupled multiphysics software used for 1D/2D/3D analyses (Li et al. 2009). COMSOL is highly flexible, enabling an analysis of complicated processes (for more information: https://www.comsol.com).

The use of a numerical model requires a detailed site conceptual model. The quality of the output generated from a numerical transport model is only as good as the accuracy and representativeness of the input parameters and conceptual model simulated. Additional information and models can be found on the U.S. Environmental Protection Agency web site (USEPA 2023).

For *unsaturated porous media*, the same fundamental transport principles as described previously are valid, except that the addition of the gas phase (e.g., air) in unsaturated soils results in an additional flow regime (e.g., vapor transport via the gas phase), partitioning between vapor, liquid, and solid phases, and a reduction in the aqueous phase through which aqueous transport occurs (resulting in reduced k, D^*, and D_m). Chemical transport in unsaturated porous media is described in detail in Jury and Horton (2004).

Nonaqueous phase liquids (NAPLs) exist as a separate phase from water and air and, therefore, are often referred to as immiscible liquids. Nonaqueous phase liquids are differentiated on the basis of liquid density relative to that for water, that is, light nonaqueous phase liquids (LNAPLs) and dense nonaqueous phase liquids (DNAPLs), where LNAPLs have densities less than that of water, and DNAPLs have densities greater than that of water. For information on the transport of DNAPLs, refer to Huling and Weaver (1991) and Pankow and Cherry (1996). For information on the transport of LNAPLs, refer to Newell et al. (2015).

1.3 FATE

Chemical and biological mass transfer processes resulting in loss or gain of solute mass (or mobile solute mass) govern the fate of solutes during migration through porous media. These processes include sorption, dissolution/precipitation, radioactive decay, acid/base reactions, complexation, hydrolysis/substitution, redox reactions, and biological transformations (National Research Council 1990, Shackelford and Rowe 1998).

Sorption is the partitioning of a contaminant between vapor or aqueous phases and mineral or organic solids of the porous media (National Research Council 1990). Sorption can be subdivided into adsorption (partitioning onto the surface of the porous media) and absorption (partitioning into the solid). Sorption mechanisms can be complex (National Research Council 1990) but generally can be differentiated between cationic (positive charged) metal ions and neutral (uncharged) organic molecules. Metal cations can sorb to minerals via ion exchange, whereby ions initially in the pore water exchange for ions initially held electrostatically to a charged mineral surface. For sorption of metals, clays are particularly important because of the large surface area and net negative structural charge characteristic of clay minerals. Other minerals (specifically mineral surfaces) such as metal oxides or oxyhydroxides can bind metal-containing compounds to the mineral surface (surface complexation). Neutral organic molecules sorb to solid organic matter, or more specifically to organic carbon, primarily because of the hydrophobic nature of these chemicals via the process referred to as hydrophobic partitioning.

Numerous empirical and semiempirical models have been developed to describe sorption reactions. These models are incorporated in the ADRE through partition coefficients, K_p [L^3M^{-1}], which relate the solid-mass or sorbed (S) concentration of the contaminant to the aqueous-phase (pore-water) concentration (c) of the contaminant at equilibrium. General shapes of sorption equilibrium isotherms (a plot of S versus c at equilibrium and

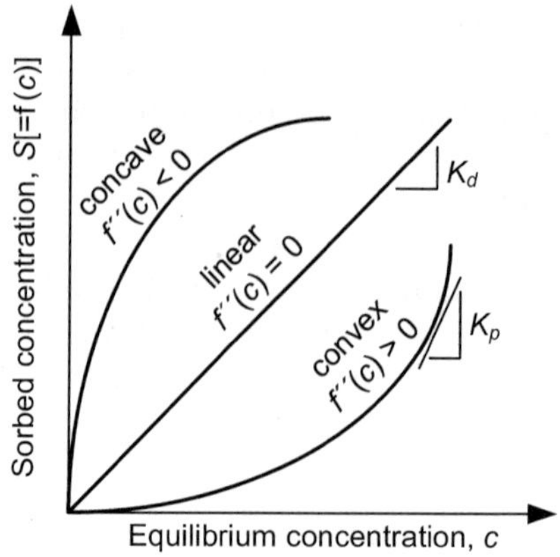

Figure 1-6. Conceptualization of sorbed versus equilibrium aqueous-phase concentration isotherms.
Source: Melnyk (1985), Shackelford and Daniel (1991).

constant temperature) are shown schematically in Figure 1-6. There are two general types of nonlinear adsorption isotherms (shown schematically in Figure 1-6), namely, concave isotherms, in which there is a limiting sorbed concentration, and convex isotherms, in which little sorption occurs at low concentrations, but no relevant sorption threshold exists (Melnyk 1985, Shackelford and Daniel 1991). Concave isotherms commonly are referred to as favorable isotherms, whereas convex isotherms are referred to as unfavorable, because for a given equilibrium concentration (c), the concave isotherm results in a greater sorbed concentration (S). Also, note that the definitions of *concave* and *convex*, as indicated in Figure 1-6, are mathematical based on the sign of the second derivative of the functional relationship [i.e., $f''(c)$]. Nonlinear sorption isotherms for soils are generally concave (favorable) because of a finite sorption capacity. In general, K_p represents the tangential slope of a nonlinear sorption isotherm evaluated at a specified equilibrium concentration, c.

In most applications, nonlinear sorption behavior is described by either the Freundlich sorption isotherm equation or the Langmuir sorption isotherm equation (Stumm and Morgan 1996). The Freundlich equation and corresponding partition coefficient are given in as follows:

$$S = K_f c^N; \quad K_p = K_f N c^{N-1} \tag{1-17}$$

where K_f [L^3M^{-1}] and N [dimensionless] are the Freundlich isotherm equation parameters. The Langmuir equation and corresponding partition coefficient are given as follows:

$$S = \frac{K_L M c}{1 + K_L c}; \quad K_p = \frac{K_L M}{(1 + K_L c)^2} \tag{1-18}$$

where K_L [L^3M^{-1}] and M [dimensionless] are the Langmuir isotherm equation parameters. When N is unity in the Freundlich sorption isotherm equation, the equation defaults to a liner expression, where $K_p = K_f$, and K_p is referred to more specifically as the distribution coefficient, K_d.

The sorption of hydrophobic organic contaminants at relatively low concentrations is often assumed to be linear, and for this condition, K_d is defined as follows:

$$K_d = f_{oc} K_{oc} \tag{1-19}$$

where f_{oc} [dimensionless] is the mass fraction of organic carbon in the soil defined as the mass of organic carbon per unit mass of dry soil solids, and K_{oc} [L^3M^{-1}] is the organic carbon partition coefficient (Karichoff et al. 1979). Often, K_{oc} is empirically approximated to the octanol–water partition coefficient, K_{ow} [L^3M^{-1}], of the organic compound (Griffin and Roy 1985). Octanol–water partition coefficients for simple organic compounds can be found in Sangster (1989). Sorption reactions are often partially or completely reversible and, thus, only act to store a contaminant onto or into the solid phase but do not permanently remove the contaminant.

Retardation factor values, reflecting the effect of sorption on transport, typically are determined using column or batch tests (Shackelford 1993). For batch tests, the value for R_d can be determined from the following relationship:

$$R_d = 1 + \frac{\rho_d}{n} K_p \tag{1-20}$$

The effects of material properties (R_d, D_m, D^*, k, and n) on the transport of sorbing solutes in porous media are illustrated in terms of concentration breakthrough curves in Shackelford (1993).

Dissolution/precipitation is the transformation of a contaminant from/to the solid to/from liquid phases (Stumm and Morgan 1996). Unlike sorption, dissolution/precipitation reactions are less reversible, and some precipitation reactions are effectively irreversible if the chemical environment remains generally constant (National Research Council 1990). Therefore, precipitation plays an important role in contaminant attenuation. Dissolution can be an important source of contaminants, as the chemical environment of the porous media is altered.

Radioactive decay is the irreversible transformation of a radionuclide (radioactive element) into a different atom through the loss of ionizing radiation from the nucleus of the radionuclide (National Research Council 1990), which results in a reduction in the concentration of the original radionuclide and an increase in the concentration of the decay by-product. Decay of radionuclide solutes can be represented by a simple rate law describing an exponential decrease in concentration with time. The half-life of a radionuclide caused by radioactive decay is the time required for the concentration of the radionuclide to reduce to half of the original value and is frequently used to quantify radioactive decay. When the half-life is less than or equal to the duration of transport from a source to a receptor, then radioactive decay may have a significant effect on contaminant migration. Otherwise, the impact of radioactive decay on contaminant transport may be minimal. Radioactive decay also results in by-products that may be of concern.

Radioactive decay is generally represented as a source/sink term in the ADRE. For example, R [$ML^{-3}T^{-1}$] for radioactive decay for a saturated porous medium (van Genuchten and Alves 1982) can be defined as follows:

$$R = -\lambda_w nc - \lambda_s \rho_d S \tag{1-21}$$

where λ_w [T^{-1}] is the first-order rate constant to describe radioactive decay of the contaminant in the aqueous phase, and λ_s [T^{-1}] is the first-order rate constant to describe radioactive decay of the contaminant on the solid phase.

Acid/base reactions are reactions that involve the loss or gain of protons (H^+); acids give up (donate) protons, whereas bases accept protons (Stumm and Morgan 1996). Acid/base reactions can affect chemical speciation and the susceptibility of a chemical to dissolution/precipitation because of changes in pH (National Research Council 1990).

Complexation reactions are the combination of anions and cations into more complex chemical species (Stumm and Morgan 1996). More complex chemical species can behave differently with respect to sorption and dissolution/precipitation reactions than the component ions. For example, complexation reactions can increase the mobility of metal ions over the range of pH common to natural groundwaters, because the complexed metal ions may not have the same affinity for sorption reactions (National Research Council 1990).

Hydrolysis/substitution reactions are abiotic transformation reactions that affect organic contaminants in groundwater (National Research Council 1990). Hydrolysis is the substitution of water (or a component of water) for a halogen (F^-, Cl^-, Br^-, I^-, At^-). Substitution reactions involve substitution of a halogen for another anion. The kinetics of hydrolysis/substitution reactions are complex. However, in some cases, hydrolysis/substitution reactions can be modeled as first-order degradation reactions (analogous to radioactive decay). Reaction products are often more susceptible to biodegradation and also more soluble (National Research Council 1990).

Redox, or oxidation/reduction, reactions refer to the transformation of the valence or an element caused by oxidation or reduction, and in dissolved elements, result in a change in valence (Stumm and Morgan 1996). Some redox reactions are reversible, whereas others are not (Lindberg and Runnels 1984). Redox reactions can play important roles in both the release and the attenuation (usually through precipitation) of metals.

Biological transformations are reactions controlled or facilitated by microorganisms; these reactions can result in the attenuation (destruction) of some organic contaminants. Biological transformations often play a key role in natural attenuation or enhanced natural attenuation remedies for contaminated groundwater. The rate of biological transformations is controlled by the abundance of microorganisms and the redox conditions within the porous media. Microorganisms that degrade both natural and man-made (xenobiotic) chemicals exist. Unsubstituted aromatics (phenols, benzenes, benzoates), nitrosubstituted aromatics, halogen-substituted aromatics, polycyclic aromatics, polychlorinated biphenyl (PCB), many pesticides, and phthalate esters are known to be biodegraded by aerobic bacteria (National Research Council 1990). Halogenated aliphatic solvents, most unsubstituted aromatics, some PCBs, and some pesticides are known to be biodegraded by anaerobic bacteria (National Research Council 1990). Further information on biological transformation can be found in Alexander (1985), Atlas (1981), Rittmann et al. (1988), Surampalli et al. (2004), and Scow and Hicks (2005), and broad information on fate, transport, and transformation guidelines is provided by the Environmental Protection Agency (USEPA 2008). Biological transformations are important processes for contaminant reduction, but similar to radioactive decay, can lead to undesirable by-products as contaminants are broken down.

REFERENCES

Anderson, M. 1984. "Movement of contaminants in groundwater: Groundwater transport-advection and dispersion." In *Groundwater contamination, studies in geophysics*, 37–45. Washington, DC: National Academy Press.

Alexander, M. 1985. "Biodegradation of organic chemicals." *Environ. Sci. Technol.* 19 (2): 106–111.

Anderson, M., and J. Cherry. 1979. "Using models to simulate the movement of contaminants through groundwater flow systems." *Crit. Rev. Environ. Control* 9 (2): 97–156.

Anderson, M., and W. Woessner. 2002. *Applied groundwater modeling, simulation of flow and advective transport*. San Diego: Academic Press.

Atlas, R. 1981. "Microbial degradation of petroleum hydrocarbons: An environmental perspective." *Microbiol. Rev.* 45 (1): 180–209.

Bear, J. 1972. *Dynamics of fluids in porous media*. New York: Elsevier.

Bear, J. 1979. *Hydraulics of groundwater*. New York: McGraw-Hill.

Bear, J., and A. Verruijt. 1987. *Theory and application of transport in porous media*. Dordrecht, Netherlands: D. Reidel.

Bohnhoff, G. L., and C. D. Shackelford. 2013. "Improving membrane performance via bentonite polymer nanocomposite." *Appl. Clay Sci.* 86 (2013): 83–98.

Clement, T. P. 1997. "A modular computer code for simulated reactive multi-species transport in 3-dimensional groundwater systems." *Tech. Rep. PNNL-11720*. Washington, DC: US Department of Energy.

Daniel, D. E. 1989. "In situ hydraulic conductivity tests for compacted clays." *J. Geotech. Eng.* 115 (9): 1205–1226.

Daniel, D. E. 1994. "State-of-the art: Laboratory hydraulic conductivity tests for saturated soils." In *Hydraulic conductivity and waste contaminant transport in soil*, edited by D. Daniel and S. Trautwein, 111–168. West Conshohoken, PA: ASTM.

Daniel, D. E., and C. D. Shackelford. 1988. "Disposal barriers that release contaminants only by molecular diffusion." *Nucl. Chem. Waste Manage.* 8 (4): 299–305.

Domenico, P., and F. Schwartz. 1990. *Physical and chemical hydrogeology*. New York: Wiley.

Dominijanni, A., M. Manassero, and S. Puma. 2013. "Coupled chemical–hydraulic–mechanical behaviour of bentonites." *Géotechnique* 63 (3): 191–205.

Freeze, R., and J. Cherry. 1979. *Groundwater*. Englewood Cliffs, NJ: Prentice-Hall.

Fried, J. 1975. *Groundwater pollution*. Amsterdam, Netherlands: Elsevier.

Griffin, R., and W. Roy. 1985. *Interaction of organic solvents with saturated soil-water systems*. Tuscaloosa, AL: Environmental Institute for Waste Management Studies, University of Alabama.

Hong, C., and C. Shackelford. 2017. "Long-term column testing of zeolite-amended backfills. II: Solute transport properties." *J. Geotech. Geoenviron. Eng.* 143 (9): 04017051.

Horton, R., M. Thompson, and J. McBride. 1985. "Estimating transit times of noninteracting pollutants through compacted soil materials." In *Proc., 11th Annual Solid Waste Research Symp.*, 275–282. EPA/600/9-85/013. Ames, IA: Iowa State University, Dept. of Agronomy.

Huling, S., and J. Weaver. 1991. *Ground water issue, dense nonaqueous phase liquids*. EPA/540/4-91-002. Washington, DC: Environmental Protection Agency.

Jury, W., and R. Horton. 2004. *Soil physics*. New York: Wiley.

Karichoff, S., D. Brown, and T. Scott. 1979. "Sorption of hydrophobic pollutants on natural sediments." *Water Res.* 13 (3): 241–248.

Kruseman, G., N. de Ridder, and A. Verweij. 1994. *Analysis and evaluation of pumping test data.* ILRI Publication 47. Wageningen, Netherlands: International Institute for Land Reclamation and Improvement.

Li, Q., K. Ito, Z. Wu, C. S. Lowry, et al. 2009. "COMSOL multiphysics: A novel approach to ground water modeling." *Groundwater* 47 (4): 480–487.

Lindberg, R., and D. Runnels. 1984. "Ground water redox reactions: An analysis of equilibrium state applied to Eh measurements and geochemical modeling." *Science* 225 (4665): 925–927.

Malusis, M., J. Scalia, A. Norris, and C. Shackelford. 2017. "Quantifying the significance of chemico-osmotic counter advection on solute transport through semipermeable clay barriers." In *Proc., 2nd Int. Symp. on Coupled Phenomena in Environmental Geotechnical.* Leeds, UK: University of Leeds.

Malusis, M., and C. Shackelford. 2002. "Theory for reactive solute transport through clay membrane barriers." *J. Contam. Hydrol.* 59 (3–4): 291–316.

Malusis, M., C. Shackelford, and J. Maneval. 2012. "Critical review of coupled flux formulations for clay membranes based on nonequilibrium thermodynamics." *J. Contam. Hydrol.* 138–139: 40–59.

Malusis, M., C. Shackelford, and H. Olsen. 2003. "Flow and transport through clay membrane barriers." *Eng. Geol.* 70 (3–4): 235–248.

Manassero, M., and A. Dominijanni. 2003. "Modelling the osmosis effect on solute migration through porous media." *Géotechnique* 53 (5): 481–492.

McDonald, M., and A. Harbaugh. 1988. "A modular three-dimensional finite-difference ground-water flow model." In Chap. A1 in *Techniques in Water-Resources Investigations*, Book 6. Reston, VA: US Geological Survey.

Melnyk, T. W. 1985. *Effects of sorption behavior on contaminant migration. Rep. No. AECL-8390.* Pinawa, MB: Atomic Energy of Canada Ltd., Whiteshell Nuclear Research Establishment.

Mitchell, J., and K. Soga. 2005. *Fundamentals of soil behavior.* 3rd ed. New York: Wiley.

National Research Council. 1990. *Ground water models: Scientific and regulatory applications.* Washington, DC: National Academy Press.

Neuzil, C., and A. Provost. 2009. "Recent experimental data may point to a greater role for osmotic pressures in the subsurface." *Water Resour. Res.* 45 (3): 1–14.

Newell, C., S. Acree, R. Ross, and S. Huling. 2015. *Ground water issue, light nonaqueous phase liquids.* EPA/540/S-95/500. Washington, DC: US Environmental Protection Agency.

Pankow, J. F., and J. A. Cherry. 1996. *Dense chlorinated solvents and other DNAPLs in groundwater*. Portland, OR: Waterloo Press.

Parker, B. L., J. A. Cherry, and P. Wanner. 2022. "Determining effective diffusion coefficients of chlorohydrocarbons in natural clays: Unique results from highly resolved controlled release field experiments." *J. Contam. Hydrol.* 250: 104075.

Perkins, T. K., and O. C. Johnson. 1963. "A review of diffusion and dispersion in porous media." *Soc. Pet. Eng. J.* 3 (1063): 70–84.

Peyton, G. R., J. P. Gibb, M. H. LeFaivre, J. D. Richey, et al. 1986. "Land disposal, remedial action, incineration, and treatment of hazardous waste." In *Proc., 12th Ann. Research Symp.*, 21–28. EPA/600/9-86/022. Washington, DC: US Environmental Protection Agency.

Rittmann, B., D. Jackson, and S. Storck. 1988. "Potential for treatment of hazardous organic chemicals with biological processes." In Vol. III of *Biotreatment systems*, edited by D. Wise. 15–64, Boca Raton, FL: CRC Press.

Robinson, R., and R. Stokes. 1959. *Electrolyte solutions*. 2nd ed. London: Butterworths Scientific.

Rowe, R. 1987. "Pollutant transport through barriers." In *Geotechnical practice for waste disposal*, edited by R. Woods, 159–181. Reston, VA: ASCE.

Sale, T., B. Parker, C. Newell, and J. Devlin. 2013. State-of-the-science review, management of contaminants stored in low permeability zones. *SERDP Project Rep. ER-1740*. Washington, DC: US Dept. of Defense.

Sangster, J. 1989. "Octanol–water partition coefficients of simple organic compounds." *J. Phys. Chem. Ref. Data* 18 (3): 1111–1229.

Schwartz, F., and H. Zhang. 2003. *Fundamentals of ground water*. New York: Wiley.

Scow, K., and K. Hicks. 2005. "Natural attenuation and enhanced bioremediation of organic contaminants in groundwater." *Curr. Opin. Biotechnol.* 16 (3): 246–253.

Shackelford, C. D. 1988. "Diffusion as a transport process in fine-grained barrier materials." *Geotech. News* 6 (2): 24–27.

Shackelford, C. D. 1989. *Diffusion of contaminants through waste containment barriers*, 169–182. Transportation Research Record 1219. Washington, DC: Transportation Research Board, National Research Council.

Shackelford, C. D. 1990. "Transit-time design of earthen barriers." *Eng. Geol.* 29 (1): 79–94.

Shackelford, C. D. 1991. "Laboratory diffusion testing for waste disposal—A review." *J. Contam. Hydrol.* 7 (3): 177–217.

Shackelford, C. D. 1993. "Contaminant transport." In Chap. 3 in *Geotechnical practice for waste disposal*, edited by D. Daniel, 33–65. London: Chapman & Hall.

Shackelford, C. D. 1994. "Waste–soil interactions that alter hydraulic conductivity." In *Hydraulic conductivity and waste contaminant transport in soil*, edited by D. Daniel and S. Trautwein, 111–168. West Conshohocken, PA: ASTM.

Shackelford, C. D. 2021. "Fundamental considerations for column testing of engineered, clay-based barriers." In Vol. 9 of *Proc., 3rd Int. Symp. on Coupled Phenomena in Environmental Geotechnics*, 441–460. Japanese Geotechnical Society Special Publication, International Society of Soil Mechanics and Geotechnical Engineering (ISSMGE), Kyoto.

Shackelford, C. D., and D. E. Daniel. 1991. "Diffusion in saturated soil. I: Background." *J. Geotech. Eng.* 117 (3): 467–484.

Shackelford, C. D., and C. Hong. 2020. "Comparative analyses of alternative breakthrough curves from cumulative mass column testing of soil–bentonite backfills." *Can. Geotech. J.* 57 (8): 1197–1214.

Shackelford, C. D., N. Lu, M. Malusis, and K. Sample-Lord. 2019. "Research challenges involving coupled flows in geotechnical engineering." In Chap. 9 in *Geotechnical fundamentals for addressing new world challenges*, edited by N. Lu and J. K. Mitchell, 237–274. Cham, Switzerland: Springer.

Shackelford, C. D., and S. Moore. 2013. "Fickian diffusion of radionuclides for engineered containment barriers: Diffusion coefficients, porosities, and complicating issues." *Eng. Geol.* 152 (1): 133–147.

Shackelford, C. D., and P. Redmond. 1995. "Solute breakthrough curves for processed kaolin at low flow rates." *J. Geotech. Eng.* 121 (1): 17–32.

Shackelford, C. D., and R. K. Rowe. 1998. "Contaminant transport modeling." In Vol. 3 of *Proc., 3rd Int. Congress on Environmental Geotechnics*, edited by P. Seco e Pinto, 939–956. Netherlands: Balkema-Rotterdam.

Shackelford, C. D., and J. Scalia. 2016. "Semipermeable membrane behavior in bentonite-based barriers: Past, present, and future." In *Proc., GeoVancouver 2016, 69th Canadian Geotechnical Conf.*, Vancouver, Canadian Geotechnical Society, Richmond, British Columbia.

Šimůnek, J., K. Huang, and M. T. van Genuchten. 1998. *The HYDRUS code for simulating the one-dimensional movement of water, heat, and multiple solutes in variably-saturated media*. Version 6.0. Research Rep. No. 144. Riverside, CA: US Salinity Laboratory, USDA, ARS.

Šimůnek, J., D. Jacques, G. Langergraber, S. Bradford, et al. 2013. "Numerical modeling of contaminant transport with HYDRUS and its specialized modules." *J. Indian Inst. Sci.* 93 (2): 265–284.

Šimůnek, J., and M. T. van Genuchten. 2008. "Modeling nonequilibrium flow and transport processes using HYDRUS." *Vadose Zone J.* 7 (2): 782–797.

Stephens, D. B., K.-C. Hsu, M. A. Prieksat, M. D. Ankeny, et al. 1998. "A comparison of estimated and calculated effective porosity." *Hydrogeol. J.* 6 (1): 156–165.

Stumm, W., and J. Morgan. 1996. *Aquatic chemistry, chemical equilibria and rates in natural waters*. 3rd ed. New York: Wiley.

Surampalli, R., S. Kee Ong, E. Seagren, J. Nuno, et al. 2004. *Natural attenuation of hazardous wastes*. Reston, VA: ASCE.

USEPA (US Environmental Protection Agency). 2008. "Test guidelines for pesticides and toxic substances: Series 835—Fate, transports and transformation test guidelines." Accessed February 21, 2024. https://www.epa.gov/test-guidelines-pesticides-and-toxic-substances/series-835-fate-transport-and-transformation-test.

USEPA. 2023. "EPA science models and research tools (SMaRT) search." Accessed February 21, 2024. https://www.epa.gov/land-research/models-tools-and-databases-land-and-waste-management-research.

van Genuchten, M. T. 1981. "Analytical solutions for chemical transport with simultaneous adsorption, zero-order production, and first-order decay." *J. Hydrol.* 49 (3–4): 213–233.

van Genuchten, M. T., and W. J. Alves. 1982. "Analytical solutions of the one-dimensional convective–dispersive solute transport equation." *Tech. Bulletin No. 1661*. Washington, DC: US Department of Agriculture.

Visudmedanukul, P., M. Kamon, and T. Katsumi. 2007. "Chloride transport through cement–bentonite barriers." *J. Geotech. Geoenviron. Eng.* 133 (2): 175–185.

Yaws, C. 2009. *Transport properties of chemicals and hydrocarbons: Viscosity, thermal conductivity, and diffusivity of C1 to C100 organics and Ac to Zr inorganics*. Norwich, NY: William Andrew Inc.

Yeung, A. 1990. "Coupled flow equations for water, electricity and ionic contaminants through clayey soils under hydraulic, electrical, and chemical gradients." *J. Non-Equilib. Thermodyn.* 15 (3): 247–268.

Zheng, C. 1990. *MT3D a modular three-dimensional transport model for simulation of advection, dispersion and chemical reactions of contaminants in groundwater systems*. Ada, OK: US Environmental Protection Agency, Robert S. Kerr Environmental Research Laboratory.

Zheng, C., and P. P. Wang. 1999. "MT3DMS: A modular three-dimensional multispecies transport model for simulation of advection, dispersion, and chemical reactions of contaminants in groundwater systems; documentation and user's guide." *Contract Rep. SERDP-99-1*. Washington, DC: US Army Corps of Engineers.

PART 2

WASTE AND SITE CHARACTERIZATION

CHAPTER 2

FIELD TESTING FOR GEOTECHNICAL CHARACTERIZATION OF MUNICIPAL SOLID WASTE

Neven Matasovic

2.1 INTRODUCTION

Recovery, processing, and testing of representative "undisturbed" or "intact" samples of municipal solid waste (MSW) is challenging. Therefore, in situ testing of MSW is the most reliable method for evaluation of its material properties, including unit weight, shear strength, stiffness, and hydraulic and thermal conductivities. In situ testing techniques are subject to limitations that may significantly affect their applicability and, consequently, the reliability of MSW parameters derived using these techniques. Because in situ testing and measurements can only be made after the waste is in place, they are directly applicable to the analysis of existing landfills. However, parameters derived from testing at existing landfills are used for design of new landfills.

In situ measurements of MSW behavior may be divided into direct and indirect measurements. Indirect measurements rely on engineering correlations and/or on back-analyses of observed behavior. In general, however, it is difficult to categorize a certain in situ test method as direct or indirect because each test method can directly measure certain waste characteristics but only indirectly measure other characteristics. For example, wave propagation surveys are a direct measurement of wave velocity but only an indirect measurement of stiffness; similarly, soundings with the standard penetration test (SPT), cone penetration test (CPT), and dilatometer test (DMT) can be used directly to identify variability in waste stratigraphy, perched liquid zones, and temperature and indirectly to assess the shear strength of MSW.

Field measurement techniques can be broadly grouped into nonintrusive, intrusive, and hybrid techniques. Nonintrusive techniques, as the name implies, do not mechanically penetrate the waste. They are mostly limited to seismological and geophysical measurements of wave propagation velocities, gross measurements of relatively large landfill surface deformations, and observations of landfill stability (or instability) for back-calculation of material properties. Intrusive measurements typically require drilling or probing and include in situ geophysical measurements, in situ testing, and internal measurements of waste mass deformations. Hybrid measurement techniques, as the name implies, combine aspects of nonintrusive and intrusive techniques.

This chapter builds on the previous work of Kavazanjian (2003) and Matasovic et al. (2011) and aims to provide a critical review of the current state-of-the practice of field methods as applied to the testing of MSW.

2.2 NONINTRUSIVE FIELD MEASUREMENT TECHNIQUES

2.2.1 General

Common nonintrusive techniques for field measurement of MSW characteristics include wave propagation velocity measurements, electrical resistivity surveys, surface measurements of settlement and lateral deformation, records of earthquake strong ground motions, ambient vibration (microtremor) measurements, and load testing. Nonintrusive techniques are generally classified as indirect methods for evaluation of MSW properties, as they generally rely on either correlation (e.g., electrical measurements for moisture content evaluation) or back-analyses (e.g., settlement measurements for compressibility evaluation). Although nonintrusive wave propagation velocity measurements yield a direct measurement of wave propagation velocity, they still require "inversion" (back-analysis) to develop a shear wave velocity profile and also require a knowledge of total unit weight to convert shear wave velocity to small strain shear stiffness (or shear modulus, or Poisson's ratio).

2.2.2 Wave Propagation Velocity Measurements

Nonintrusive wave propagation velocity measurement techniques include surface wave measurements, seismic refraction tests (SRT), and the measurement of ambient vibrations [e.g., microtremor analysis method (MAM), refraction microtremor (ReMi)]. The most common surface wave measurement technique currently in use in the United States is the spectral analysis of surface waves, or SASW (Stokoe et al. 1994). In this technique, two geophones are placed in-line at known spacings from a source of surface wave energy. The measurements of the phase of the surface waves at the two geophones are processed mathematically to develop a plot of surface wave velocity versus wavelength. The calculated surface wave velocity is assumed, on the basis of analytical studies, to correspond to the average surface wave velocity of the subsurface material between the two geophones over a depth of one-half wavelength. Based on an estimated value of Poisson's ratio, the plot of average surface wave velocity versus wavelength is then converted to average shear wave velocity versus depth. By looking at progressively longer wavelengths, a plot of shear wave velocity versus depth is developed from the data.

The surface wave energy source used with SASW [and its derivative, multichannel analysis of surface waves (MASW)] (Park et al. 1997) may simply be a falling weight or a hammer striking a board or plate. However, to generate higher energy levels and thus achieve deeper penetration, heavy equipment such as a bulldozer, a backhoe, or a triaxial truck-mounted shaker can be used to generate excitation. The University of Texas, Austin-owned 30-t shaker (T-Rex) is shown in Figure 2-1. This and similar shakers were used at several Southern California's landfills, including the Operating Industries, Inc. (OII), BKK landfill, and Lamb Canyon landfill, to measure shear wave velocity to a depth of more than 90 m (Matasovic and Kavazanjian 1998, Matasovic et al. 2018).

A more sophisticated (although not necessarily more accurate) application of the SASW technique is the use of an electromechanical vibrator (shaker) to generate surface waves of controlled frequency content. In general, a sinusoidal excitation is used and a sweep through the frequency range of interest is conducted, simplifying the processing and analysis of the data. This type of excitation has been referred to by some

Figure 2-1. High-force, three-axis shaker T-Rex (testing at Lamb Canyon Landfill).

authors as controlled surface wave or continuous surface wave (CSW) sounding (e.g., Kavazanjian et al. 1994, Bouazza and Kavazanjian 2000).

An important aspect of surface wave testing is that the "representative" volume to which the measurement applies increases with depth. In principle, the average surface wave velocity measured using this technique may be conceptualized as applying to a two-dimensional (2D) slice along a plane passing through the geophones with a width of one wavelength (i.e., approximately twice as long as the reference depth). Thus, the "deeper you go," the greater the "representative" volume. This may be advantageous if an attempt is made to characterize the gross behavior of the waste mass for site response calculations but may be disadvantageous when attempting to use the surface wave measurements to try to identify localized zones of anomalous stiffness within the waste mass.

Other limitations of the evaluation of stiffness using the surface wave measurement technique include the need to estimate a total unit weight for the waste material. A shear stiffness calculated from a shear wave velocity is directly proportional to the estimated mass density. Therefore, a 25% error in estimated unit weight would induce a 25% error in the calculated small strain stiffness. Poisson's ratio must also be estimated to convert surface wave velocity to shear wave velocity. However, uncertainty with respect to Poisson's ratio becomes a factor only when the value of Poisson's ratio exceeds 0.4 and approaches 0.5, and this occurs only when the waste approaches saturation. Thus, except for bioreactor landfills, high moisture waste (e.g., sludge and or sediment mixed with MSW) landfills, and unlined landfills where the waste is below the water table, estimating a value for Poisson's ratio should not introduce significant uncertainty into the derived values of shear wave velocity and shear stiffness.

Attempts were also made by Lai et al. (2002) to evaluate small strain damping from the results of surface wave velocity measurements. However, as damping evaluated in this manner is of limited use in engineering applications (e.g., can be used to calibrate strain-dependent damping within a small-strain range), this technique is not widely used.

In principle, SRT and reflection surveys can also be used to evaluate the shear wave velocity of MSW. One of the first attempts to use SRT surveys was by Woodward Clyde Consultants (WCC 1987) at the OII landfill. Subsequent SASW measurements by Kavazanjian et al. (1996, 2013) were consistent with these measurements within the (refraction) depth of penetration. However, the SRT only gives an average velocity over the full depth of the waste mass (often misses low-velocity zones and requires a knowledge of the depth of waste) unless there is a distinct (and defined) layering within the waste mass (i.e., a distinct impedance contrast, where impedance is the mass density times shear wave velocity). Furthermore, SRT surveys are generally more complex to set up and more expensive to conduct than surface wave measurements. For these reasons, SRT and reflection measurements are not generally used at MSW and hazardous landfill sites to evaluate shear wave velocity or shear stiffness.

2.2.3 Electrical Resistivity and Conductivity Surveys

Nonintrusive measurements of characteristic parameters (e.g., resistivity, conductivity, and/or inductance surveys) are sometimes employed to evaluate the homogeneity, consistency, and limits of a waste mass. Although these techniques are more commonly used at hazardous waste sites, they can also be used at MSW landfills. The primary benefit of these techniques is to identify anomalous zones (i.e., wet zones or areas with buried drums or metallic objects) for further investigation and not to evaluate the material (mechanical) properties of waste. In principle, characteristic electrical parameter measurements can be correlated to moisture content and waste composition. Such correlations are dependent on moisture content and composition (of both the waste and the pore fluid) and thus are highly site-specific. Furthermore, waste composition often changes spatially because of waste mass heterogeneity and temporally because of decomposition and deposition and/or dissolution of salts and metals within the waste mass and pore fluid (leachate). Therefore, these methods are best used to provide an indication of changes in waste composition (and hence properties) and moisture content in time and/or space. Once a change of interest is identified, supplemental investigation using another measurement technique is generally required to evaluate the nature of the change and its impact on material properties.

Electrical resistivity and conductivity surveys have been used in mapping numerous waste disposal sites. Soupios et al. (2007a) reported on the use of electrical resistance tomography (ERT) and electromagnetic measurements (EM) at the Fodele MSW landfill on the island of Crete. The measurements were used to define the depth and extent of the landfill as well as to monitor the spatial distribution of perched leachate zones within the landfill.

An example from the author's files of interpreted ERT measurements at a confidential site in Northern California is shown in Figure 2-2. The thick dashed line indicates the interpreted boundary between refuse (MSW), native materials, and landfill cover. These boundaries were confirmed by drilling, sampling, and laboratory testing. The resistivity of refuse materials ranged from 20 (green) to 50 (red) Ohm-meters.

Mondelli et al. (2007) used ERT in conjunction with resistivity piezocone (RCPTu) to identify contamination plumes and contaminant saturated zones at an MSW landfill in Bauru, Brazil. Guérin et al. (2004) tracked leachate recirculation at two sites by studying resistivity variations versus time using electrical surveys. A similar use of electrical resistivity and conductivity methods at landfill sites was also reported by Bernstone and

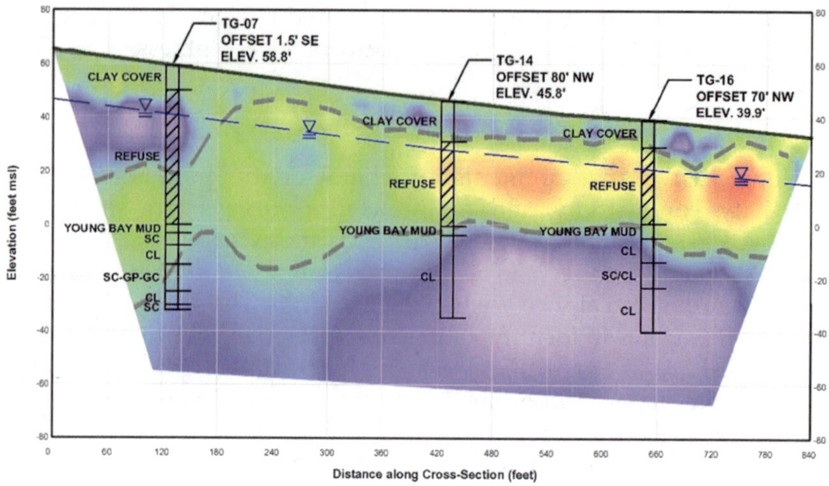

Figure 2-2. Extent of MSW as identified by ERT and confirmed by drilling. Source: Author's project files.

Dahlin (1996), Aristodemou and Thomas-Betts (2000), Stanton and Schrader (2001), Batayneh and Barjous (2005), Chambers et al. (2006), and Soupios et al. (2007b), among others.

2.2.4 Surface Deformation Measurements

Surface (and near-surface) deformation measurements can be conducted on macro and micro scales. Micro-scale deformation measurements are performed with a measurement tape. Macro-scale deformation measurement techniques include surveys of settlement monuments and settlement plates, light detection and ranging (LiDAR) imaging, and interpretation of unmanned aerial vehicle (UAV), aerial photography, and even interpretation of InSAR, satellite imagery. The LiDAR imaging conducted by using a computer-controlled laser beam over preset time intervals is presently considered the most accurate measurement applicable at the macro scale. This technique along with settlement surveys conducted with traditional methods (level) theoretically may be used to evaluate the mechanical properties of MSW through a back-calculation of settlement parameters and/or from an observed bulging of landfill slopes. The application of LiDAR is hampered by its relatively high cost. An alternate approach is the development of DEMs using digital imagery and the results have been shown to be comparable to aerial LiDAR scanning. Details about the application of UAV for measurement of waste settlement using imagery are provided in Champagne et al. (2020).

2.2.5 Strong Ground Motion Measurements

A site-specific measurement of earthquake-induced strong ground motions is a specialized type of nonintrusive field measurement. The results of such a measurement, when available, can be used to evaluate, by back-analysis, the dynamic stiffness and damping of the waste mass. In general, two strong motion stations are required: one on the waste mass and one on (level) ground far enough from the landfill not to be influenced by the presence of the waste mass but close enough to be representative of the ground motions that would have occurred at the site if the landfill was not there. A back-analysis of low-intensity ground motions (from low-intensity nearby earthquakes or larger but distant earthquakes) recorded at the site will yield information on small strain shear wave velocity and stiffness (which can be compared to surface wave and other types of wave propagation velocity surveys) and, to some extent, on layering within the landfill. A back-analysis of high(er)-intensity ground motions will yield information on strain-dependent modulus reduction, damping, and Poisson's ratio within the waste mass, provided that waste thickness and unit weight profile are known or can be reasonably estimated.

Although strong motion instruments have now been installed at several landfills in Southern California, usable strong ground motion records are available only from the OII and Puente Hills landfills. As reported by Matasovic and Kavazanjian (1998), a back-analysis of a series of small nearby earthquakes and a relatively large but distant earthquake (the 1992 Moment Magnitude M 7.3 Landers Earthquake) recorded at the OII landfill helped identify the presence of a continuous layer of interim soil cover at a depth beneath the strong motion station located on the waste mass. A back-analysis of strong ground motion data recorded on the top deck of OII landfill (Matasovic et al. 1995, Matasovic and Kavazanjian 1998) yielded valuable information on strain-dependent modulus reduction and damping of the landfilled waste.

2.2.6 Shear Strength Estimation Based on Stability Observations

Observations of waste fill side slopes, along with observations of trenches and excavations in waste (Figure 2-3), are useful for back-calculation of design MSW shear strength. Back-analysis may be performed on the basis of observations of "unstable" (i.e., failed or near incipient failure) and "stable" (i.e., no visible signs of instability, such as bulging at the toe and/or cracking) waste fill slopes. Studies documenting this approach include Kavazanjian et al. (1995), Matasovic et al. (1998a), Eid et al. (2000), Merry et al. (2005), Blight (2008), Stark et al. (2009), Jafari et al. (2013), and Bonaparte et al. (2020). Depending on the intended use, back-analysis results may be combined with other results from the same site or

Figure 2-3. Excavation in MSW (Azusa Landfill, Southern California).

combined with and/or compared to the results of other studies (e.g., Kavazanjian et al. 1995) to develop a low-bound envelope or an average value for forward design.

A back-analysis of "unstable" slopes is generally a more reliable approach (versus a back-analysis of "stable" slopes) to estimate the shear strength of waste. This is because the factor of safety is assumed to be near 1 (FS = 1.0) (i.e., imminent failure), the shape of the failure surface can be reasonably inferred (from field observations supplemented with drone and/or LiDAR surveys), and the unit weight and cohesion of waste, as described in Chapter 5, are reasonably well defined. In some cases, more detailed assumptions for waste fill properties may be warranted (e.g., use of unit weight and cohesion distributions for back-analysis of deep-seated failures that reach old waste) (Kavazanjian et al. 2013). Care should be taken to ensure that the assumptions used in the back-analysis are adequately representative of field conditions.

Although an "unstable" slope suitable for back-analysis may be available in some cases (e.g., for remedial grading and/or buttressing an existing waste slope failure), such a condition may not be available in other cases. For these cases, historical failures (e.g., see failure in MSW shown in Figure 2-4) and/or failures at comparable sites (e.g., landfill within the same region, with waste fill of a similar age and nature) may be convenient for back-analysis. Bonaparte et al. (2020) estimated that waste fill failures

Figure 2-4. 2022 Failure in MSW at the Novy Svet (New World) Landfill near Saint Petersburg, Russian Federation.

occur in the United States at a frequency of "several per year" and noted that there might be additional stability incidents they are not aware of.

Alternatively, a back-analysis of "stable" landfill waste slopes is occasionally performed as well. Applications may include landfill sites where stability analysis using literature shear strength values and/or other conservative assumptions (e.g., high liquid levels and gas pressures) suggests impending failure despite observations to the contrary (e.g., calculated FS values below unity in conflict with the absence of any visible signs of instability in the field). This analysis may be performed, for example, for old, unlined waste piles with relatively steep side slopes and relatively thick waste piles. A back-analysis of "stable" landfill waste slopes is also applicable in areas of high seismicity, where past seismic loading can be estimated, and observational data are available. However, observational data on key parameters for such assessments (i.e., the maximum permanent seismic displacement) are rare and may not be reliable (e.g., Matasovic et al. 1998a).

2.3 INTRUSIVE FIELD MEASUREMENT TECHNIQUES

2.3.1 General

Intrusive techniques for field measurement of MSW characteristics are in situ procedures that require physical penetration through the landfill cover into waste. Available intrusive techniques include borehole (and test pit) sampling for compositional analysis and moisture content evaluation, borehole (and test pit) testing, in-hole density (i.e., unit weight) testing, down-hole, cross-hole, and in-hole wave propagation velocity surveys, pressuremeter and SPT soundings, CPT and DMT probes, in situ direct shear tests, various down-hole and/or internal electrical and nuclear measurements, and internal deformation measurements using in-hole geotechnical instrumentation such as inclinometers, tilt meters, and extensometers. These are referred to as "indirect" tests because parameters measured must be related back to the material properties of the representative volumes through either engineering correlation or by a back-analysis of observed behavior.

2.3.2 Borehole Sampling and Testing

2.3.2.1 Types of Borehole Tests. A variety of different types of field measurements can be made on samples recovered from boreholes or on the waste mass surrounding a borehole. Compositional classification of bulk samples recovered from a borehole is perhaps the most common type of field characterization test. In situ MSW moisture content and temperature are also commonly evaluated on recovered bulk samples. If the bulk sample is recovered from a relatively wet zone within the waste mass, it is not known whether the measured moisture content is representative of the waste mass field capacity. Often, perched liquid conditions are present within waste mass. Moisture content of samples recovered from such areas are likely close to or at saturation, (i.e., well above the field capacity of MSW).

In situ tests that can be conducted within a borehole advanced into a waste mass include density (total unit weight), standard (and nonstandard) SPT soundings, Becker penetration testing (BPT), plate load testing, (self-boring) pressuremeter tests, and various geophysical soundings including down-hole and in-hole velocity measurements, electrical conductivity and capacitance measurements, and nuclear measurements of gamma radiation and neutron absorption and scattering. In addition to boreholes advanced for the purpose of sampling and in situ testing, boreholes may also be required to install internal measurement devices such as inclinometers, tilt meters, extensometers, piezometers, temperature probes, and/or specialty devices such as in-hole strong motion instruments (accelerometers).

2.3.2.2 Complications Inherent to Borehole Sampling and Testing. Obstructions and large pieces of solid waste may necessitate special drilling equipment and can, in the worst case, prevent drillers from reaching target drilling depth. If a recovery of relatively intact pieces of solid waste is required (e.g., for compositional analysis or for reconstitution of specimens in the laboratory), a large-diameter bucket auger (e.g., 750 mm diameter) may be required. In situ density testing also requires a relatively large-diameter boring to provide sufficient accuracy. However, a relatively small-diameter hole (e.g., 100 to 150 mm diameter) may be required for down-hole pressuremeter or wave propagation velocity testing and for installation of inclinometers, tilt meters, and extensometers. Drilling mud is rarely used when drilling in waste because of concerns about excessive fluid loss and the introduction of additional liquids into the waste mass. Therefore, if borehole stability is a concern, methods such as air rotary casing hammer (ARCH) drilling may be required to advance a borehole in waste. All these considerations can significantly increase the cost of borehole drilling and sampling and down-hole testing at a landfill site.

2.3.2.3 Health and Safety Considerations. Intrusive investigation at landfill sites (e.g., advancement of boreholes for sampling, testing, and/or instrumentation) is complicated by various health and safety concerns posed by the presence of potentially hazardous substances within the waste mass. Potential safety concerns include a combination of acute hazards (e.g., fire, explosion, asphyxiation) and/or chronic hazards (e.g., longer-term health impacts). Appropriate protective measures are required, and these can significantly increase the cost of borehole sampling and testing compared to "conventional" (nonenvironmental) geotechnical testing.

Landfill gas components mobilized during investigation may include methane, hydrogen sulfide, and volatile organic compounds, which may pose explosion and health hazards. Continuous air quality monitoring at the head of the borehole and in the breathing zone is generally required during intrusive investigation, at a minimum. If hazardous emissions are present, the borehole may require ventilation and workers in the vicinity of the borehole may require additional personal protective equipment (PPE) (e.g., respiratory protection).

Dermal contact or ingestion of waste materials may also pose health hazards. Protective clothing (e.g., a combination of gloves, boot covers, and/or Tyvek coveralls) may be required in addition to standard field attire. Appropriate decontamination procedures should be followed to minimize the risk of ingestion (e.g., transference to food items, etc., during breaks) or off-site transport.

2.3.2.4 Disposition of Investigation Derived Waste. Another consideration associated with drilling (and trenching) in waste is the disposition of investigation-derived waste (IDW). The IDW includes drill cuttings, liquid, and drilling mud. If investigation is performed at a landfill with active landfilling operations, IDW may be disposed of on-site at the working face. However, if off-site disposal is required, a suite of environmental testing is required. Depending on the test results, material may be classified as "hazardous waste" and require special manifests, shipping, and disposal in an appropriate landfill facility.

2.3.2.5 Borehole and Test Pit Sampling. Bulk samples of MSW recovered from conventional and bucket-auger boreholes, and test pits are often visually classified and tested for moisture content in the field. Although standard visual classification and correlation schemes do not exist for MSW, several firms have their own formal in-house classification systems, whereas engineers and researchers sometimes employ their own informal correlations to relate waste composition to material properties (e.g., Geosyntec 1996; Matasovic et al. 1998b; Dixon and Langer 2006, 2008; Zekkos et al. 2010). The objective of waste classification, and thus the basis for development of a waste classification scheme, is to capture the characteristics of the waste that are expected to influence its behavior (e.g., its mechanical and physical properties). These characteristics are generally considered to include the types and percentages of material present (e.g., paper, plastic, wood, fibers, soil, soil-like material), the state of decomposition of the material, and the moisture content (on a relative basis, from very dry to very wet). The detailed description of a scheme for field classification of MSW, including field forms, can be found in Matasovic et al. (2011).

2.3.2.6 Borehole and Test Pit Unit Weight Testing. To measure in situ unit weight at four MSW landfill sites in California, Geosyntec (1996) developed a borehole unit weight (density) test. This test was patterned after the sand cone density test (ASTM D1556) (ASTM 2007) that is commonly used in earthwork engineering. In this borehole density test, a bulk sample of waste is recovered from a borehole using a large-diameter bucket auger. Upon removal, the sample is carefully weighed, and the corresponding borehole interval is backfilled with a calibrated gravel or rock. The gravel (or rock) is tremied through a pipe of fixed length into the bottom of the borehole to minimize intrusion of waste from the borehole sides. The gravel (or rock) is calibrated by tremieing it into a 227 L drum through the same fixed-length pipe used in the field. Several trials are used to establish the expected value and variability of the calibration factor. The weight of gravel tremied into the borehole is monitored [by before and after measurements of the weight of the bin(s) from which the gravel is taken] and then divided by the calibration factor to calculate the volume of the sampling interval. The weight of sample removed from the sampling interval is divided by the sampling interval volume to obtain the in situ unit weight. Several investigators have used a similar procedure to obtain the unit weight of MSW from a test pit, except that the test pit is generally lined with a plastic sheet and filled with a measured volume of water to determine test pit volume similarly to the method used to determine the density of rock and soil in place according to ASTM D5030 (ASTM 2004). The procedures and results from several such investigations are summarized in Zekkos et al. (2006).

When it was first used at the OII landfill (Matasovic and Kavazanjian 1998), the in situ unit weight test yielded, at the time, unexpectedly large values for unit weight. An average value measured across 45 m test interval was 15.7 kN/m^3. This value was further confirmed within top 6 m of the profile by an in situ density test conducted in a large test pit that was 170 m^3 in volume. The total unit weight of the test pit was approximately 15 kN/m^3 (i.e., within about 95% of the value measured in the borehole). On the basis of postfailure large-scale field testing of refuse debris at the Crossroads Landfill, Maine, Reynolds (2015) reported an average in situ unit weight of MSW of 15.1 kN/m^3. The high unit weight of the waste was attributed to a high percentage of cover soil in the waste coupled with high moisture content caused by the lack of leachate collection.

2.3.2.7 Borehole Hydraulic Conductivity Testing. Several in situ testing methods have been used to assess the hydraulic conductivity of waste utilizing boreholes, wells, and/or temporary piezometers. These tests are generally carried out by a monitored injection or pumping of fluids into or out of the borehole. The hydraulic conductivity of waste is then back-calculated using measured hydraulic heads and flow rates. Hence, such tests are considered to provide an indirect measurement of hydraulic conductivity. In situ permeability tests that have been used in measuring the hydraulic conductivity of waste include slug tests, pumping tests, and "constant head" and "falling head" borehole permeameter tests.

Slug tests and pumping tests are generally used to evaluate the hydraulic properties of aquifers. Therefore, these tests may be conducted on waste if the waste is submerged. This condition may be encountered in unlined landfills if portions of the waste are below the groundwater table or in well-functioning lined landfills with high rates of leachate recirculation. The presence of standing liquid in a landfill borehole (or in a landfill gas well) is not necessarily an indication of the presence of a continuous phreatic surface within the waste mass.

In a slug test, water is rapidly added or removed from the screened well installed in a borehole, followed by a monitoring of the change in hydraulic head with time. Results of slug tests are representative of the area within proximity to the screened zone and are generally affected by the packing material in the well adjacent to the screen. Pumping tests are conducted at either a constant rate or at multiple-step rates. Draw down in the borehole and at surrounding observation wells at a given pumping rate is used to back-calculate the hydraulic conductivity. Ettala (1987) and Oweis et al.

(1990) used the Jacob and Theis methods, respectively, to calculate the saturated hydraulic conductivity of MSW.

The borehole permeameter (e.g., Reynolds 1993, Jain et al. 2006) may be used for in situ measurement of the saturated hydraulic conductivity of both soils and waste. Water is injected into the borehole and is maintained at a constant level above the ambient phreatic surface in the soil or waste. Steady-state conditions are reached when the infiltration rate becomes constant. Analytical methods are used to back-calculate the saturated hydraulic conductivity of the soil, or waste, on the basis of the measured head and infiltration rate.

Shank (1993) estimated the saturated hydraulic conductivity of MSW from a series of slug tests performed in flooded vertical landfill gas wells (waste age > 10 years; the site was in Florida; calculations were performed using the Bouwer and Rice 1976 method). Oweis et al. (1990) conducted pump tests at an unlined MSW landfill and used standard techniques for analyzing aquifer pump test data to estimate the transmissivity of the landfilled waste and then calculated the hydraulic conductivity assuming saturated waste thickness. Wysocki et al. (2003) performed a variety of pumping tests (variable rate, constant rate, and recovery tests) at an MSW landfill in New York and reported the hydraulic conductivity of the landfilled waste. However, the pumping test analysis method, density, and the age of the waste were not provided.

Jain et al. (2006) provided the most comprehensive summary of relevant past studies and, building on that experience, performed borehole permeameter tests at 23 representative locations at an MSW landfill in central Florida designed as a bioreactor. At each location, three to four injection wells were constructed for a total of 77 injection wells. The locations were considered representative of relatively "fresh" (i.e., up to 8 years old) mixed residential and commercial waste. The maximum waste thickness was 22 m. Elapsed time to achieve a steady-state condition varied widely, with some wells not reaching steady state after 10 weeks of injection. Hydraulic conductivities were back-calculated using three different methods at different depths (confining stresses). The values ranged from 5.4×10^{-6} to 6.1×10^{-5} cm/s, with the lowest value associated with the highest confining stress. The values reported by Jain et al. (2006) are on the low end of the range of values reported by some of the other studies cited in this chapter. In situ hydraulic conductivities as high as 0.25 cm/s were reported by Ettala (1987).

A common limitation of the testing methods discussed herein is that the analytical procedures used for back-analysis of hydraulic conductivity assume a homogeneous waste mass with equal vertical and horizontal hydraulic conductivities. As described in Chapter 5, this is not generally the case as the waste is anisotropic. Therefore, the back-calculated hydraulic conductivity is considered an approximate value influenced by three-dimensional variations in hydraulic conductivity over the volume of material tested. Benson and Wang (2000) used a two-stage borehole permeameter (TSB) to evaluate the in situ hydraulic conductivity of paper sludge (the TSB is also called the *Boutwell permeameter*). The tests allow for back-calculation of both vertical and horizontal hydraulic conductivities. However, TSB tests have not been reported on MSW, and many argue that it would be impossible to perform this test in MSW.

2.3.2.8 Borehole Wave Propagation Velocity Testing. Wave propagation velocity testing can be conducted in a borehole using cross-hole, down-hole, and/or in-hole methods. Cross-hole testing can be conducted if there are two or more boreholes spaced close enough together for a down-hole hammer to generate a signal of sufficient strength to be detected in the adjacent borehole (typically limited to a spacing of about 5 to 10 m and requires borehole survey for verticality). Down-hole also measures average velocity and testing is limited to the depth at which a surface excitation can be detected within the boring, generally on the order of 30 m, whereas cross-hole testing and the OYO P-S suspension logging method (OYO) are not depth-limited. ASTM and EUROCODE 7 standards are available for down-hole and cross-hole testing. The standard guide for planning and conducting borehole geophysical logging is available as ASTM D5753 (ASTM 2018).

In both down-hole and cross-hole methods, the distance between the signal generation point and the signal detection point is divided by the time interval between signal generation and signal detection to obtain the wave propagation velocity. In the case of cross-hole testing, three borings can be used, and the measurement interval may then also be between two different detection points. Down-hole testing is generally restricted to shear wave velocity testing, whereas cross-hole testing can yield both shear and compressional wave velocities.

OYO is the best known and most widely used in-hole wave propagation velocity testing method. The OYO calls for a borehole to be filled with borehole fluid and an in-hole source applied as a rapid load against the casing. The method can be used in either cased or uncased boreholes. For best results, the uncased borehole must be between 10 and 20 cm in diameter. A cased borehole may be as small as 75 mm in diameter, if properly grouted. Geophones suspended in the borehole fluid sense the (grouted casing) deformations as the wave propagates upward in the soil (waste) around the casing. Thus, by suspending two geophones at a known distance apart (typically 1 m), wave propagation velocity can be measured. This method can yield both shear and compressional wave velocity. In-hole testing is not depth-limited (testing in boreholes as deep as 600 m has been reported by a specialty contractor).

The seismic dilatometer test (SDMT) can be conducted in-hole as well (no casing should be used). This test is, however, rarely used.

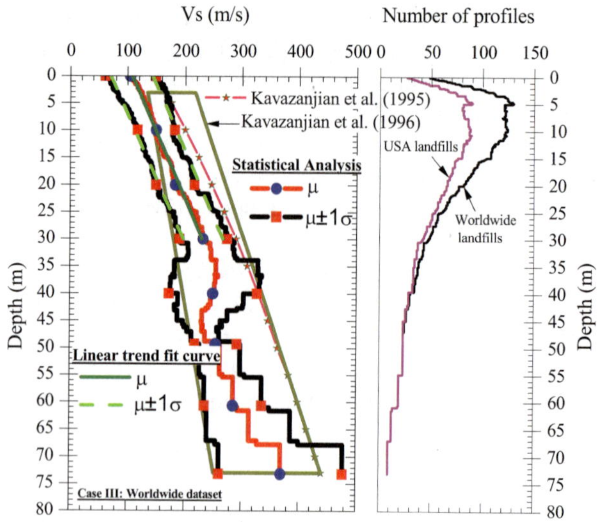

Figure 2-5. Shear wave velocity in MSW: Worldwide data.
Source: Ramaiah et al. (2016).

Ramaiah et al. (2016) and Zekkos et al. (2014) compiled the results of worldwide measurements of shear wave velocity in MSW. The processed and interpreted results, shown in Figure 2-5, combine the SASW, CSW, MASW, MAM, ReMi, and SRT measurements with their intrusive counterparts, including OYO suspension logging, SDMT, down-hole, and cross-hole soundings. With the exception of measurements performed close to the ground surface (top 20 m) and on relatively young waste, the results fall within the range of values recommended by Kavazanjian et al. (1996) (also included in Figure 2-5) that are still commonly used for seismic design of landfills and landfill covers. Zekkos et al. (2014) pointed out the similarities of shear wave velocity profiles in different regions (e.g., Michigan versus Southern California).

As both shear and compressional wave velocities can be recorded by in-hole and cross-hole measurements, the in situ small-strain Poisson's ratio can be computed. This is advantageous over the nonintrusive wave propagation measurements presented in Section 2.2.2. Using the OYO suspension logs from the OII Landfill, Matasovic and Kavazanjian (1998) estimated that Poisson's ratio of MSW ranges from 0.33 to 0.4 (a constant value of more than 90 m of the MSW profile).

2.3.2.9 Borehole Electrical and Nuclear Testing. A variety of electrical and nuclear measurements can be conducted in a borehole advanced through MSW, including "conventional" (i.e., such as those used in the oil exploration industry) down-hole electrical resistivity logging, and gamma radiation and neutron probe logging. The advantages and disadvantages of these measurements when made down-hole are similar to those associated with nonintrusive surface measurements using the same or similar techniques: because of "calibration" issues, they are most reliable as an indicator that something is "different" (e.g., they may indicate that there is a change in waste composition) or that conditions have changed over time (e.g., moisture content change) rather than as a means of measuring the absolute value of a parameter (i.e., of moisture content or waste composition). "One-time" measurements in a borehole or series of boreholes are generally used to indicate spatial differences in composition (i.e., as an index of stratigraphy), whereas measurements taken at the same location at different times are generally used as an index of temporal change (e.g., an increase in moisture content).

Yuen et al. (2000) reported on the use of neutron probes for monitoring moisture content changes with time in a bioreactor landfill. On the basis of a survey of available methods, they concluded that despite calibration difficulties and other limitations of the method, neutron probe measurements through "permanent" access tubes installed in the waste was the best method to monitor moisture content changes over time within the waste mass.

2.3.2.10 Borehole Pressuremeter Testing. The use of a pressuremeter to evaluate the mechanical properties of MSW was reported by Dixon et al. (1999, 2006). These authors employed an 83 mm diameter, 1.2 m long self-boring pressuremeter in a "pre-drilled" borehole. A "Chinese lantern" consisting of stainless-steel strips bonded to a rubber membrane was placed around the cylindrical probe to protect the 1.25 mm polyurethane inner membrane from damage during probe advancement in MSW. Once the probe fully penetrated the bottom of the borehole and reached the desired test depth, the inner membrane was expanded radially and radial deformation versus expansion pressure was measured and monitored. It should be noted that, in principle, this type of testing could be performed entirely in the self-boring mode (i.e., without the aid of a borehole) and thus avoid possibly "poor" contact between the membrane and MSW. However, it seems unlikely that a relatively small device such as a self-boring pressuremeter could penetrate any appreciable distance in a typical MSW landfill without reaching "refusal."

The pressuremeter test calls for several cycles of unloading and reloading. Interpretation of the test results typically yields values for the coefficient of lateral earth pressure at rest (K_0), the shear modulus at a reference cavity strain (typically 1%) and an indication of shear strength. It should be noted that the modulus values are evaluated at a relatively high strain level and therefore are generally lower than their counterparts evaluated from shear wave velocity or by back-calculation. The interpretation of MSW shear strength from pressuremeter testing is difficult, as radial strains induced in this test, even though relatively large, rarely approach failure strains.

2.3.2.11 Borehole Standard Penetration Test and Becker Penetration Test. Both the SPT and the BPT have been conducted at MSW landfills mostly for sample recovery and visual field and/or laboratory classification. Gabr and Valero (1995) used the results of SPT sounding as indicators of waste mass consistency. The results of the BPT were used as means for evaluating the properties of buried soil layers (e.g., for evaluating the properties of intermediate and interim covers within the waste mass). At older landfills, where large-particle waste was disposed of (e.g., large pieces of concrete, appliances), multiple trials with the SPT and/or use of the BPT were required to achieve target depths.

Attempts were made to interpret the SPT and BPT test logs in MSW on the basis of standard engineering correlations. An example of such an attempt (i.e., of an attempt to estimate the shear strength of MSW using correlations) can be found in Gabr and Valero (1995). However, this attempt to estimate shear strength was not validated on a site- (or waste-) specific basis. Until (if ever) more validation studies become available, the use of the SPT and BPT for evaluation of MSW properties is not recommended for deriving strength estimates but may be useful to interpret variations in buried waste.

2.3.2.12 In Situ Direct Shear Tests. Richardson and Reynolds (1991) and Houston et al. (1995) conducted large-scale in situ direct shear tests on MSW. The general procedure for these tests consisted of isolating a "pedestal" of waste by excavation and trenching, confining the "pedestal" of waste in some type of rigid container (box) split along a horizontal plane, fixing the bottom of the box against lateral deformation, loading the top of the box with the desired normal load, and then shearing the top of the box laterally while measuring both the shear force and the lateral displacement.

The advantage of an in situ direct shear test is that a relatively large volume of MSW (typically $> 0.5\ m^3$) can be tested in its "natural" state (i.e., as deposited in and consolidated within the landfill). The main disadvantages of in situ direct shear testing include a relatively high cost (the test requires an excavation of pedestal of waste, either at the surface or in the trench), health and safety considerations, a limitation on the normal pressure that can be applied on the waste pedestal, and a restriction on testing only relatively "shallow" waste. The other limitations of the direct shear test include shearing of the waste along a horizontal plane, which may be aligned with waste layering, a nonuniform strain field throughout the sample, and the sample tested (waste pedestal) may be disturbed during preparation. However, despite these limitations, the ability to test MSW in its in situ state makes this type of field test potentially attractive and worth considering.

2.3.2.13 Internal Deformation Measurements. Internal measurements of vertical and lateral deformation within a waste mass can provide both direct and indirect information on in situ waste characteristics. Internal deformation measurements made over a relatively small interval can provide elemental information on waste compressibility. Larger-scale measurements of vertical deformation and internal measurements of lateral deformation can provide gross measures of MSW deformability that can be back-analyzed to yield elemental property data. Oftentimes, the deformation measurement device can be installed as the waste is placed and subsequently accessed from outside the waste mass, eliminating the need to drill into the waste, minimizing health and safety concerns, and making the measurement only semi-intrusive.

Internal deformations are commonly measured by using settlement platforms (plates, borehole settlement gauges), inclinometers, tiltmeters, and/or extensometers, and occasionally by using embedded crack gauges, earth pressure cells, and settlement profiler systems. Bowders et al. (2000) report on internal settlement measurements in a lined landfill in Columbia, Missouri. An internal settlement plate was constructed on top of the initial 2 m thick lift of MSW and then monitored during and after placement of an additional 5 m of MSW at that location. The results were presented as a plot of settlement versus time and were used to calculate primary and secondary compression indexes.

Sharma and De (2007) presented a landfill case history with an internal measurement of secondary settlement of MSW under self-weight. Internal deformations were measured using a Sondex device. The Sondex device is an extensiometer that consists of a corrugated pipe equipped with magnetic rings. The pipe is vertically installed in a borehole and grouted to the surrounding waste. The distance between the magnetic rings is measured using a magnetic probe lowered into the pipe. The change in distance between the magnetic rings is a measure of the compression of the waste. On the basis of a 2-year measurement period, Sharma and De (2007) back-calculated secondary compression indexes (c_α) at two monitoring locations. Sharma and De (2007) reported another case history of a landfill in Northern California where internal deformations caused by fill placement were measured using extensometers and surface measurements.

Bachus et al. (2006) described a "waste settlement profiler" for use in evaluating the settlement of liner systems for "piggy back" landfills (landfills where a liner system for a new waste unit is placed on top of an old unlined waste unit). The waste settlement profiler consists of a flexible guide tube placed horizontally and a guide wire used to pull a water-level indicator. By recording the water-level indicator data, the deformation of the guide tube is measured. These measurements can be interpreted as the gross compressibility of the underlying waste subject to the load of the newly placed waste (overburden).

2.3.2.14 Internal Moisture Content and Temperature Measurements. In situ moisture content and temperature measurements are often performed together. These measurements can provide insight into moisture distribution and leachate flow within the waste fill. These measurements can be useful for a variety of applications, including the design of leachate recirculation and bioreactor landfills and, recently, for design of temporary contact water infiltration systems at desert landfills. Such measurements may also be used to back-calculate the hydraulic properties of the waste mass by means of numerical modeling.

Moisture content measurements are "index" values (i.e., they are not necessarily meaningful in isolation). On the one hand, these values are more useful as indicators of temporal and/or spatial changes (e.g., variation of liquid content over time) than as stand-alone measurements. On the other hand, temperature is a very important measurement, particularly in recent years with the occurrence of elevated temperature landfill events (ETLEs). The presence of ETLEs can impact the integrity of the landfill cover and liner systems, leachate quality, gas composition, slope stability, differential settlement, and odor mitigation, and even impact landfill gas recovery rates (elevated temperatures may kill off mesophilic bacteria that regulate methane generation).

An example of the simplest form of moisture and temperature measurements and waste moisture assessment within waste mass is presented in Figure 2-6. Temperature measurements and qualitative (i.e., visual) moisture content assessment were performed at the ground surface, whereas laboratory moisture content was measured on 19 L samples recovered from a 760 mm diameter borehole. At this Southern California desert landfill, the three reported parameters correlate relatively well along the profile: higher qualitative moisture content is generally associated with higher laboratory moisture content and lower temperature. This correlation, however, should not be generalized. It may not be representative of the behavior in many landfills.

Internal measurement of moisture content can also be carried out in a more sophisticated manner using borehole electrical and nuclear measurements, as previously discussed. Similarly, electrical resistance and impedance sensors can be embedded in arrays within the waste mass for measurement of moisture content changes. Electromagnetic sensors, such as for time domain transmissivity (TDT) and time domain reflectometry (TDR), can also be used to measure changes in moisture content. Sensors used for measuring moisture content require calibration on representative

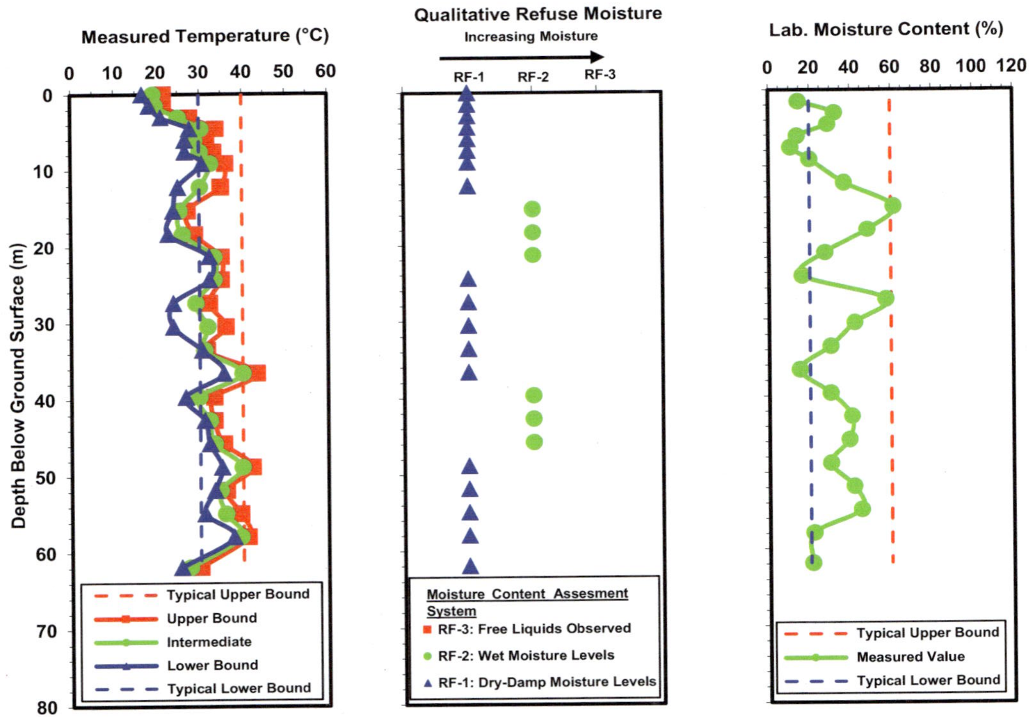

Figure 2-6. Waste temperature and moisture profiles at a Badlands Sanitary Landfill, California. Waste age: Top of Profile 2020, bottom of profile 2004.
Source: Rowe et al. (2010) for typical temperature and moisture content ranges and as measured at the BKK Landfill in Southern California.

samples of the waste mass (i.e., sensor calibrations need to be site-specific because of the wide variability in waste composition).

When constructing new landfills, measurement devices (both temperature and moisture probes) can be placed as the waste fill placement progresses. This allows for continuous probe access from outside of the waste mass. Thus, the need to drill into the waste mass to install instrumentation such as fiber optic cables, thermocouples, and vibrating wire piezometers with thermistors and fiber optic sensors is eliminated along with many health and safety concerns associated with drilling in waste (see previous discussion in Section 2.3.2.3). This approach to geotechnical instrumentation during construction is referred to as *semi-intrusive*.

Masbruch and Ferre (2003) reported on calibration of TDT sensors for measuring the moisture content of MSW. They concluded that large errors may result from the use of a calibration based on a single waste sample. They calculated a root mean square error in volumetric water content on the order of 4% on the basis of TDT calibration using nine oven-dried samples. However, they suggested that a simplified two-point calibration can be performed on each sample because of the highly linear relationship between the volumetric moisture content and the square root of the dielectric permittivity.

Khire and Haydar (2007) used TDR and electrical impedance sensors for monitoring moisture content changes within a bioreactor landfill in Jackson, Michigan. An array of sensors was used to monitor the progress of the wetting front from leachate injected into a geocomposite drainage blanket (the geocomposite drainage blanket was used for leachate recirculation within the waste mass). The impedance and TDR measurements responded to leachate arrival. A constant measurement value was reached during reflection. The constant value was interpreted to correspond to a degree of saturation of 100% on the basis of laboratory calibration tests. Khire and Haydar (2007) further performed numerical modeling of the leachate flow and back-calculated a waste hydraulic conductivity.

Partitioning gas tracer tests (PGTTs) can be used to measure the moisture content of MSW. PGTT is an intrusive method that involves injection and extraction of two tracers within the soil or MSW mass, both under steady gas flow. The arrival time of the tracers, measured using chromatography, can be related to the degree of saturation. Imhoff et al. (2007) reported good agreement between measured moisture contents using the PGTT and gravimetric measurements at two landfills on the basis of tests conducted by Han et al. (2004, 2006).

Moisture content can also be monitored in situ by measuring changes in temperature. In addition to moisture monitoring, temperature sensors can be used to measure the in situ heat transfer properties and heat generation characteristics of MSW (Oettle 2008). Temperature data can also be used to analyze frost depths, liner and cover system desiccation, and geosynthetic aging (Yesiller et al. 2008). Thermocouples are the most popular way of measuring in situ landfill temperatures because of their chemical and thermal resistance. Jafari et al. (2017) reported the use of Type T thermocouples rated for a temperature range of −185 °C to 315 °C. Imhoff et al. (2007) and Oettle (2008) used vibrating wire piezometers with thermistors and fiber optic sensors for a combined monitoring of moisture and temperature.

Temperature sensors can be placed on liners and covers (e.g., Koerner 2006). Long vertical and horizontal arrays of sensors can be installed to measure the spatial and temporal variation of temperatures at discrete locations (Hanson et al. 2008, Jafari et al. 2017). In the case of fiber optic sensors, nearly continuous data can be obtained along the length of the cable (Imhoff et al. 2007). Similar to electrical, electromagnetic, and nuclear measurements, PGTT and thermal measurements may only be indicative of changes in moisture content rather than absolute values.

2.4 HYBRID FIELD MEASUREMENT TECHNIQUES

2.4.1 General

Hybrid techniques for field measurement of MSW properties are techniques that require penetration into the waste mass but, generally, cannot be used for recovery of representative MSW samples. These techniques include penetration tests such as the CPT and DMT and a recently developed technique for direct (i.e., in situ) measurement of shear modulus reduction and damping. The recently developed technique consists of obtaining a modulus reduction curve on the basis of the procedure developed by Sahadewa (2014) and Sahadewa et al. (2015) and obtaining a damping curve through back-calculation from the modulus reduction curve using a technique developed by Matasovic et al. (2018), thus coupling nonintrusive, intrusive, and numerical modeling efforts. This technique is discussed subsequently.

No significant amount of drilling cuttings, liquid, and drilling mud is generated by implementation of the hybrid measurement techniques.

2.4.2 Cone Penetration Test Soundings

Cone penetration test (CPT) sounding has been used to characterize waste materials and/or underlying foundation conditions at hazardous waste landfills, mostly because of health and safety concerns associated with the application of other methods. An example of such an application is an attempt to evaluate the (undrained) properties of contaminated cohesive soils below liquid level. This effort involved a correlation of the results of SASW sounding, triaxial compression testing of representative samples, and CPT sounding, as presented in Matasovic et al. (2006).

Until recently, CPT and seismic CPT (SCPT) soundings were rarely used for characterization of MSW. This was not only because of an apparent mismatch between cone tip size, MSW particle size, and the corresponding lack of engineering correlations, but also because of economic reasons. Many operators required large upfront payments because many CPT cones were lost or damaged during sounding at landfills. However, with the relatively recent advent of tiltmeter-enabled cones [i.e., cones that allow for real-time monitoring of rod-cone system lateral deformation (and buckling)], performing CPT and especially SCPT soundings at MSW landfills has become more common.

Currently, CPT soundings are best used to identify the location of soft zones (e.g., sludges) and perched leachate in the waste. They are also used to assess the thickness of interim landfill cover, a task that is commonly required in the support of design of evapotranspirative cover at landfills. The results of such an exploration in Southern California are shown in Figure 2-7. SCPT soundings are used for greenfield and brownfield developments where measurement of shear wave velocity is required to establish site class for seismic and structural design. Special, dual accelerometer cones are available for direct measurement of compressional wave velocity as well.

Instrumentation can be added to the standard cone to provide additional measurements of waste characteristics. For example, the CPT with piezocone (CPTu), which allows pore-water pressure dissipation measurement through a pore-water pressure sensor, can be used to evaluate the hydraulic properties of waste. The in situ pore pressure can be estimated by extrapolation of the pressure dissipation profile at a

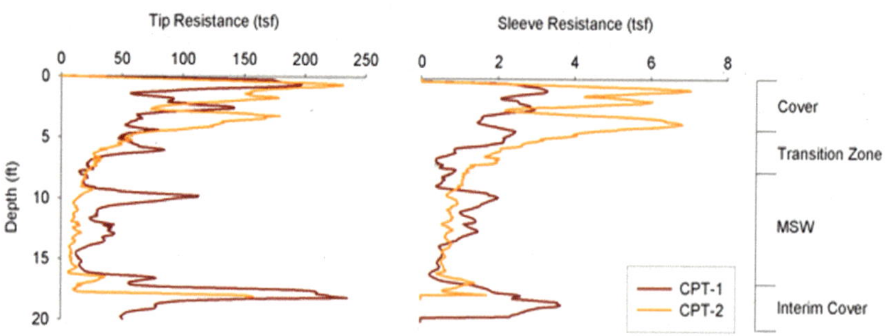

Figure 2-7. Results of CPT sounding performed in support of an evapotranspirative landfill cover design in Southern California.

specific location. CPTu soundings are being increasingly used in the assessment of MSW landfills, particularly those that practice leachate recirculation, accept special wastes such as sludges, or are otherwise relatively wet moisture content above 60% (Rowe et al. 2010) (Figure 2-6).

The electrical resistivity or conductivity cone (RCPT or CPT-EC, respectively) can measure the electrical resistivity or conductivity of the waste for monitoring moisture conditions, subject to the limitations previously mentioned for nonintrusive and borehole electrical testing methods. For example, Mondelli et al. (2007) used ERT in conjunction with resistivity piezocone (RCPTu) to identify contaminant plumes and contaminant-saturated zones at an MSW landfill in Bauru, Brazil. ERT provided an overall understanding of the shape and direction of the contaminant plume. The RCPTu results provided higher resolution and identified contaminant-saturated zones for leachate sampling.

The thermal cone penetration test (T-CPT) uses a standard cone penetrometer with the addition of a temperature sensor placed in the center of the cone tip. At the target depth of thermal measurement, the CPT sounding is interrupted, and the temperature decay is recorded. The testing procedure, including the theoretical background of the procedure suggested for evaluation of the thermal properties of the waste (i.e., thermal conductivity), is presented in Vardon et al. (2018). The same authors also presented a method for evaluation of the initial in situ temperature within the profile sounded.

2.4.3 Dilatometer and Seismic Dilatometer Testing

The dilatometer Marchetti test (DMT) has been widely used for in situ measurement of physical and mechanical soil properties such as hydraulic conductivity, undrained shear strength, friction angle, coefficient of consolidation, and constrained modulus (e.g., Marchetti 1997). The dilatometer blade is constructed of high-strength stainless steel and may be advanced into the ground using a common CPT rig or by hydraulic push capability of a drill rig. Testing can also start from the bottom of a borehole.

The latest iteration of the DMT, the seismic dilatometer (SDMT), is equipped with a seismic module for measuring wave propagation velocities. The small strain and large strain moduli back-calculated from the SDMT measurements can be combined to develop the modulus reduction curve for a specific soil (Mayne et al. 1999). Castelli and Maugeri (2014) used the SDMT to evaluate the friction angle of MSW.

2.4.4 In Situ Measurement of Shear Modulus Reduction

Mayne et al. (1999) used the SDMT to indirectly evaluate modulus reduction in soil. There is an indication that their approach may be used in

MSW as well. Sahadewa (2014) and Sahadewa et al. (2015) developed a technique for direct (i.e., in situ) measurement of shear modulus reduction. The use of this technique for development of in situ shear modulus reduction curves requires (1) a relatively powerful source of excitation (e.g., a 30 t Vibroseis truck); (2) a vertical, embedded instrumentation array; (3) specialty signal processing equipment; and (4) specialty reduction of measurement results. Figure 2-8 shows the top of an embedded instrumentation array (a round concrete foundation) where shaking is applied.

The Sahadewa (2014) and Sahadewa et al. (2015) techniques allow for evaluation of the small-strain shear modulus (G_{max}) and Poisson's ratio. This modulus is evaluated from the results of small-strain testing (cross-hole, down-hole, and/or SASW) at the same location and MSW density measurements on samples recovered from the instrument borehole. Small-strain damping and a damping curve cannot be measured directly. A damping curve can be inferred indirectly, however. Matasovic et al. (2018) used the MKZ constitutive model (Matasovic and Vucetic 1993) to back-calculate damping in MSW from an in situ-measured modulus reduction. The measured modulus reduction and corresponding back-calculated damping curves are shown in Figure 2-9.

Like the Sahadewa (2014) and Sahadewa et al. (2015) techniques, the Matasovic et al. (2018) technique is limited to evaluation of damping curves representative of waste close to the ground surface (top 10 m).

Figure 2-8. Preparation for in situ measurement of shear modulus reduction. Source: Matasovic et al. (2018).

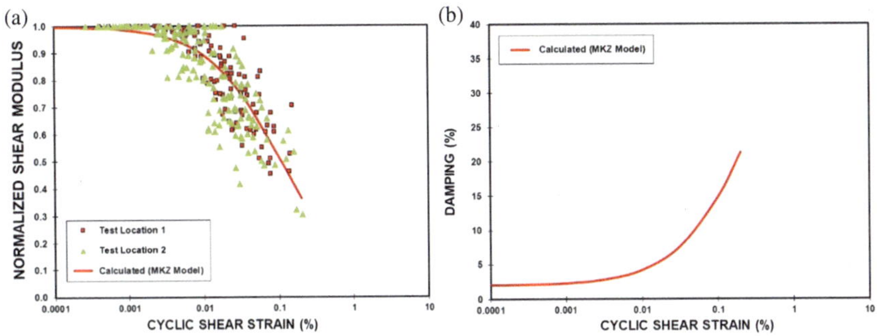

Figure 2-9. (a) Results of in situ nonlinear testing fitted with the MKZ model, (b) a damping curve calculated from nonlinear testing data by using the MKZ model. Source: Matasovic et al. (2018).

2.5 RECOMMENDATIONS

A variety of nonintrusive, intrusive, and hybrid techniques for field characterization of MSW, including measurement of MSW properties, are available. Techniques reviewed and critiqued herein are listed in Table 2-1. Techniques listed in Table 2-1 include both direct methods for measurement of MSW properties of interest and indirect methods that rely on correlation between the measurements and the properties of interest. Further categorization of these techniques with respect to test nature (intrusive versus nonintrusive) is provided in Table 2-1 with symbols.

It is not possible to single out a particular technique or test as a "recommended" procedure. Nonintrusive techniques are generally preferred from a health and safety perspective, and are generally less expensive, faster, and easier to conduct than their intrusive counterparts. However, nonintrusive tests rarely yield direct measurements of key waste properties required for stability evaluation, such as shear strength, unit weight, and modulus reduction and damping ratio. Therefore, for stability evaluation applications, direct measurements are preferable to indirect measurements. Direct measurements generally require intrusion of the waste mass, recovery, and processing of waste samples and therefore require implementation of health and safety measures. Therefore, on the one hand, direct measurements are generally more expensive than indirect measurements. On the other hand, several of these direct measurement techniques can be used to directly evaluate shear strength, unit weight, and several other engineering parameters.

All field measurement methods, including nonintrusive, intrusive, and hybrid techniques, as applied to evaluation of MSW properties, are subject to limitations. One of the most significant limitations is the need for site-specific calibrations and correlations. This is not only because of the lack

Table 2-1. Summary of Field-Testing Methods for Geotechnical Characterization of MSW.

In situ measurement technique \ Material property	Shear strength parameters	Unit weight	Shear/compressional wave velocity	Modulus reduction	Damping	Hydraulic conductivity	Moisture retention	Young's modulus	Modulus of subgrade reaction	Deformations/settlement
Nonintrusive										
SASW and MASW		▒	■					▒		
ReMi			■							
Electrical resistivity (ERT) and RCPTu							▒			
LiDAR, InSAR, and/or UAV										■
Back-analysis	▒			▒	▒			▒	▒	
Plate load test (surface)									■	
Intrusive										
TSB						■				
Extensiometers/Sondex device										▒

Settlement plate										■
OYO suspension logging		▒	■							
Cross-hole, in-hole, and/or down-hole		▒	■							
In-hole density		■								
Plate load test (in-borehole test)									■	
SPT and BPT	▒	▒								
Pressuremeter (PMT)								■		
Plate load test (in-trench)									■	
Borehole and test pit sampling		■					■			
Borehole permeability tests						■				
TDT, TDR, PGTT, and/or fiber optic sensors							▒			
In situ direct shear test	■									
Gamma radiation and neutron probe							▒			
Hybrid										
CPT, CPTu, RCPTu, CPTu-EC, SCPT, and SCPTu	▒	▒	■			▒	▒	▒		
Dilatometers (DMT and SDMT)	▒	▒	■	▒	▒			■		
Vibroseis/T-Rex + embedded instrumentation				■	▒					

Note: ■ = direct measurement of properties; ▒ = indirect measurement of properties (by correlation).

of sufficient test data for MSW required for calibration of testing and sounding discussed herein, but also because of the high variability of waste deposits and hence related material properties. Such limitations can sometimes be addressed through a coordinated field testing and large-size laboratory testing program. However, despite these limitations, the values of MSW properties obtained from measurement through field methods are, in general, considered more representative than the values obtained in a geotechnical laboratory. Whenever possible, several in situ methods should be employed, and the interpretation of the results should be combined to assess the representative properties of MSW. The assessed properties should be compared to the values published in the literature as a check for range.

ACKNOWLEDGMENTS

The author wishes to thank the individuals who reviewed the paper and provided valuable comments. The most extensive and most valuable comments were provided by Rudy Bonaparte of Geosyntec Consultants, Inc., Brookhaven, Georgia. Additional comments and suggestions were provided by Dimitrios Zekkos of University of California, Berkeley; Alan Witthoeft of Geo-Logic Associates, Inc.; Costa Mesa, California; Richard Thiel of Thiel Engineering, Oregon House, California; James K. Mitchell, Distinguished Professor Emeritus, Virginia Tech; Craig K. Benson, Distinguished Professor Emeritus, University of Wisconsin, Madison; Anirban De of Manhattan College, The Bronx, New York; R. Jeffrey Dunn, R.J. Dunn Geotechnical Consultants, Walnut Creek, California; Xunchang Fei of Nanyang Technological University, Singapore; and Tarik Hadj-Hamou of SRK Consulting, Reno, Nevada.

REFERENCES

Aristodemou, E., and A. Thomas-Betts. 2000. "DC resistivity and induced polarisation investigations at a waste disposal site and its environments." *J. Appl. Geophys.* 44 (2): 275–302.

ASTM International. 2004. *Standard test method for density of soil and rock in place by the water replacement method in a test pit*. ASTM D5030. West Conshohocken, PA: ASTM.

ASTM. 2007. *Standard test method for density and unit weight of soil in place by sand-cone method*. ASTM D1556. West Conshohocken, PA: ASTM.

ASTM. 2018. *Standard guide for planning and conducting geotechnical borehole geophysical logging*. ASTM D5753. West Conshohocken, PA: ASTM.

Bachus, R. C., T. E. Zettler, and J. E. Fleming. 2006. "Development of a settlement profiling system to assess MSW compression characteristics." In *Proc., GeoCongress 2006: Geotechnical Engineering in the Information Technology Age*. Reston, VA: ASCE.

Batayneh, A. T., and M. O. Barjous. 2005. "Resistivity surveys near a waste-disposal site in the Qasr Tuba area of Central Jordan." *Bull. Eng. Geol. Environ.* 64 (3): 285–291.

Benson, C. H., and X. Wang. 2000. "Hydraulic conductivity assessment of hydraulic barriers constructed with paper sludge." In *Geotechnics of high water content materials*, edited by T. B. Edil and P. J. Fox, 91–107. West Conshohocken, PA: ASTM.

Bernstone, C., and T. Dahlin. 1996. "2D resistivity surveying of old landfills." In *Proc., 2nd European Meeting of Environmental and Engineering Geophysics*, 188–191. The Netherlands: European Association of Geoscientists & Engineers.

Blight, G. 2008. "Slope failures in municipal solid waste dumps and landfills: A review." *Waste Manage. Res.* 26 (5): 448–463.

Bonaparte, R., R. C. Bachus, and B. A. Gross. 2020. "Geotechnical stability of waste fills: Lessons learned and continuing challenges." *J. Geotech. Geoenviron. Eng.* 146 (11): 0502000.

Bouazza, A.-M., and E. Kavazanjian, Jr. 2000. "Characterization of municipal solid waste sites using the continuous surface wave method." In *Proc., GeoEng 2000, Int. Conf. on Geotechnical and Geological Engineering*. Lancaster, England: Technomic Publishing.

Bouwer, H., and R. C. Rice. 1976. "A slug test for determining hydraulic conductivity of unconfined aquifers with completely or partially penetrating wells." *Water Resour. Res.* 12 (3): 423–428.

Bowders, J., A.-M. Bouazza, E. Loehr, and M. Russell. 2000. "Settlement of municipal solid waste landfills." In *Proc., 4th Kansai Int. Geotechnical Forum*, 101–106. Osake, Japan

Castelli, F., and M. Maugeri. 2014. "Mechanical properties of municipal solid waste by SDMT." *Waste Manage.* 34 (2): 256–265.

Chambers, J. E., O. Kuras, P. I. Meldrum, R. D. Ogilvy, et al. 2006. "Electrical resistivity tomography applied to geologic, hydrogeologic, and engineering investigations at a former waste-disposal site." *Geophysics* 71 (6): B231–B239.

Champagne, C. L., D. Zekkos, J. P. Lynch, and S. O'Loughlin. 2020. "Waste settlement measurements using unmanned aerial vehicles at a municipal solid waste landfill in Michigan." In *Proc., Geo-Congress 2020: Geo-Systems, Sustainability, Geoenvironmental Engineering, and Unsaturated Soil Mechanics*. Reston, VA: ASCE.

Dixon, N., D. R. V. Jones, and R. W. Whittle. 1999. "Mechanical properties of household waste: In-situ assessment using pressuremeters." In Vol. III of *Proc., Sardinia '99: 11th Int. Waste Management and Landfill Symp.*, 453–460. Cagliari, Italy: Environmental Sanitary Engineering Centre.

Dixon, N., and U. Langer. 2006. "Development of a MSW classification system for the evaluation of mechanical properties." *Waste Manage.* 26 (3): 220–232.

Dixon, N., and U. Langer. 2008. "Classification and mechanical behavior relationships for municipal solid waste: Study using synthetic wastes." *J. Geotech. Geoenviron. Eng.* 134 (1): 79–90.

Dixon, N., R. W. Whittle, D. R. V. Jones, and S. Ng'ambi. 2006. "Pressuremeter tests in municipal solid waste: Measurement of shear stiffness." *Géotechnique* 56 (3): 211–222.

Eid, H. T., T. D. Stark, W. D. Evans, and P. E. Sherry. 2000. "Municipal solid waste slope failure. I: Waste and foundation soil properties." *J. Geotech. Geoenviron. Eng.* 126 (5): 397–407.

Ettala, M. 1987. "Infiltration and hydraulic conductivity at a sanitary landfill." *Aqua Fennica* 17 (2): 231–237.

Gabr, M. A., and S. N. Valero. 1995. "Geotechnical properties of municipal solid waste." *Geotech. Test. J.* 18 (2): 241–251.

Geosyntec. 1996. "Waste mass field investigation, Operating Industries, Inc. landfill, Monterey Park, California." *Rep. No. SWP-2.* Huntington Beach, CA: Geosyntec Consultants.

Guérin, R., M. L. Munoz, A. Christophe, C. Laperrelle, et al. 2004. "Leachate recirculation: Moisture content assessment by means of a geophysical technique." *Waste Manage.* 24 (8): 785–794.

Han, B., V. N. Gallagher, P. T. Imhoff, R. Yazdani, et al. 2004. "Measuring water in bioreactor landfills Abstract GC54A-06." *EOS Trans. Am. Geophys. Union* 85 (47): 344–355.

Han, B., B. Jafarpour, V. N. Gallagher, P. T. Imhoff, et al. 2006. "Measuring seasonal variations of moisture in a landfill with the partitioning gas tracer test." *Waste Manage.* 26 (4): 344–355.

Hanson, J. L., N. Yesiller, and N. K. Oettle. 2008. "Spatial variability of waste temperatures in MSW landfills." In *Proc., 1st Global Waste Management Symp.*, 1–11. International Solid Waste Association, Indian Wells, California.

Houston, W. N., S. L. Houston, J. W. Liu, and A. Elsayed. 1995. "In-situ testing methods for dynamic properties of MSW landfills." In *Earthquake design and performance of solid waste landfills*, Geotechnical Special Publication 54, edited by E. Yegian and W. Liam Finn, 73–82. Reston, VA: ASCE.

Imhoff, P. T., D. R. Reinhart, M. Englund, and R. Guérin. 2007. "Review of state of the art methods for measuring water in landfills." *Waste Manage.* 27 (6): 729–745.

Jafari, N. H., T. D. Stark, and S. Merry. 2013. "The July 10, 2000 Payatas landfill slope failure." *Int. J. Geoeng. Case Hist.* 2 (3): 208–228.

Jafari, N. H., T. D. Stark, and T. Thalhamer. 2017. "Progression of elevated temperatures in municipal solid waste landfills." *J. Geotech. Geoenviron. Eng.* 143 (8): 05017004.

Jain, P., J. Powell, T. G. Townsend, and D. R. Reinhart. 2006. "Estimating the hydraulic conductivity of landfilled municipal solid waste using the borehole permeameter test." *J. Environ. Eng.* 132 (6): 645–652.

Kavazanjian, E., Jr. 2003. "Evaluation of MSW properties using field measurements." In *Proc., 17th GSI/GRI Conf.: Hot Topics in Geosynthetics—IV*, edited by R. M. Koerner, G. R. Koerner, Y. Hsuan, and M. V. Ashley, 74–113. Foisom, PA: Geosynthetic Institute.

Kavazanjian, E., Jr., N. Matasovic, and R. C. Bachus. 2013. "11th Peck lecture: Pre-design geotechnical investigation for the OII superfund site landfill." *J. Geotech. Geoenviron. Eng.* 139 (11): 1849–1863.

Kavazanjian, E., Jr., N. Matasovic, R. Bonaparte, and G. R. Schmertmann. 1995. "Evaluation of MSW properties for seismic analysis." In *Proc., Specialty Conf. on Geotechnical Practice in Waste Disposal*, Geotechnical Special Publication 46, 1126–1141. Reston, VA: ASCE.

Kavazanjian, E., Jr., N. Matasovic, K. H. Stokoe, II, and J. D. Bray. 1996. "In-situ shear wave velocity of solid waste from surface wave measurements." In *Proc., 2nd Int. Congress Environmental Geotechnics*, 97–104. London: Balkema.

Kavazanjian, E., Jr., M. S. Snow, N. Matasovic, C. Poran, et al. 1994. "Non-intrusive Rayleigh wave investigations at solid waste landfills." In *Proc., 1st Int. Congress on Environmental Geotechnics*, 707–712. Edmonton, Alberta: International Society of Soil Mechanics and Foundation Engineering and Canadian Geotechnical Society.

Khire, M. V., and M. M. Haydar. 2007. "Leachate recirculation in bioreactor landfills using geocomposite drainage material." *J. Geotech. Geoenviron. Eng.* 133 (2): 166–174.

Koerner, G. G. 2006. "Long-term temperature monitoring of geomembranes at dry and wet landfills." *Geotext. Geomembr.* 24 (1): 72–77.

Lai, C. G., G. J. Rix, S. Foti, and V. Roma. 2002. "Simultaneous measurement and inversion of surface wave dispersion and attenuation curves." *Soil Dyn. Earthquake Eng.* 22 (9–12): 923–930.

Marchetti, S. 1997. "The flat dilatometer: Design applications." In *Proc., 3rd Geotechnical Engineering Conf.* Cairo: Cairo University.

Masbruch, K., and T. P. A. Ferre. 2003. "A time domain transmission method for determining the dependence of the dielectric permittivity on volumetric water content: An application to municipal landfills." *Vadose Zone J.* 2 (2): 186–192.

Matasovic, N., R. El-Sherbiny, and E. Kavazanjian, Jr. 2011. "In-situ measurements of MSW properties." Chap. 6 in *Geotechnical characterization, field measurement, and laboratory testing of municipal solid waste*, edited by D. Zekkos, 153–194. Geotechnical Special Publication 209. Reston, VA: ASCE.

Matasovic, N., and E. Kavazanjian, Jr. 1998. "Cyclic characterization of OII landfill solid waste." *J. Geotech. Geoenviron. Eng.* 124 (3): 197–210.

Matasovic, N., E. Kavazanjian, Jr., and F. Abourjeily. 1995. "Dynamic properties of solid waste from field observations." In *Proc., 1st Int. Conf. on Earthquake Geotechnical Engineering*, 549–554. Tokyo: Balkema-Rotterdam.

Matasovic, N., E. Kavazanjian, Jr., and R. L. Anderson. 1998a. "Performance of solid waste landfills in earthquakes." *Earthquake Spectra* 14 (2): 319–334.

Matasovic, N., E. Kavazanjian, Jr., A. De, and J. Dunn. 2006. "CPT-based seismic stability assessment of a hazardous waste site." *Soil Dyn. Earthquake Eng.* 26 (2–4): 201–208.

Matasovic, N., and M. Vucetic. 1993. "Cyclic characterization of liquefiable sands." *J. Geotech. Eng.* 119 (11): 1805–1822.

Matasovic, N., T. A. Williamson, and R. C. Bachus. 1998b. "Cyclic direct simple shear testing of OII landfill solid waste." In *Proc., 11th European Conf. on Soil Mechanics and Foundation Engineering*, 441–448. Porec, Croatia: Balkema-Rotterdam.

Matasovic, N., D. Zekkos, A. Sahadewa, and C. P. Carson. 2018. "In-situ testing for evaluation of modulus reduction and damping of municipal solid waste at hazardous waste landfill." In *Geotechnical Earthquake Engineering and Soil Dynamics V*, Geotechnical Special Publication 293, 452–460. Reston, VA: ASCE.

Mayne, P. W., J. A. Schneider, and G. K. Martin. 1999. "Small- and large-strain soil properties from seismic flat dilatometer tests." In *Proc., Int. Symp. on Pre-failure Deformation Characteristics of Geomaterials*, 419–426.

Merry, S. M., E. Kavazanjian, Jr., and W. Fritz. 2005. "Reconnaissance of the July 10, 2000 Payatas landfill failure." *J. Perform. Constr. Facil.* 19 (2): 100–107.

Mondelli, G., H. Giacheti, M. Boscov, and V. Elis. 2007. "Geoenvironmental site investigation using different techniques in a municipal solid waste disposal site in Brazil." *Environ. Geol.* 52 (5): 871–887.

Oettle, N. K. 2008. "Thermal analysis of landfill barriers." M.S. thesis, California Polytechnic State University, Dept. of Civil and Environmental Engineering.

Oweis, I., D. Smith, R. Ellwood, and D. Greene. 1990. "Hydraulic characteristics of municipal refuse." *J. Geotech. Eng.* 116 (4): 539–553.

Park, C. B., R. D. Miller, and J. Xia. 1997. "Multi-channel analysis of surface waves (MASW)." A summary report of technical aspects, experimental results, and perspective. *Kansas Geological Survey Open-File Rep. No. 97-10*. Lawrence, KS: University of Kansas.

Ramaiah, B. J., G. V. Ramana, E. Kavazanjian, Jr., N. Matasovic, et al. 2016. "Empirical model for shear wave velocity of municipal solid waste in situ." *J. Geotech. Geoenviron. Eng.* 142 (1): 06015012.

Reynolds, R. T. 2015. "25-Years of loading an old Presumpscot formation landslide: A case history." In *Proc., 2015 Symp. on the Presumpscot Formation*. Orono, ME: University of Maine.

Reynolds, W. D. 1993. "Saturated hydraulic conductivity: Field measurement." In *Soil sampling and methods of analysis*, edited by M. R. Carter, 599–613. Boca Raton, FL: Lewis Publishers.

Richardson, G., and D. Reynolds. 1991. "Geosynthetic considerations in a landfill on compressible clays." In *Proc., Geosynthetics '91*. St. Paul, MN: Industrial Fabrics Association International.

Rowe, R. K., A. Hoor, and A. Pollard. 2010. "Numerical examination of a method for reducing the temperature of municipal solid waste landfill liners." *J. Environ. Eng.* 136 (8): 794–803.

Sahadewa, A. 2014. "In-situ assessment of the linear and nonlinear dynamic properties of municipal solid waste." Ph.D. thesis, University of Michigan, Ann Arbor, Dept. of Civil and Environmental Engineering.

Sahadewa, A., D. Zekkos, R. D. Woods, and K. H. Stokoe. 2015. "Field testing method for evaluating the small-strain shear modulus and shear modulus nonlinearity of solid waste." *Geotech. Test. J.* 38 (4): 427–441.

Shank, K. L. 1993. "Determination of the hydraulic conductivity of the Alachua County southwest landfill." Master's thesis, University of Florida, Dept. of Civil Engineering.

Sharma, H. D., and A. De. 2007. "Municipal solid waste landfill settlement: Postclosure perspectives." *J. Geotech. Geoenviron. Eng.* 133 (6): 619–629.

Soupios, P., I. Papadopoulos, M. Kouli, I. Georgaki, et al. 2007b. "Investigation of waste disposal areas using electrical methods: A case study from Chania, Crete, Greece." *Environ. Geol.* 51 (7): 1249–1261.

Soupios, P., N. Papadopoulos, I. Papadopoulos, M. Kouli, et al. 2007a. "Application of integrated methods in mapping waste disposal areas." *Environ. Geol.* 53 (3): 661–675.

Stanton, G. P., and T. P. Schrader. 2001. "Surface geophysical investigation of a chemical waste landfill in northwestern Arkansas." In *Proc., U.S. Geological Survey Karst Interest Group*, 107–115. Water-Resources Investigations Report 01-4011. Reston, VA: US Geological Survey.

Stark, T. D., N. Huvaj-Sarihan, and G. Li. 2009. "Shear strength of municipal solid waste for stability analyses." *Environ. Geol.* 57 (8): 1911–1923.

Stokoe, K. H. I. I., S. G. Wright, J. A. Bay, and J. M. Roesset. 1994. "Characterization of geotechnical sites by SASW method." In *Geophysical characterization of sites*, edited by R. D. Woods, 15–26.

Vardon, P. J., D. Baltoukas, and J. Peuchen. 2018. "Thermal cone penetration test (T-CPT)." In *Proc., Cone Penetration Testing 2018*. Delft, Netherlands: Delft University of Technology.

WCC (Woodward Clyde Consultants). 1987. *Seismic stability evaluations: Operating industries landfill*. Santa Ana, CA: Woodward Clyde Consultants.

Wysocki, E. J., T. R. Nabavi, and S. Djafari. 2003. "The use of leachate recovery wells to evaluate municipal solid waste hydraulic characteristics at fresh kills landfill." In *Proc., 8th Annual Landfill Symp.* Silver Spring, MD: Solid Waste Association of North America.

Yeşiller, N., J. L. Hanson, N. K. Oettle, and W. Liu. 2008. "Thermal analysis of cover systems in municipal solid waste landfills." *J. Geotech. Geoenviron. Eng.* 134 (11): 1655–1664.

Yuen, S. T. S., T. A. McMahon, and J. R. Styles. 2000. "Monitoring in situ moisture content of municipal solid waste landfills." *J. Environ. Eng.* 126 (12): 1088–1095.

Zekkos, D., J. D. Bray, E. Kavazanjian, Jr., N. Matasovic, et al. 2006. "Unit weight of municipal solid waste." *J. Geotech. Geoenviron. Eng.* 132 (10): 1250–1261.

Zekkos, D., E. Kavazanjian, Jr., J. D. Bray, N. Matasovic, et al. 2010. "Physical characterization of municipal solid waste for geotechnical purposes." *J. Geotech. Geoenviron. Eng.* 136 (9): 1231–1241.

Zekkos, D., A. Sahadewa, R. D. Woods, and K. H. Stokoe. 2014. "Development of model for shear-wave velocity of municipal solid waste." *J. Geotech. Geoenviron. Eng.* 140 (3): 04013030.

CHAPTER 3

LABORATORY TESTING FOR CHEMICAL CHARACTERIZATION OF SOLIDS, LIQUIDS, AND GASES

Marisa Chrysochoou and Nefeli Bompoti

3.1 INTRODUCTION

This chapter describes laboratory methods to conduct chemical characterization of solids, liquids, and gases related to geoenvironmental practice. In addition, the analysis of solids extends to testing techniques that determine mineralogical composition, such as diffraction, thermogravimetric analyses, and spectroscopy, as these are highly relevant to engineering practice. Specifically, the chapter covers the following methods:

- Digestion, extraction, and spectroscopic methods to analyze bulk elemental composition and inorganic and organic contaminants in solids;
- Mineralogical analysis methods (X-ray diffraction, scanning electron microscopy, thermal analyses, infrared spectroscopy);
- Liquid analysis methods for index properties [pH, conductivity, redox potential, dissolved oxygen (DO)], metals, anions, and organic compounds; and
- Gas analysis methods for organic contaminants and trace gases (CO_2, CH_4, and so on).

It should be noted that the terms solid, liquid, and gas used in this chapter generally refer to the types of media encountered in geoenvironmental practice, namely,

1. Solids: soils, sediments, rocks, municipal and industrial solid waste, as well as by-products (fly ash, cement kiln dust, blast furnace slag, and so on);

2. Liquids: aqueous solutions (groundwater, surface water, stormwater, leachate), digestion solutions; and
3. Gases: soil vapor, landfill gas.

This distinction is important, as analytical methods are dependent on the type of matrix and the range of concentration of the target parameter of element.

A general resource for laboratory analysis of geoenvironmental media is the SW-846 compendium of the Environmental Protection Agency (USEPA 2014), which consists of more than 200 laboratory methods for sampling and analysis of waste and other matrixes. The SW-846 manual also offers guidance on sampling, quality assurance and quality control, and selection of appropriate methods to comply with Resource Recovery and Conservation Act regulations, but which can also be used as general guidance for geoenvironmental media. The EPA methods stipulated in the Clean Water Act (CWA), "Part 136" methods (USEPA 2024) are also relevant to chemical and biological pollutants in media, such as wastewater, ambient water, sediment, and biosolids (sewage sludge). In addition, ASTM International offers a wide array of standard methods, which can substitute or supplement EPA methods. Finally, the Soil Science Society of America's Methods of Soil Analysis series (Parts 1 to 5) is a comprehensive reference for specialty methods and applications. This chapter presents the most relevant methods for geoenvironmental practice.

3.2 SOLID ANALYSIS METHODS

3.2.1 Bulk Chemical Analysis

In general, analysis of the chemical composition of solids in geoenvironmental practice is performed through a two-step process: digestion of the solid into the liquid phase, followed by analysis of the resulting digestion solution. The strength of the digestion solution will ultimately determine the recovery of individual elements and the accuracy of the analysis. Some knowledge of the sample matrix is required to select an appropriate digestion method and users are advised to carefully consider both the matrix and the application to select the appropriate method. Digestion methods can be divided into two categories: (1) Methods that target total recoverable elements that utilize a combination of medium-strength acids such as nitric and hydrochloric acid. (2) Methods that target complete dissolution of all elements including refractory ones (e.g., Si, Ti, and Ag). A four-acid digestion method including hydrofluoric and fusion methods with flushing agents such as lithium tetraborate ($Li_2B_4O_7$) followed by dilute acid dissolution are included in this category.

The digestion method also depends on the selection of the subsequent method used to analyze the extracted liquid. The title of a method is not always instructive in this respect, and therefore, users are advised to examine the scope and application notes of a particular method. An overview of ASTM International and EPA digestion methods is given in Table 3-1. It should be noted that these methods cover only inorganic target analytes; for organic target analytes, the EPA 3500 series should be referenced.

Table 3-1. Overview of Digestion Methods for Targeting Inorganic Analytes in Geoenvironmental Samples.

Method number	Method name	Notes
ASTM E2941-14	Standard Practices for Extraction of Elements from Ores and Related Metallurgical Materials by Acid Digestion	Nitric and four-acid versions for mine soil, waste rock, and tailings
ASTM D5198-17	Standard Practice for Nitric Acid Digestion of Solid Waste	Total recoverable elements
ASTM D6349-13	Standard Test Method for Determination of Major and Minor Elements in Coal, Coke, and Solid Residues from Combustion of Coal and Coke by Inductively Coupled Plasma—Atomic Emission Spectrometry	Fusion and digestion methods for total recovery included in the method
ASTM D6357-11	Test Methods for Determination of Trace Elements in Coal, Coke, and Combustion Residues from Coal Utilization Processes by Inductively Coupled Plasma Atomic Emission, Inductively Coupled Plasma Mass, and Graphite Furnace Atomic Absorption Spectrometry	Includes a description of 4-acid digestion
EPA 3050B	Acid Digestion of Sediments, Sludges, and Soils	Total recoverable elements, separate procedure for flame and graphite AA analysis
EPA 3051A	Microwave-Assisted Acid Digestion of Sediments, Sludges, Soils, and Oils	Alternative to EPA 3050
EPA 3052	Microwave-Assisted Acid Digestion of Siliceous and Organically Based Matrixes	Total sample decomposition

All the acid digestion methods described previously require that the use of acids be conducted under a fume hood using appropriate safety equipment. In addition to these general digestion methods, specialty methods are used to target particular compounds of interest, such as hexavalent chromium (EPA 3060A), cyanide (EPA 9013A), sulfides (EPA 9031), mercury (EPA 3200), halides (EPA 9022 and 9023), and others. EPA 9060A is used to determine total organic carbon (TOC) in water and solid waste. Analysis of light elements such as C, O, S, and P is performed via other methods, including the use of elemental (CHNS) analyzers.

An alternative to wet chemical methods to determine the inorganic composition of geoenvironmental samples quantitatively is X-ray fluorescence (XRF). Laboratory wavelength dispersive XRF equipment requires borate fusion and pelletizing of samples. It should be noted that specialized X-ray training is required for the use of XRF. Method ASTM E1621-13 provides guidance on the development of analytical procedures for WDXRF analysis of solids. Portable XRF relies on energy-dispersive technology and thus is generally less accurate than WDXRF. However, recent advances in portable XRF equipment have rendered it powerful enough to determine even lighter elements such as Mg, Al, and Si. Most geoenvironmental applications utilize portable XRF as a screening tool for in situ and on-site measurements. EPA method 6200 provides guidance on portable XRF applications. ASTM International has a variety of XRF standard methods that cover different matrixes, including coal ashes (D4326-04). XRF users are cautioned that matrix effects may render built-in calibrations inapplicable to a material. XRF analyzes the composition in terms of equivalent oxides, which can be converted to elemental composition by taking into account the molecular weight ratio of the oxide over the elemental molecular weight.

3.2.2 Mineralogical Analysis

The composition of solid materials can also be described in terms of the minerals present. Minerals are natural solid phases with defined crystal structures. In geoenvironmental practice, the identification of mineralogical properties can aid in the following ways:

- Assist in positive material identification: Forensic work often requires the identification of the origin of waste materials such as ashes and slags. Although chemical composition is useful in this respect, it is not a unique material "signature," especially for siliceous matrixes. Mineralogy, however, offers a higher level of information; for example, the mineral mullite ($3Al_2O_32SiO_2$) is commonly associated with fly ashes and not with natural soils (Winburn et al. 2000, Fernández-Jiménez et al. 2006, Ward and French 2006).

- Identify retardation and immobilization mechanisms for contaminants, i.e., heavy metal precipitates and potential sorption surfaces (iron oxides, clays). D'Amore et al. (2005) provide a comprehensive review of methods to determine metal speciation in soils and waste.
- Assist in the evaluation of soil origin, fabric, and reactivity.

A wide variety of methods is available for mineralogical studies and only the most common ones are discussed here: X-ray diffraction, scanning electron microscopy with energy-dispersive X-ray (SEM/EDX), and thermal methods [thermogravimetric analysis (TGA) and differential thermal analysis (DTA)]. An overview of these is given in Table 3-2. All these methods are commercially available.

XRD and SEM are by far the most widely applied techniques in geoenvironmental practice, both to study the properties of cementitious by products, as described previously and to study clay mineralogy (Grim 1968, Mitchell and Soga 2005, Cui and Jia 2013, Chrysochoou 2014, Lin and Cerato 2014). There is no standard method for sample preparation and qualitative and quantitative analyses of geoenvironmental media, except for quantitative XRD for portland cement clinker. Bish and Post (1989) provided both a theoretical overview and practical guidance with respect to XRD application in soils and rocks. Interpretation of XRD results requires acquisition and knowledge of specialized software such as MDI JADE (version 7.7) or others. For SEM analysis, the ASTM standard for hardened concrete (ASTM C1723-16) can serve as a useful reference for most geoenvironmental media as well.

Thermogravimetric analysis and differential thermal analysis analyses are among the most popular methods to study the hydration of portland cement (Lothenbach and Winnefeld 2006) as well as of other lime–mineral mixtures (Klimesch and Ray 1998). Differential thermal analysis locates the ranges corresponding to thermal decompositions of different phases in paste, whereas TGA simultaneously measures the weight loss because of the decomposition. Typical thermal decompositions are dehydration (usually around 100 °C), dehydroxylation, and ultimately, melting. Thermogravimetric analysis and differential thermal analysis analyses are particularly useful to monitor the evolution of reactions involving hydration/dehydration of minerals.

3.3 LIQUID ANALYSIS METHODS

Liquid samples including aqueous and digestion solutions can be characterized by utilizing wet chemistry methods for testing general parameters, metals and metalloids, organics, and other compounds.

Table 3-2. Overview of Basic Mineralogical Analysis Methods.

Method	Information provided	Quantitative?	Standard methods	Min. Sample quantity (g)
XRD	Crystalline phases >1 wt.%	Usually not, but possible	ASTM C1365-06(2011) (cements)	1
SEM/EDX	Size, morphology, and composition of individual particles	EDX semiquantitative for composition	ASTM C1723-16 (concrete) E2809-13 (paint) E2142-08 (steel)	Approximately 0.1
TGA/DTA	Water content, hydration, and mineralogy using thermal decomposition patterns	Usually not, but possible	ASTM E2105-00(2016) (with IR)	0.1–10

3.3.1 General Parameters

The most common parameters measured in aqueous solutions include the following:

Color: The color of an aqueous sample or waste can be determined by EPA method 110.1 or 110.3 using a spectrophotometer (and Tristimulus filters in case of EPA method 110.1). This method provides mostly a qualitative characterization of the sample.

Temperature: Temperature measurements may be made with any good mercury-filled Celsius thermometer, a dial-type centigrade thermometer, or a thermistor according to EPA method 170.1.

Turbidity: The turbidity of an aqueous sample can be measured using a turbidimeter (EPA 180.1 or method M5171) or a slit turbidimeter (ASTM D1889-00). All methods cover a range of 0–40 NTU, whereas ASTM method D6855-12 is more applicable in low-turbidity samples (< 5 NTU).

pH: The pH of an aqueous solution is usually measured with a pH meter equipped with a glass electrode (EPA method 150.1, EPA-SW 9040c, or

ASTM E70-07). A calibration of the measuring device is required before the measurement using a series of standards solutions of known pH. If no accuracy is needed or the previous methods are not applicable, the pH paper method (EPA-SW 9040c) can be used as an alternative. Specifically, for water samples, ASTM methods 1293-99 and D5464-16 can be applied for samples of high and low conductivity, respectively.

Redox Potential: The oxidation–reduction potential is measured electrometrically with ASTM method 1498-14. The instrumentation required is a pH meter equipped with a calomel, silver–silver chloride, or another reference electrode of constant potential.

Hardness: Hardness in water can be determined either colorimetrically (EPA 130.1) or titrimetrically (EPA 130.2 or ASTM D1126-17).

Electrical Conductivity or Conductance: Electrical conductance is estimated with a conductivity meter (EPA methods 120.1 and 9050A). Both methods are applicable to aqueous solutions; however, the 9050A method is not applicable to organic samples. ASTM method D1125-14 describes the methodology of measuring the electrical conductivity in water samples (Method A).

Electrical conductivity is strongly correlated with the ionic strength of the solution and consequently with the total dissolved solids (TDS). An empirical relationship to estimate the TDS from conductivity is given by way of Equation (3-1):

$$\text{TDS (mg L}^{-1}) \approx 640 \times EC(\text{dS m}^{-1}) \tag{3-1}$$

Dissolved Oxygen: Mainly there are two categories of methods for measuring the DO in aqueous solutions: (1) instrumental methods that measure DO using a probe [EPA 360.1 or ASTM D888-12 (Methods B and C)], and (2) titrimetric methods [(EPA 360.2 and ASTM D888-12 (Method A)]. The main difference between these methods is that in the titrimetric methods, the sample is titrated with a series of manganous sulfate, potassium hydroxide, and potassium iodide, and finally, sulfuric acid, and is therefore subject to several interferences such as oxidizing and reducing agents, nitrate ion, ferrous iron, and organic matter. In the presence of interfering materials, the probe method should be applied. For low concentrations of DO in water (< 100 μg/L), ASTM method D5543-12 is more applicable. The two methods by ASTM are applicable to water samples.

A summary of the methods is given in Table 3-3.

3.3.2 Analysis of Metals and Metalloids

The most common methods to determine metals in solution are atomic absorption (AA) spectrometry and inductive coupled plasma (ICP), both

Table 3-3. Summary of Methods for Estimation of General Parameters in Liquid Samples.

Parameter	Method	Quantitative/ qualitative	Special safety precautions
Color	EPA 110.1 and EPA 110.3	Qualitative	No
Temperature	EPA 170.1	Quantitative	No
Turbidity	EPA 180.1, ASTM D1889-00, and D6855-12	Quantitative	No
pH	EPA 150.1 and ASTM E70-07	Quantitative	No
	EPA-SW 9040c	Semiquantitative	No
Redox potential	ASTM 1498-14	Quantitative	No
Hardness	EPA 130.2 and ASTM D1126-17	Quantitative	No
Electrical conductivity	EPA 120.1, EPA 9050A, ASTM D1125-14	Quantitative	No
Dissolved oxygen	EPA 360.1, ASTM D888-12, EPA 360.2, and ASTM D5543-12	Quantitative	No

of which have different variations. Specifically, AA is available as flame (AAFL) and graphite furnace (AAGF) spectroscopy, which have different working ranges and detection limits. AAFL methods typically have detection limits higher than 1 part per billion (ppb), whereas GFAA detection limits can be as low as 0.01 ppb, depending on the element (Wright and Stuczynski 1996). Inductively coupled plasma is coupled with either atomic emission or mass spectrometry (ICP-AES or ICP-MS). Inductively coupled plasma-based methods are generally preferred as they have a wider working range of concentrations, lower detection limits, and are free of matrix-related interferences. Inductively coupled plasma-mass spectrometry (ICP-MS) is the superior method in terms of all these considerations. However, AA techniques can also provide comparable results, with proper sample preparation and method development. It should be noted that digestion of aqueous samples may or may not be necessary prior to analysis via any of these methods. For relatively clean, filtered groundwater samples, the EPA methods of the CWA may be used, whereas for digested samples, the EPA methods in SW-846 should be used along with the appropriate digestion pretreatment. The digestion method is specific to the type of analysis that follows, except for EPA method 3015A (microwave-assisted acid digestion of aqueous samples and extracts), which is applicable to all four spectroscopies. The target metals to be analyzed also influence the decision on the digestion process and spectroscopy method to adopt. Table 3-4 summarizes the methods applicable to common metals in geoenvironmental practice. All metal and metalloids analysis methods provide quantitative results. Additional safety precautions should be taken when handling certain metals such as arsenic, cadmium, chromium, lead, mercury, and thallium.

3.3.3 Analysis of Organics

The analysis of organic compounds in solution may be performed through the determination of index parameters such as TOC or total petroleum hydrocarbons (TPH), or through the determination of the presence of individual compounds. This chapter focuses on general parameters and on the most frequent organic compounds of interest to geoenvironmental practice, namely, VOCs (volatile organic compounds), PAHs (polycyclic aromatic hydrocarbons), PCBs (polychlorinated biphenyls), and PFAS (poly- and per-fluoroalkyl substances). The SW-846 5000 series includes methods for sample preparation for VOC analysis and the 8000 series chromatography-based methods for a wide range of specific organic compounds. In principle, the analysis of organic mixtures requires first the separation of individual compounds using a chromatographic method [e.g., gas chromatography (GC) or high-performance liquid chromatography (HPLC)], followed by detection of the compounds using detectors such as infrared, thermal ionization, or mass spectrometry. All the methods described subsequently provide quantitative results.

Total Organic Carbon: The methods for determination of TOC in waters and wastes include methods with combustion and oxidation using a carbonaceous analyzer (EPA 415.1 and EPA SW-846 9060a), methods using UV-persulfate oxidation (EPA 415, ASTM D5904-02(2017) and ASTM D6317-15 for low level of carbon). ASTM methods D5904-02(2017) and ASTM D6317-15 can also measure the inorganic carbon in the sample. ASTM method D4129-05(2013) is applicable for a wide range of concentrations (2 to 20,000 mg/L of carbon) by high-temperature oxidation and coulometric detection.

Table 3-4. Applicable Analytical Methods for Common Metals and Metalloids in Water.

Metals	Method	Description
Aluminum	EPA 200.1	FLAA
	EPA 200.2	GFAA
	ASTM D857-17	FLAA
	EPA 200.7	ICP-AES
	EPA 200.8	ICP-MS
Antimony	EPA 204.1	FLAA
	EPA 204.2	GFAA
	ASTM D3697-17	FLAA
	EPA 200.7	ICP-AES
	EPA 200.8	ICP-MS
Arsenic	EPA 206.2	GFAA
	EPA 206.3	FLAA (gaseous hydride)
	EPA 206.4	Spectrophotometric
	ASTM D2972-15	Colorimetric, FLAA (hydride), GFAA
	EPA 200.7	ICP-AES
	EPA 200.8	ICP-MS
Barium	EPA 208.1	FLAA
	EPA 208.2	GFAA
	ASTM D4382-12	FLAA, GFAA
	EPA 200.7	ICP-AES
	EPA 200.8	ICP-MS

(*Continued*)

Table 3-4. (*Continued*)

Metals	Method	Description
Beryllium	EPA 210.1	FLAA
	EPA 210.2	GFAA
	ASTM D3645-15	FLAA, GFAA
	EPA 200.7	ICP-AES
	EPA 200.8	ICP-MS
Cadmium	EPA 213.1	FLAA
	EPA 231.2	GFAA
	ASTM D3557-17	FLAA, FLAA (chelation–extraction), Differential pulse anodic stripping voltammetry, GFAA
	EPA 200.7	ICP-AES
	EPA 200.8	ICP-MS
Calcium	EPA 215.1	FLAA
	EPA 215.2	Titrimetric
	ASTM D511-14	Complexometric titration, FLAA
	EPA 200.7	ICP-AES
Chromium	EPA 218.1	FLAA (total Cr)
	EPA 218.2	GFAA (total Cr)
	EPA 218.3	FLAA chelation extraction (total Cr)
	EPA 218.4	FLAA chelation extraction (hexavalent Cr)
	EPA 1636	Ion chromatography (hexavalent Cr)
	ASTM D1687-17	Photometric, FLAA, GFAA (total Cr)
	ASTM D5257-17	Ion chromatography (hexavalent Cr)

(*Continued*)

Table 3-4. (*Continued*)

Metals	Method	Description
	EPA 200.7	ICP-AES (total Cr)
	EPA 200.8	ICP-MS (total Cr)
Cobalt	EPA 219.1	FLAA
	EPA 219.2	GFAA
	ASTM D3558-15	FLAA, FLAA (chelation–extraction), GFAA
	EPA 200.7	ICP-AES
	EPA 200.8	ICP-MS
Copper	EPA 220.1	FLAA
	EPA 220.2	GFAA
	ASTM D1688-17	FLAA, FLAA (chelation–extraction), GFAA
	EPA 200.7	ICP-AES
	EPA 200.8	ICP-MS
Iron	EPA 236.1	FLAA
	EPA 236.2	GFAA
	ASTM D1068-15	FLAA, GFAA, photometric Bathophenanthroline
	EPA 200.7	ICP-AES
Lead	EPA 239.1	FLAA
	EPA 239.2	GFAA
	ASTM D3559-15	FLAA, FLAA (chelation–extraction), differential pulse anodic stripping voltammetry, GFAA

(*Continued*)

Table 3-4. (*Continued*)

Metals	Method	Description
	EPA 200.7	ICP-AES
	EPA 200.8	ICP-MS
Magnesium	EPA 242.1	FLAA
	ASTM D511-14	Complexometric titration, FLAA
	EPA 200.7	ICP-AES
Manganese	EPA 243.1	FLAA
	EPA 243.2	GFAA
	ASTM D858-17	FLAA, FLAA (chelation–extraction), GFAA
	EPA 200.7	ICP-AES
	EPA 200.8	ICP-MS
Mercury	EPA 245.1, 245.2	Cold vapor AA
	EPA 245.7	Cold vapor atomic fluorescence spectrometry
	ASTM D3223-17	Cold vapor atomic absorption
	EPA 200.7	ICP-AES
	EPA 200.8	ICP-MS
Molybdenum	EPA 246.1	FLAA
	EPA 246.2	GFAA
	ASTM D3372-17	FLAA
	EPA 200.7	ICP-AES
	EPA 200.8	ICP-MS
Nickel	EPA 249.1	FLAA

(*Continued*)

Table 3-4. (*Continued*)

Metals	Method	Description
	EPA 249.2	GFAA
	ASTM D1886-14	FLAA, FLAA (chelation–extraction), GFAA
	EPA 200.7	ICP-AES
	EPA 200.8	ICP-MS
Potassium	EPA 258.1	FLAA
	ASTM D4192-15	FLAA
	EPA 200.7	ICP-AES
Selenium	EPA 270.2	FLAA (gaseous hydride)
	EPA 270.3	GFAA
	ASTM D3859-15	FLAA (gaseous hydride), GFAA
	EPA 200.7	ICP-AES
	EPA 200.8	ICP-MS
Silver	EPA 272.1	FLAA
	EPA 272.2	GFAA
	ASTM D3866-12	FLAA, FLAA (chelation–extraction), GFAA
	EPA 200.7	ICP-AES
	EPA 200.8	ICP-MS
Sodium	EPA 273.1	FLAA
	EPA 273.2	GFAA
	ASTM D4191-15	FLAA
	EPA 200.7	ICP-AES

(*Continued*)

Table 3-4. (*Continued*)

Metals	Method	Description
Thallium	EPA 279.1	FLAA
	EPA 279.2	GFAA
	EPA 200.7	ICP-AES
	EPA 200.8	ICP-MS
Tin	EPA 282.1	FLAA
	EPA 282.2	GFAA
	EPA 200.7	ICP-AES
Vanadium	EPA 286.1	FLAA
	EPA 286.2	GFAA
	ASTM D3373-17	GFAA
	EPA 200.7	ICP-AES
	EPA 200.8	ICP-MS
Zinc	EPA 289.1	FLAA
	EPA 289.2	GFAA
	ASTM D1691-17	FLAA, FLAA (chelation–extraction)
	EPA 200.7	ICP-AES
	EPA 200.8	ICP-MS

Total Petroleum Hydrocarbons: Water and wastewater samples can be analyzed for TPH spectrophotometrically (EPA method 418.1) or with an IR laser spectrometer (ASTM D7678-17). EPA method 8440 can be used for measurements of leachates and extracts from solid samples. The extraction of total recoverable petroleum hydrocarbons from soils, sediments, and other solid materials is achieved with the use of supercritical fluids (EPA 3560) or microwave heating (ASTM D5765-16).

Volatile Organic Compounds: EPA method 8260B (VOCs by GC/mass spectrometry) is applicable for nearly any geoenvironmental matrix, including groundwater, surface water, wastes, and soils. The method to extract VOCs from the matrix depends on the material. Method 8260B includes a list of 5000-series extraction methods for different matrixes. Method 8270d is the corresponding GC/MS method for semivolatile organic compounds. ASTM method 5790-95 is also a GC/MS specification for purgeable organic compounds. Screening in the field is also frequently performed using a photoionization detector (EPA method 3815).

Polycyclic Aromatic Hydrocarbons: EPA SW-846 specifies method 8100 to determine 24 PAH compounds in appropriate extracts, whereas the CWA stipulates method 610 for 16 PAH compounds using GC/MS or HPLC. ASTM method E2143-01 provides a rapid screening tool for a field-portable spectrometer to quantify PAH and other aromatics in the field, and various kits are available for this purpose.

Polychlorinated Biphenyls: Several method combinations both by SW-846 and by the CWA are available and these are used to determine the presence of PCBs. Three general methods exist that target different types of PCBs (Coots 2014):

- EPA 1668—HRGC/HRMS (high-resolution gas chromatography/high-resolution mass spectrometry): This method determines all 209 individual PCB congeners.
- EPA 8270D—GC/LRMS (gas chromatography/low-resolution mass spectrometry): This reports the total concentration of PCB homolog groups and is of intermediate resolution.
- 8082A—GC/ECD (gas chromatography/electron capture detector): This is the lowest resolution and most frequently used PCB analysis method, determining the concentration for each Aroclor.

Poly- and Per-Fluoroalkyl Substances : For nonpotable water (groundwater, surface water, and wastewater matrixes), EPA SW-846 method 8327 can provide a targeted analysis of 24 PFAS compounds using liquid chromatography/tandem mass spectrometry (LC/MS/MS). Nondrinking aqueous samples can also be analyzed with the ASTM D7979-19 method using LC/MS/MS and isotopic dilution. Solid matrixes can be analyzed with ASTM D7968-17 using LC/MS/MS and isotopic dilution, similar to ASTM D7979-19.

3.3.4 Analysis of Other Compounds and Parameters

Analyses are often conducted for other compounds and parameters to fully characterize the liquid phase of the samples. These include chloride, chlorine, fluoride, nitrogen, phosphorus, sulfate/sulfide, and oxygen demand. All the methods described subsequently provide quantitative results.

Chloride: Chloride concentrations in aqueous solutions can be determined colorimetrically by forming a colored ferric thiocyanate solution (EPA 325.1

and 325.2) or by titrating with mercuric nitrate (EPA 325.3 and ASTM D512-12 method A) or silver nitrate ASTM D512-12 method B). Chloride can also be measured potentiometrically using a chloride ion–selective electrode (EPA SW-846 9212 and ASTM D512-12 method C).

Chlorine: Most methods of measuring chlorine in solution use a titrimetric amperometric method (EPA 330.1 or ASTM D1253-14), back iodometric titration, especially for wastes (EPA 330.2), iodometric titration (EPA 330.3), or *N*-diethyl-*p*-phenylene diamine—the ferrous ammonium sulfate titration method (EPA 330.4). For chlorine concentrations in the range of 0.2 to 4 mg/L, the D-colorimetric method can be applied (EPA 330.5).

Cyanide: Cyanide complexes can be measured with titrimetric methods (EPA 335.1, EPA 335.2, or EPA SW-846 9014), with colorimetric methods (EPA 335.3, EPA 335.4, EPA SW-846 9012b, or EPA SW-846 9014), and with anion exchange chromatography methods (ASTM D6994-15 and EPA SW-846 9015).

Fluoride: Determination of fluoride concentrations in water and wastes can be estimated colorimetrically (EPA 340.1 and EPA 340.3) or with an ion-selective electrode [EPA 340.2, EPA SW-846 9214, and ASTM D1179-16 (Method B)]. In case of method ASTM D1179-16 (Method A), a distillation step is required to isolate the fluoride before measurement by using the ion-selective electrode.

Nitrogen–Ammonia. Kjeldahl, Nitrate, Nitrite: Ammonia as nitrogen (NH_3–N) can be measured colorimetrically [EPA 350.1, 350.2, ASTM D1426-15 (Method A)], titrimetrically (EPA 350.2), potentiometrically with an ion-selective electrode [EPA 350.3 and ASTM D1426-15 (Method B)].

Total Kjeldahl nitrogen is measured with colorimetric methods (EPA 351.1, EPA 351.2, and EPA 351.3), titrimetric (EPA 351.3), or potentiometric methods (EPA 351.4 and ASTM D8083-16).

Nitrates in water and wastes can be measured colorimetrically (EPA 352.1) and potentiometrically with an ion-selective electrode (EPA SW-846 9210a). In the same way, nitrites can be measured with methods EPA 354.1 and EPA SW-846 9216. The total nitrate–nitrite nitrogen is measured spectrophotometrically with cadmium reduction (EPA 353.2, EPA 353.3, and ASTM D3867-16) or with hydrazine reduction (EPA 353.1). To avoid cadmium toxicity, ASTM method ASTM D7781-14 uses nitrate reductase to measure the total of nitrate and nitrite.

Phosphorus: Several colorimetric methods are available that measure phosphorus in aqueous samples (EPA 365.1, EPA 365.2, EPA 365.3, and EPA 365.4). Methods EPA 365.1, EPA 365.2, and EPA 365.3 can be used to determine phosphorus concentrations (all forms), dissolved and total orthophosphate, and hydrolyzable phosphate. With all these methods, the sample is pretreated to select the phosphorus forms of interest and these forms are then converted to orthophosphate (which is amenable to the color chemistry of the method). EPA method 365.4 (colorimetric) and

ASTM D8001-16e1 (ion chromatography) can determine the total phosphorus content in aqueous samples and wastes.

Sulfate and Sulfide: Determination of sulfate concentrations in aqueous solutions and wastes is achieved colorimetrically by adding barium chloranilate (EPA 375.1 and EPA SW-846 9035) or barium chloride and methylthymol blue (EPA 375.2 and EPA SW-846 9036). Other methods involve turbidimetry (EPA 375.4 or EPA SW-846 9038, and ASTM D516-16) and gravimetry (EPA 375.3).

Concentrations of acid-soluble sulfide ions can be determined by titrimetric methods (EPA 376.1), colorimetric methods (EPA 376.2), or ion-selective methods (EPA SW-846 9215 and ASTM D4658-15). Concentrations of acid-insoluble sulfides can be measured by EPA method SW-846 9034.

Total Oxygen Demand: The total oxygen demand (TOD) in water and waste samples can be estimated by method ASTM D6238-98(2011) and can be correlated to both biochemical oxygen demand (BOD) and chemical oxygen demand (COD).

Biochemical Oxygen Demand: Biochemical oxygen demand is a measure of the DO consumed by microbial life while assimilating and oxidizing the organic matter present. The standard EPA method 405.1 involves incubation of the samples for 5 days at 20 °C in the dark. The reduction in DO concentration during the incubation period yields a measure of the BOD. The DO is measured with one of the methods as described.

Chemical Oxygen Demand: Chemical oxygen demand is a representation of the amount of oxygen required to oxidize the organic matter in a waste sample, under specific conditions of oxidizing agent, temperature, and time. Titrimetric methods for low (5 to 50 mg/L) (EPA 410.2), mid-level (50 to 2,000 mg/L) (EPA 410.1), and high level (> 2,000 mg/L) (EPA 410.3) use potassium dichromate and ferrous ammonium as the reagents to facilitate complete oxidation. EPA method 410.4 measures COD colorimetrically for a range of concentrations 3 to 900 mg/L.

3.4 GAS ANALYSIS

Analysis of gases in geoenvironmental practice is mostly required for in situ or on-site applications such as VOC detection in contaminated sites and gas analysis in landfill collection systems including methane emissions. EPA's Air Emission Measurement Center provides information on measuring pollutants from smokestacks. Methods for monitoring emissions from the surface of landfills are also available on their website (EMC 2024). RCRA requires monitoring of methane around the landfill perimeter; however, several other parameters are typically included in landfill gas monitoring programs. A complete guide on landfill gas

Table 3-5. Overview of Some Common Methods for Landfill Gas Analysis.

Standard	Standard description	Analysis method
EPA 2E	Determination of Landfill Gas Production Flow Rate	Network of 3–5 extraction wells
EPA 3C	Method 3C—Carbon Dioxide, Methane, Nitrogen, and Oxygen Concentrations—Thermal Conductivity Detector (TCD)	GC with TCD
ASTM D5504-12	Standard Test Method for Determination of Sulfur Compounds in Natural Gas and Gaseous Fuels by Gas Chromatography and Chemiluminescence	GC with chemiluminescence—applicable to landfill gas as well
EPA 18	Volatile Organic Compounds by Gas Chromatography	GC/MS
EPA 25C	Nonmethane Organic Compounds Landfill Gases	GC with flame ionization detector

evaluation is provided by the EPA guidance document EPA-600/R-05/123a (USEPA 2005). All the methods described in Table 3-5 are quantitative methods.

For in situ measurements of VOCs, EPA method 8265 can be used, which targets "direct sampling ion trap mass spectrometry for the rapid quantitative measurement, continuous real-time monitoring, and qualitative and quantitative preliminary screening of VOCs in water, soil, soil gas, and air." An additional in situ method for CH_4 estimated using a time correlation tracer with a Fourier-transform infrared spectroscopy have also been reported in the literature (Galle et al. 2001, De la Cruz et al. 2016).

3.4.1 Questionnaire

The following questionnaire serves as an analysis guide for target parameters, categorized by the sample phase (solid, liquid, or gas). The purpose of the questionnaire is to provide a checklist of the parameters of interest and instrumentation needed for each method.

1. What is the phase of the sample?
   Solid
   - Are you interested in mineralogy?
     - X-ray diffraction, SEM microscopy, and TGA analysis;
   - Are you interested in chemical composition?
     - X-ray fluorescence (XRF);
     - Wet chemical methods: digestion and liquid sample analysis;

   Liquid sample or dissolved leachate
   - Are you interested in analyzing general parameters?
     - Color: spectrophotometer;
     - Temperature: thermometer;
     - pH: pH meter;
     - Turbidity: turbidimeter;
     - Oxidation–reduction potential: pH meter with a silver chloride electrode;
     - Hardness: spectrophotometer or titration;
     - Electrical conductivity: conductivity meter;
     - Dissolved oxygen: oxygen probe or titration;
   - Are you interested in analyzing metals and metalloids?
     - Atomic adsorption (FLAA, GFAA), inductive coupled plasma (ICP-MS, ICP-AES), or other methods.
   - Are you interested in analyzing organic compounds?
     - Total organic carbon (TOC): carbonaceous analyzer or UV;
     - Petroleum hydrocarbons (TPH): spectrophotometer or IR;
     - Volatile organic compounds: GC/MS;
     - Polycyclic aromatic hydrocarbons (PAHs): GC/MS;
     - Polychlorinated biphenyls (PCBs): HRGC/HRMS, GC/LRMS, GC/ECD;
     - Poly- and per-fluoroalkyl substances (PFAS): LC/MS/MS;
   - Are you interested in analyzing other compounds?
     - Chloride: spectrophotometer or titration;
     - Chlorine: titration;
     - Cyanide: spectrophotometer, titration, or anion exchange chromatography;
     - Fluoride: spectrophotometer or titration;
     - Nitrogen–Ammonia. Kjeldahl, Nitrate, Nitrite: spectrophotometer, titration, or ion-selective electrode;
     - Phosphorus: spectrophotometer
     - Sulfate, sulfide: spectrophotometer, t;urbidimetry, titration;
     - Total oxygen demand: oxygen probe or titration;
     - Biochemical oxygen demand: oxygen probe or titration;
     - Chemical oxygen demand: oxygen probe or titration;

   Gas
   - Are you analyzing soil vapor or landfill gas?

- Carbon dioxide, methane, nitrogen, and oxygen concentrations: GC with thermal conductivity detector (TCD);
- Sulfur compounds in natural gas and gaseous fuel: GC with chemiluminescence;
- Volatile organic compounds: GC/MS; and
- Nonmethane organic compounds in landfill gases: GC with flame ionization detector.

REFERENCES

Bish, D. L., and J. E. Post. 1989. "Modern powder diffraction." In Vol. 20, *Reviews in mineralogy*. Washington, DC: Mineralogical Society of America.

Chrysochoou, M. 2014. "Investigation of mineral dissolution rate and strength development in stabilized soils using quantitative x-ray diffraction." *J. Mater. Civ. Eng.* 26 (2): 288–295.

Coots, R. 2014. *PCB method comparison of high and low resolution sediment analysis*. Publication No. 14-03-009. Olympia, WA: Washington State Department of Ecology, Toxics Studies Unit, Environmental Assessment Program.

Cui, Z.-D., and Y. J. Jia. 2013. "Analysis of electron microscope images of soil pore structure for the study of land subsidence in centrifuge model tests of high-rise building groups." *Eng. Geol.* 164: 107–116.

D'Amore, J. J., S. R. Al-Abed, K. G. Scheckel, and J. A. Ryan. 2005. "Methods for metal speciation in soils." *J. Environ.Qual.* 34 (5):1707–1745.

De la Cruz, F., R. Green, G. Hater, J. Chanton, et al. 2016. "Comparison of field measurements to methane emissions models at a new landfill." *Environ. Sci. Technol.* 50 (17): 9432–9441.

EMC (Air Emission Measurement Center). "Air emission measurement center: The EPA hub for stationary source air emission test methods and procedures." Accessed April 29, 2024. https://www.epa.gov/emc.

Fernández-Jiménez, A. G., A. de la Torre, G. Palomo, and M. M. López-Olmo. 2006. "Quantitative determination of phases in the alkali activation of fly ash. Part I. Potential ash reactivity." *Fuel* 85 (5–6): 625–634.

Galle, B., J. Samuelsson, B. H. Svensson, and G. Borjesson. 2001. "Measurements of methane emissions from landfills using a time correlation tracer method based on FTIR absorption spectroscopy." *Environ. Sci. Technol.* 35 (1): 21–25.

Grim, R. E. 1968. Clay mineralogy. 2nd ed. New York: McGraw-Hill.

Klimesch, D. S., and A. Ray. 1998. "DTA ± TGA of unstirred autoclaved metakaolin ± lime ± quartz slurries. The formation of hydrogarnet." *Thermochim. Acta* 316 (2): 149–154.

Lin, B., and A. B. Cerato. 2014. "Applications of SEM and ESEM in microstructural investigation of shale-weathered expansive soils along swelling-shrinkage cycles." *Eng. Geol.* 177: 66–74.

Lothenbach, B., and F. Winnefeld. 2006. "Thermodynamic modeling of the hydration of Portland cement." *Cem. Concr. Res.* 36 (2): 209–226.

Mitchell, J. K., and K. Soga. 2005. *Fundamentals of soil behavior*. New York: Wiley.

USEPA (US Environmental Protection Agency). 2005. *Guidance for evaluating landfill gas emissions from closed or abandoned facilities*. EPA-600/R-05/123a. Washington, DC: USEPA.

USEPA. 2014. "The SW-846 compendium." Accessed February 26, 2024. https://www.epa.gov/hw-sw846/sw-846-compendium.

USEPA. 2024. "Approved CWA chemical test methods." Accessed February 26, 2024. https://www.epa.gov/cwa-methods/approved-cwa-chemical-test-methods.

Ward, C. R., and D. French. 2006. "Determination of glass content and estimation of glass composition in fly ash using quantitative x-ray diffractometry." *Fuel* 85 (16): 2268–2277.

Winburn R. S., D. G. Grier, G. McCarthy, and R. Peterson. 2000. "Rietveld quantitative x-ray diffraction analysis of NIST fly ash standard reference materials." *Powder Diffract.* 15 (3): 163–172.

Wright R. J., and T. Stuczynski. 1996. "Atomic absorption and flame emission spectrometry." In *Methods of soil analysis,* "Part 3—Chemical methods," edited by D. L. Sparks, A. L. Page, P. A. Helmke, R. H. Loeppert, et al. New York: Wiley.

PART 3
WASTE MATERIAL PROPERTIES

CHAPTER 4
HAZARDOUS WASTE

Prabir K. Kolay, Ph.D. P.E., F.ASCE and Rudolph Bonaparte, Ph.D., P.E., NAE

4.1 INTRODUCTION

In the United States, hazardous waste is defined in, and regulated by, the Resource Conservation and Recovery Act (RCRA). The Resource Conservation and Recovery Act, passed by Congress in 1976 and updated numerous times since then, including the substantive changes in the Hazardous and Solid Waste Amendments (HSWA) of 1984, describes the program mandated by Congress for management of the country's solid waste. The RCRA statute gives the Environmental Protection Agency (EPA) authority to develop the RCRA regulatory framework and to administer the RCRA program. Regulations promulgated by EPA under the RCRA statute are contained in Title 40 of the Federal Code of Regulations, Parts 239 to 272 (40 CFR §239 to §272). Subtitle C of RCRA focuses on hazardous solid waste. Subtitle D of RCRA focuses on nonhazardous solid waste [e.g., municipal solid waste (Chapter 5) and coal combustion residuals (Chapter 6)]. Note too that RCRA allows EPA to authorize individual states to take the lead in administering portions of the RCRA hazardous waste program if the state meets certain conditions. A state can impose regulatory requirements in addition to those in federal regulations, but in no case can the state's regulations be less stringent than the federal regulations.

The remainder of this chapter addresses only federal regulations and describes (1) the RCRA definition of and identification process for hazardous waste; (2) the cradle-to-grave regulatory philosophy and framework for hazardous waste management in the United States; (3) regulations governing the permitting, design, and operation of hazardous waste treatment, storage, and disposal facilities (TSDFs); (4) RCRA restrictions on the land disposal (i.e., landfilling) of hazardous wastes; (5)

remediation of contamination (soil, groundwater, surface water) at permitted RCRA hazardous waste TSDFs; (6) remediation of contamination (containing hazardous waste, hazardous substances, and/or hazardous constituents, defined subsequently) at non-RCRA sites under the Comprehensive Environmental Response, Compensation and Liability Act (CERCLA); (7) special regulatory requirements for managing and disposing one class of hazardous chemicals regulated separately from RCRA hazardous wastes, those being polychlorinated biphenyls (PCBs); and (8) special requirements for mixed waste [i.e., hazardous waste and low-level radioactive waste (LLRW)].

4.2 DEFINITION AND IDENTIFICATION PROCESS FOR HAZARDOUS WASTE

Under the RCRA statute, hazardous waste is defined as any waste material in solid, liquid, or gaseous form that "because of its quantity, concentration, or physical, chemical or infectious characteristic may cause or significantly contribute to an increase in mortality, serious irreversible illness, or incapacitating reversible illness; or pose a substantial present or potential hazard to human health or the environment when improperly treated, stored, transported or disposed of, or otherwise managed."

In implementing RCRA, EPA developed a regulatory framework and process to identify whether a substance is a regulated hazardous waste. The framework has several steps, as shown in Figure 4-1, and it can be quite complex to achieve full implementation. Recognizing this, EPA provides extensive online guidance and information to help determine whether a substance is a hazardous waste.

Referring to Figure 4-1, for a material to be classified as a hazardous waste, it must first be a solid waste. Therefore, the first step in the

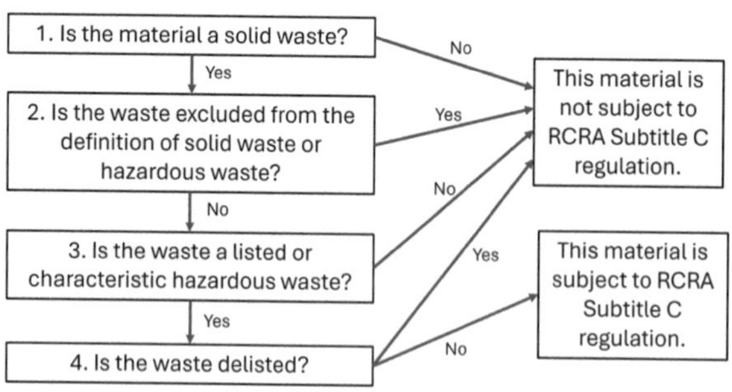

Figure 4-1. The hazardous waste identification process.
Source: USEPA (2024a).

hazardous waste identification process is determining whether a material is a solid waste. Although identifying a material as a solid waste seems intuitively straightforward, EPA provides a specific regulatory definition for a solid waste (USEPA 2005, 2024b).

The second step in the process examines whether the solid waste identified in the first step is specifically excluded by EPA from regulation as a solid or hazardous waste. Several materials are excluded from the definition of solid waste for a variety of reasons, including public policy, economic impacts, regulation by other laws, lack of data, or impracticability of regulating the waste. Thus, wastes that are excluded from the definition of solid waste are not subject to RCRA Subtitle C hazardous waste regulations. A list of wastes that are excluded from the definition of solid waste is contained in 40 CFR §261.4(a). The list contains 25 wastes, to name just a few: domestic sewage, radioactive waste, spent sulfuric acid, spent wood preservatives, coke by-product wastes, and excluded scrap metal.

Once it is determined that a waste meets the definition of a solid waste and it is not an excluded waste, the third step is to investigate whether the waste satisfies the criteria to be classified in either of the two regulatory categories of hazardous waste, namely, a *listed* hazardous waste or a *characteristic* hazardous waste. Listed wastes are wastes or discarded commercial products from common manufacturing and industrial processes. Characteristic wastes are wastes that exhibit any one or more of the following characteristic properties: ignitability, corrosivity, reactivity, or toxicity.

4.2.1 Listed Hazardous Waste

A nonexcluded solid waste is a listed hazardous waste if it is on one of four lists given in 40 CFR §261. The four categories of listed waste are as follows:

1. F-list (40 CFR §261.31): These are wastes from common manufacturing and industrial processes. F-listed wastes can be subdivided into seven groups depending on the type of manufacturing or industrial process that created them: (1) spent solvent wastes, (2) electroplating and other metal finishing wastes, (3) dioxin-bearing wastes, (4) chlorinated aliphatic hydrocarbon production wastes, (5) wood-preserving wastes, (6) petroleum refinery wastewater treatment sludges, and (7) multisource leachate.
2. K-list (40 CFR §261.32): These are wastes from specific sectors of industry and manufacturing and are considered source-specific wastes. To qualify as a K-listed hazardous waste, a waste must fit into one of the 13 categories on the list and the waste must match one of

the 148 detailed K-list waste descriptions in 40 CFR §261.32. A few examples of the industries that generate K-listed wastes are: (1) organic chemical manufacturing, (2) inorganic chemical manufacturing, (3) pesticide manufacturing, (4) explosives manufacturing, (5) secondary lead processing, and (6) coking (processing coal to produce coke).

3. P-list and U-list (40 CFR §261.33): These are discarded commercial chemical products, off-specification products, container residues of these products, and spill residues of these products. In comparison with U-listed wastes, P-listed wastes are considered by EPA to be acutely hazardous. A table identifying 239 P-listed wastes and 485 U-listed wastes is given in 40 CFR §261.33(e).

4.2.2 Characteristic Hazardous Waste

According to EPA, a hazardous waste characteristic is a property that, when present in a waste, indicates that the waste poses enough threat to merit regulation as hazardous. EPA has established four hazardous waste characteristics: ignitability, corrosivity, reactivity, and toxicity.

- Ignitability (40 CFR §261.21): Wastes that are hazardous because of the ignitability characteristic include liquids with flash points below 60 °C, nonliquids that cause fire through specific conditions, ignitable compressed gases, and oxidizers. EPA has assigned D001 as the characteristic waste code for ignitable hazardous wastes. Test methods for ignitability include the Flash Point by Pensky-Martens Closed-Cup Tester (SW-846Test Method 1010A) (USEPA 2014b), the Setaflash Closed-Cup Method for Determining Ignitability (SW-846 Test Method 1020B) (USEPA 2004), and the Ignitability of Solids test method (SW-846 Test Method 1030) (USEPA 2014a).
- Corrosivity (40 CFR §261.22): Wastes that are hazardous because of the corrosivity characteristic include aqueous wastes with a pH of less than or equal to 2 or greater than or equal to 12.5, as determined using a pH meter (SW-846 Test Method 9040C). Wastes with the corrosivity characteristic include those that corrode plain carbon steel (Society of Automotive Engineers [SAE] grade 1020 steel) at a rate faster than 6.35 mm (0.250 in.) per year at a test temperature of 55°C (130°F), as determined using SW-846 Test Method 1110A. EPA has assigned D002 as the characteristic waste code for ignitable hazardous wastes.
- Reactivity (40 CFR §261.23): Wastes that are hazardous because of the reactivity characteristic may be unstable under normal conditions, may react with water, may give off toxic gases, and may be capable of detonation or explosion under normal conditions or when heated. EPA assigned D003 as the waste code for reactive hazardous wastes.

40 CFR §261.23 does not provide specific test methods to determine reactivity. Instead, EPA uses narrative criteria to define reactive wastes and allows waste generators to use their judgment in determining whether a waste is sufficiently reactive to be regulated. EPA indicates that this is acceptable because reactive hazardous wastes are relatively uncommon and the risks they pose are well-known to the few waste generators who deal with them.

- Toxicity (40 CFR §261.24): Wastes that are hazardous because of the toxicity characteristic are harmful when ingested or absorbed by humans. Toxic wastes present a concern as they may be able to leach from waste and pollute groundwater and surface water. The toxicity of a waste is determined by the Toxicity Characteristic Leaching Procedure (TCLP) (SW-846 Test Method 1311, USEPA 1992). EPA assigned wastes codes D004 through D043 for the suite of chemicals for which toxicity characteristic TCLP concentrations have been developed.

4.2.3 Hazardous Wastes, Constituents, Chemicals, and Substances

In practice, it is not uncommon to encounter the terms hazardous waste, hazardous substance, and hazardous constituent. Each has a specific definition and the terms are not interchangeable. Moreover, based on the authors' experience, the terms are not always used correctly. The correct definitions are as follows:

- Hazardous wastes are defined in 40 CFR §261 as explained in the preceding parts of this chapter.
- Hazardous constituents are defined in 40 CFR §268.2 as those constituents listed in Appendix VIII of 40 CFR §261.
- Hazardous chemicals are defined by US Occupational Safety and Health Administration (OSHA) in 29 CFR §1910.1200 as any chemical which is a physical hazard or a health hazard, a simple asphyxiant, combustible dust, pyrophoric gas, or hazard not otherwise classified. OSHA illustrates the definition by saying that antifreeze, oil, and diesel fuel are all considered hazardous chemicals.
- Hazardous substance is a term that has found a variety of uses and definitions, for example, for use in real estate contracts. The definitions can be quite broad. Perhaps the most important and comprehensive definition is given in OSHA regulations at 29 CFR §1910.120, which defines a hazardous substance as any:
  - RCRA hazardous wastes as defined previously;
  - Substance listed by the US Department of Transportation (USDOT) as a hazardous material under 49 CFR §172.101 and its appendixes;
  - Substance defined under Section 103(14) of the Comprehensive Environmental Response Compensation and Liability Act (CERCLA);

- Biologic agents and other disease-causing agents which after release into the environment and upon exposure, ingestion, inhalation, or assimilation into any person, either directly from the environment or indirectly by ingestion through food chains, will or may reasonably be anticipated to cause death, disease, behavioral abnormalities, cancer, genetic mutation, physiological malfunctions (including malfunctions in reproduction) or physical deformations in such persons or their offspring.

4.3 CRADLE-TO-GRAVE REGULATORY FRAMEWORK

As already discussed, with RCRA and its amendments, Congress established a framework for the management of hazardous waste from "cradle to grave," meaning from the time the waste is created (cradle), through when it is transported and stored, and then through final recycling, treatment, or disposal (grave). This cradle to grave regulatory framework is briefly described subsequently.

4.3.1 Hazardous Waste Generation and Transportation

Under RCRA, hazardous waste generators are the first link in the hazardous waste management system. All generators must determine whether their waste is hazardous and they must oversee the fate of the waste; that is, generators must ensure and fully document that the hazardous waste that they produce is properly identified and managed prior to final recycling, treatment, or disposal. The degree of regulation that applies to each generator depends on the amount of waste they produce. EPA provides extensive guidance for hazardous waste generators (USEPA 2024c).

After generators produce a hazardous waste, transporters may move the waste to a facility that can use, recycle, treat, store, or dispose of the waste. Because such transporters are moving regulated wastes on public roads, highways, rails, and waterways, transporters must follow not only EPA hazardous waste regulations, but also the US Department of Transportation (USDOT) hazardous materials regulations. The EPA also provides guidance for hazardous waste transporters (USEPA 2024d). USDOT also has regulations applicable to the transport of hazardous materials (USDOT 2024).

4.3.2 Hazardous Waste Recycling, Treatment, Storage, and Disposal

Recycling hazardous waste has a variety of benefits, including reducing the consumption of raw materials and the volume of waste materials that must be treated and/or disposed. Recycling can reduce emissions of greenhouse gases (GHGs). When hazardous wastes are recycled, less energy is needed to obtain raw materials to manufacture new products. When energy demand decreases, fewer fossil fuels are burned and less GHGs are emitted into the atmosphere, which can help lessen the impacts of climate change and decrease air pollution. Information on EPA's hazardous waste recycling policies, regulations, and guidance can be found at USEPA (2024e).

RCRA treatment storage, and disposal facilities (TSDFs) are used to treat, temporarily store, and/or permanently dispose of hazardous wastes. Treatment storage, and disposal facilities are highly regulated by EPA through the RCRA program. RCRA TSDF regulations are administered through a facility permitting process (USEPA 2024). The RCRA permit will describe establish minimum facility operating, monitoring, and reporting requirements, minimum performance and design requirements for the various types of hazardous waste management units at the facility, and additional requirements designed to protect soil, groundwater, surface water, and air resources.

RCRA regulations covering hazardous waste landfills, surface impoundments, and waste piles are presented in, respectively, 40 CFR §264.300 to §317, §264.220 to §232, and §264.250 to §259. Because of their specific relevance to geoenvironmental practice, hazardous waste landfills are discussed in more detail in the next section of this chapter. As applicable, the requirements for surface impoundment and waste pile TSDFs generally follow those for landfills.

4.4 REQUIREMENTS FOR HAZARDOUS WASTE LANDFILLS

Geoenvironmental engineers often lead projects involving the permitting, design, performance monitoring, and closure of RCRA hazardous waste landfills. The regulatory requirements for these landfills start at 40 CFR §264.300. The regulations cover: (1) design and operating requirements, (2) action leakage rate, (3) monitoring and inspection, (4) response actions, (5) closure and postclosure care, and other requirements. Several of these requirements are discussed subsequently.

4.4.1 Double-Liner System

RCRA regulations governing the design and operation for hazardous waste landfills are intended to prevent leakage of landfill leachate from the landfill into the surrounding environment. Final cover systems placed on top of the landfill function to minimize leachate generation by preventing precipitation infiltration into the landfill. Final cover system requirements are discussed subsequently. Complementing the final cover system, the liner system beneath the landfill functions to contain landfill leachate, while providing systems to remove the leachate from the landfill as it is generated. For RCRA Subtitle C hazardous waste landfills, the entire landfill footprint must be underlain by a double-liner system. The double-liner system requirements are as follows (Figure 4-2):

1. Top (primary) liner designed and constructed of materials (e.g., a geomembrane) to prevent the migration of hazardous constituents into such liner during the active life and post-closure care period.
2. Composite bottom (secondary) liner, consisting of at least two components: (1) The upper component must be designed and constructed of materials (e.g., a geomembrane) to prevent the migration of hazardous constituents into this component during the active life and post-closure care period, and (2) the lower component must be designed and constructed of materials to minimize the migration of hazardous constituents if a breach in the upper component were to occur. The lower component must be constructed of at least 3 ft (91 cm) of compacted soil material with a hydraulic conductivity of no more than 1×10^{-7} cm/s.
3. *Leachate collection and removal system* (LCRS) immediately above the top liner must be designed, constructed, operated, and maintained to collect and remove leachate from the landfill during the active life and post-closure care period. The Regional Administrator will specify design and operating conditions in the permit to ensure that the leachate depth over the liner does not exceed 30 cm (one foot). Further, the LCRS must be
   - Constructed with a bottom slope of 1% or more,
   - Constructed of granular drainage materials with a hydraulic conductivity of 1×10^{-2} cm/s or more and a thickness of 12 in.

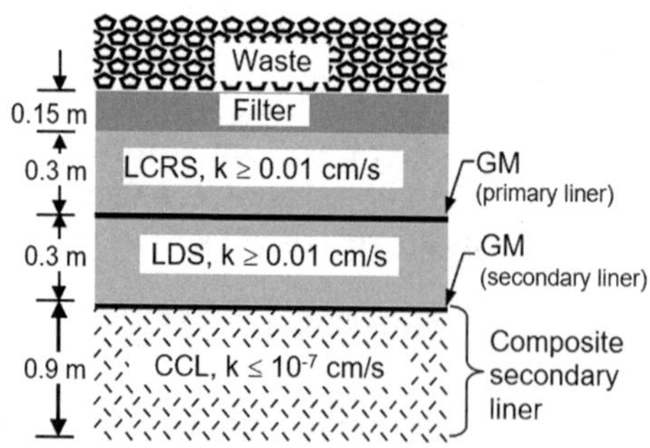

Figure 4-2. Example double-liner system meeting the minimum prescriptive requirements of 40 CFR §264.
Source: Bonaparte et al. (2002).

(30.5 cm) or more; or constructed of synthetic or geonet drainage materials with a transmissivity of 3×10^{-5} m^2/s or more,

- Constructed of materials that are chemically resistant to the waste managed in the landfill and the leachate expected to be generated, and of sufficient strength and thickness to prevent collapse under the pressures exerted by overlying wastes, waste cover materials, and equipment used at the landfill, and
- Designed and operated to minimize clogging during the active life and post-closure care period.

4. "*Leachate collection and removal system* between the liners, and immediately above the bottom composite liner in the case of multiple leachate collection and removal systems, is also a *leak detection system*. This leak detection system must be capable of detecting, collecting, and removing leaks of hazardous constituents at the earliest practicable time through all areas of the top liner likely to be exposed to waste or leachate during the active life and post-closure care period. The leak detection system must satisfy the bottom slope, chemical resistance, and clogging resistance requirements for LCRSs as described, and in addition it must be
   - Constructed of granular drainage materials with a hydraulic conductivity of 1×10^{-2} cm/s or more and a thickness of 12 in. (30.5 cm) or more; or constructed of synthetic or geonet drainage materials with a transmissivity of 3×10^{-5} m^2/s or more,
   - Constructed of materials that are chemically resistant to the waste managed in the landfill and the leachate expected to be generated, and of sufficient strength and thickness to prevent collapse under the pressures exerted by overlying wastes, waste cover materials, and equipment used at the landfill, and
   - Constructed with sumps and liquid removal methods (e.g., pumps) of sufficient size to collect and remove liquids from the sump and prevent liquids from backing up into the drainage layer.

Detailed information on the field performance of operating and closed landfills with double-liner systems can be found in Bonaparte et al. (2002), NRC (2007), and USEPA (2017).

4.4.2 Action Leakage Rate and Response Action Plan

To assure that any leachate leakage into the leak detection system between the top and the bottom liners is detected at the earliest practicable time, the concepts of an action leakage rate (ALR) and response action plan (RAP) were added to 40 CFR §264 in 1987 (USEPA 1987). A technical background document used by EPA to establish these regulatory requirements is presented in Bonaparte et al. (1987). The idea behind the ALR and RAP is to require the landfill owner/operator to address the cause of leakage through the top (primary) liner as soon as liquid flows in the leak detection system (LDS) exceed a certain rate, the ALR, and before there is time for any leakage (at a rate above the ALR) to migrate through the bottom liner should there be a leakage pathway through it (i.e., a defect in the bottom liner). The ALR is defined by EPA as the "maximum design flow rate that the LDS can remove without the fluid head on the bottom liner exceeding 1 ft. The action leakage rate must include an adequate safety margin to allow for uncertainties in the design (e.g., slope, hydraulic conductivity, thickness of drainage material), construction, operation, and location of the LDS, waste and leachate characteristics, and likelihood and amounts of other sources of liquids in the LDS." The ALR must also consider decreases in the flow capacity of the system over time resulting from siltation and clogging, rib layover and creep of geosynthetic drainage layers, overburden pressures, and other potential mechanisms for reduction in flow capacity. Flow rates from the LDS are monitored on a routine basis to assess whether the ALR has been exceeded.

The performance of primary liners at double-lined landfills in limiting leakage of leachate is generally inferred by comparing LCRS and LDS flow rate and chemical constituent data. Studies of operational landfill data by Bonaparte and Gross (1990) and Gross et al. (1990) identified that not all flows in an LDS are typically caused by top liner leakage. In determining whether the ALR has been exceeded, the total flow rate for the LDS is used and potential nonleakage sources of flow in the LDS are not considered. However, in evaluating the need and scope for response actions (discussed subsequently), an assessment is made as to whether the observed flow in the LDS is caused by top liner leakage or other sources. Bonaparte and Gross (1990) identified four potential sources of liquids contributing to LDS flow (Figure 4-3), these being

1. Top (primary) liner leakage (Source A in Figure 4-3);
2. Construction water and compression water (Source B), comprising water (mostly rainwater) that infiltrates the LDS during construction and continues to drain to the LDS sump after the start of the facility operation;
3. Consolidation water expelled into the LDS from the CCL/GCL components of a composite top liner because of clay consolidation under the weight of the waste (Source C); and
4. Groundwater (including perched water) that percolates vertically or laterally through the bottom liner from outside the landfill or other external sources of water (e.g., condensation of water vapor in any gases that seep into the LDS) (Source D).

The RAP is an EPA-approved document that describes the actions the facility owner or operator must take if the ALR is exceeded. These actions include (1) submit a preliminary assessment of the cause(s)

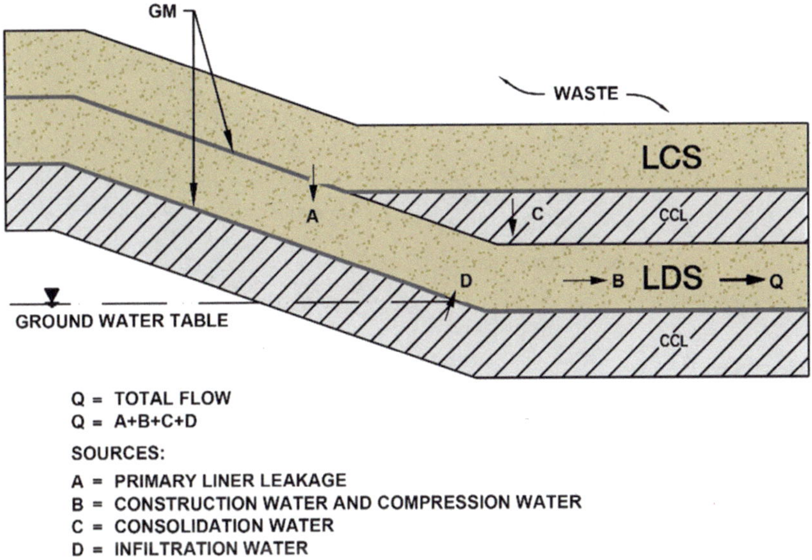

Figure 4-3. Possible sources of liquids in the LDS of double-liner systems.
Source: Bonaparte and Gross (1990).
Note: GM = geomembrane.

of the exceedance to EPA within 14 days of determination that the ALR has been exceeded; (2) evaluate, as practicable, the location, size, and cause of any leak; (3) evaluate whether waste receipt into the affected landfill cell or area should be temporarily or permanently stopped, and whether any waste need to be removed for inspection or to repair the liner system; and (4) take other measures as necessary to mitigate the leakage and other actions as spelled out in 40 CFR §264.304(c). The evaluation under (1) will consider the timing of flows in the LDS, an evaluation of the chemical constituents in the flows and comparison to the constituents in the LCRS flows, and quantitative analyses, to assess whether the LDS flows are caused by top liner leakage or, instead, by another source.

4.4.3 Closure and Postclosure Care

As discussed, final cover systems placed on top of the landfill function to minimize leachate generation by preventing precipitation infiltration into it. The effect of final cover system placement on leachate generation potential in a landfill is illustrated in Figure 4-4.

The regulatory requirements for closure of RCRA Subtitle C hazardous waste landfills are contained in 40 CFR §264.310. The closure performance standards are as follows:

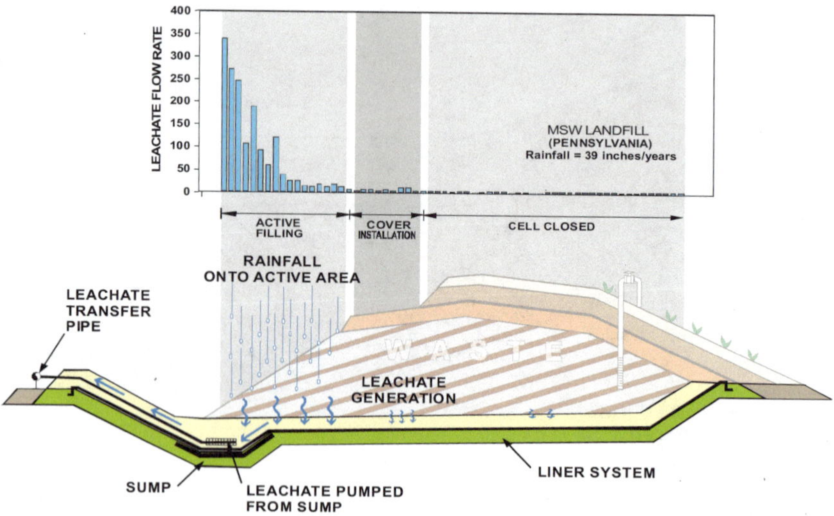

Figure 4-4. Relationship between leachate generation and placement of the final cover system. Time units for leachate flow histogram are months. Units of flow measurements are gallons per acre per day (multiply times 10 for liters per hectare per day).
Source: Bonaparte (1995).

- Provide long-term minimization of migration of liquids through the closed landfill,
- Function with minimum maintenance,
- Promote drainage and minimize erosion or abrasion of the cover system,
- Accommodate settling and subsidence so that the cover system's integrity is maintained, and
- Achieve a cover system permeability less than or equal to the permeability of any bottom liner system and natural subsoils present.

As described in Bonaparte et al. (2004), the regulations of 40 CFR §264.310 provide for a performance-based design of the final cover system; there are no prescriptive minimum design requirements in the regulations. However, USEPA (1989) has issued minimum technology guidance for a final cover system that they have interpreted as likely meeting, for many project-specific conditions, the performance standards of §264.310. The final cover system given in the guidance consists of the following (Figure 4-5):

- Top layer containing two components: (1) either a vegetated or an armored surface layer, selected to minimize erosion and, to the extent possible, promote drainage off the cover; and (2) a protection layer,

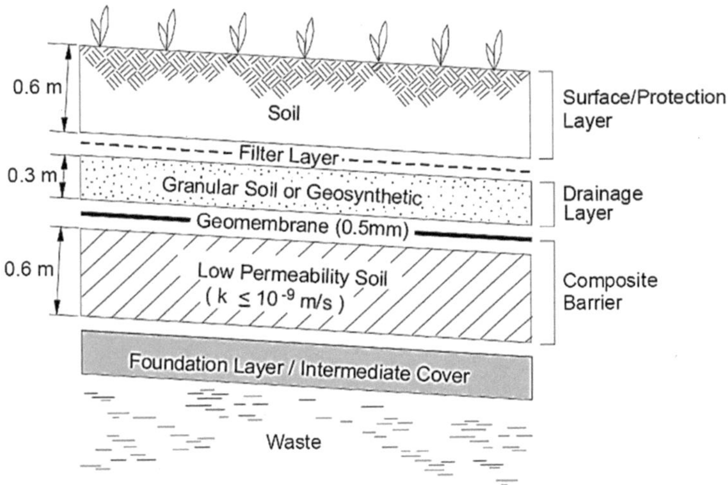

Figure 4-5. Hazardous waste landfill final cover system based on EPA (1989) minimum technology guidance.
Source: Bonaparte et al. (2004).

comprising topsoil and/or fill soil, as appropriate; the recommended top layer surface slope is 3% to 5%; the 1989 EPA guidance notes that the top layer soil component should be at least 0.6 m thick, and that a greater thickness may be required to assure that the underlying hydraulic barrier is below the frost zone,

- Soil drainage layer with a minimum thickness of 0.3 m and a minimum hydraulic conductivity of 1×10^{-4} m/s that will effectively "minimize water infiltration into the underlying low-permeability barrier" and have a final slope of at least 3% after settlement and subsidence, or a drainage layer consisting of a geosynthetic material with performance characteristics equivalent to the soil drainage layer, and
- Composite hydraulic barrier consisting of (1) a geomembrane of a minimum thickness of 0.5 mm; underlain by (2) a compacted clay liner with a minimum thickness of 0.6 m and a maximum hydraulic conductivity of 1×10^{-9} m/s; the 1989 EPA guidance notes that the entire hydraulic barrier should lie below the frost zone.

The requirements for postclosure care (for the duration of the postclosure care period, which is a minimum of 30 years) are as follows:

1. Maintain the integrity and effectiveness of the final cover system, including making repairs to it as necessary to correct the effects of settling, subsidence, erosion, or other events;
2. Continue to operate the LCRS system until leachate is no longer detected;

3. Maintain and monitor the LDS and comply with all applicable requirements;
4. Maintain and monitor the landfill's groundwater monitoring system and comply with all applicable requirements;
5. Prevent runon and runoff from eroding or otherwise damaging the final cover system; and
6. Protect and maintain surveyed benchmarks.

4.5 RESOURCE CONSERVATION AND RECOVERY ACT LAND DISPOSAL RESTRICTIONS

4.5.1 Basic Requirements and Treatment Standards

The 1984 RCRA HSWA placed severe limitations on which hazardous wastes can be disposed of in hazardous waste landfills. These limitations are codified in the EPA Land Disposal Restriction (LDR) contained in 40 CFR §268.1. The RCRA HSWA for the first time prohibited the land disposal of untreated hazardous wastes. The LDR program identifies treatment standards for hazardous wastes and specifies requirements that generators, transporters, and owners or operators of TSDFs that manage hazardous wastes destined for land disposal must meet. The LDRs do not apply to wastes discharged to surface waters [regulated under EPA's National Pollutant Discharge Elimination System (NPDES)] or to groundwater [where drinking water maximum contaminant levels (MCLs) or other regulatory criteria apply].

The LDR program is based on three overall principles:

1. Disposal prohibition: Requires that waste-specific treatment standards must be met before a waste can be land-disposed;
2. Dilution prohibition: Prohibits achievement of the disposal requirements of a waste through the process of dilution (e.g., mixing clean soil with a contaminated soil to reduce the average concentration of the hazardous constituents in the soil to below the treatment standards); and
3. Storage prohibition: Prohibits the indefinite storage of hazardous wastes at a TSDF, or anywhere else, instead of treating the waste promptly.

Treatment in the LDR program is defined as a process that destroys, removes, or permanently immobilizes hazardous constituents. Treatment standards are established on the basis of best-demonstrated available technology (BDAT). The LDR program includes three types of treatment standards: (1) maximum allowable concentration levels for land disposal; (2) use of a specified treatment technology (e.g., vitrification); and (3) a "no land disposal" treatment standard when a specific waste is no longer generated, is completely destroyed through treatment, or no residuals are produced from treatment. One example of an LDR is the treatment standard for a waste that is characteristically hazardous because of arsenic. The treatment standard is 1.4 mg/L for wastewaters and 5.0 mg/L for nonwastewaters as measured in a TCLP test. Detailed information on the treatment standards for hazardous wastes under the LDR program can be found in 40 CFR §268.40 (USEPA 2002a, 2024g).

4.5.2 Alternative Land Disposal Restrictions for Contaminated Soils

Geoenvironmental engineers are often involved in the design of remediation projects where the management and disposal of contaminated soils must be addressed. On this topic, in 1998, EPA promulgated LDR treatment standards specific to contaminated soils. These are referenced as "LDR Alternative Soil Treatment Standards" and they are contained in 40 CFR §268.49. They are also described in USEPA (2002a). The agency promulgated the standards to provide more flexible treatment and disposal requirements for contaminated soils given the heterogeneous nature of soil matrixes, the potentially very large volumes of remediation wastes, and the lack of effectiveness of available treatment technologies developed for industrial hazardous wastes. Under 40 CFR §268.49, there are several requirements to comply with the LDR alternative soil treatment standards when a soil has hazardous constituents above the applicable treatment standards and it cannot be treated to meet them:

1. Total concentrations of *nonmetal* hazardous constituents in the soil must be reduced by at least 90% through treatment so that no more than 10% of their initial concentration remains;
2. TCLP concentrations of *metal* hazardous constituents in the soil must be reduced by at least 90% through treatment so that no more than 10% of their initial concentration remains;
3. Alternatively, hazardous constituents must not exceed 10 times the universal treatment standards (UTS) contained in 40 CFR §268.49; and
4. In addition, contaminated soils that exhibit the characteristic of ignitability, corrosivity, or reactivity must be treated to eliminate these characteristics before the soils can undergo landfill disposal.

The purpose of the LDR treatment standards is to ensure that prohibited hazardous wastes are properly treated before disposal (i.e., treated so that short- and long-term threats to human health and the environment posed by landfill disposal are minimized). The alternative soil treatment standards, like other LDR treatment standards, are intended to achieve this goal. Potential treatment technologies include chemical treatment (e.g., neutralization, oxidation, reduction), thermal treatment (e.g., incineration, thermal desorption, smoldering), and physical treatment (e.g., solidification, separation, evaporation, filtration).

4.6 REMEDIATION OF CONTAMINATION AT RESOURCE CONSERVATION AND RECOVERY ACT HAZARDOUS WASTE TREATMENT, STORAGE, AND DISPOSAL FACILITIES

Geoenvironmental engineers are often involved in the investigation and remedial design of sites with soil, groundwater, and surface water contamination, including those at permitted RCRA Subtitle C TSDFs. The 1984 RCRA HSWA established a regulatory framework (known as the RCRA corrective action program) for the investigation and remediation of RCRA Subtitle C TSDFs that may have caused contamination of soil, groundwater, or surface water. The RCRA permit for each hazardous waste TSDF in the United States contains requirements for the facility owner or operator to implement a corrective action process if the facility detects a release of contaminants above permitted levels. EPA enforces the corrective action program along with the 44 states and territories that the agency has authorized to do so. Nearly 4,000 US facilities have undergone investigation and cleanup under the RCRA program, including former chemical manufacturing plants, oil refineries, lead smelters, wood preservers, steel mills, commercial landfills, federal facilities, and a variety of other types of facilities. A typical RCRA corrective action process includes: (1) initial site assessment, (2) site characterization, (3) interim actions to address immediate or short-term risks, (4) evaluation of remedial alternatives and selection of a specific site remedy, (5) implementation of the selected site remedy, (6) tracking cleanup progress, and (7) long-term monitoring and care. Regulations for the RCRA corrective action program are contained in 40 CFR §264.100. Additional information on the RCRA corrective action program can be found in USEPA (1998, 2024h, and 2024i).

It is important to note that not all contaminated sites are subject to RCRA corrective action requirements. These requirements apply only to hazardous waste TSDFs. Cleanup of contamination at sites and facilities outside of the RCRA Subtitle C permit program is discussed subsequently.

4.7 REMEDIATION OF CONTAMINATION AT NON-RESOURCE CONSERVATION AND RECOVERY ACT SITES UNDER COMPREHENSIVE ENVIRONMENTAL RESPONSE COMPENSATION AND LIABILITY ACT

4.7.1 Comprehensive Environmental Response Compensation and Liability Act Basics

As just noted, remediation of non-RCRA-Subtitle C contaminated sites may be regulated under one of any number of federal or state programs. These include, for example, 40 CFR §258 Subpart E for corrective actions at RCRA Subtitle D municipal solid waste landfills and 40 CFR §257 Subpart D for electric power coal combustion residual (CCR) landfills and surface impoundments. Perhaps the most far-ranging and important of these federal and state programs is CERCLA, more commonly known as Superfund. CERCLA was established by Congress in 1980 in response to national attention given to human health impacts and risks associated with several toxic waste dumps around the country, most notably Love Canal in Niagara Falls, New York (USEPA 2024j) and Valley of the Drums in Brooks, Kentucky (USEPA 2024k). Some of the most contaminated sites in the country are being investigated under the CERCLA program. CERCLA includes not only upland sites with facility, soil, groundwater, and/or surface-water contamination, but also contaminated sediment sites in rivers, lakes, ports, and harbors where the contamination creates risks for humans (e.g., from swimming, ingesting, fishing) and also for the ecological community (e.g., food chain effects, from benthic organisms to birds and mammals). These sediment sites can be very large and the names of some of them are well-known, for example: New Bedford Harbor (Massachusetts), Hudson River (New York), Gowanus Canal and Newtown Creek (New York City), Lower Passaic River Study Area and Berry's Creek Study Area (New Jersey), Portland Harbor (Oregon), and Lower Duwamish River (Washington). A photograph of a CERCLA remedial action (defined subsequently) at the Gowanus Canal Superfund site is shown in Figure 4-6.

CERCLA is of importance to geoenvironmental engineers because many of the approximately 1,200 current (and 41 proposed) sites on the CERCLA list will involve geoenvironmental engineers in the investigation and design of the cleanups at these sites. In addition to the current and proposed sites, as of 2022, investigation and cleanup (as needed) have been completed at 431 CERCLA sites. A map showing the locations of CERCLA sites around the United States is presented in Figure 4-7.

An important part of the CERCLA statute (§104 and §106) is the enforcement authority it gives EPA. Through CERCLA, EPA has the authority to seek out those parties responsible for any release and assure

Figure 4-6. Mechanical dredging of contaminated sediment at the Gowanus Canal superfund site, Brooklyn, New York City, 2021.
Source: Courtesy of Geosyntec Consultants.

their cooperation in the cleanup. EPA obtains private party cleanup through various enforcement tools, such as orders, consent decrees, and others. EPA itself contracts for the design and implementation of site cleanups when potentially responsible parties cannot be identified or located (i.e., orphan sites), or when they fail to act. In the case where a responsible party fails to act and EPA implements the remedy, the agency will undertake legal cost recovery proceedings from financially viable parties.

4.7.2 Comprehensive Environmental Response Compensation and Liability Act Removal Actions and Remedial Actions

CERCLA regulatory requirements are contained in 40 CFR §300, a section of the federal code of regulations titled "National Oil and Hazardous Substances Pollution Contingency Plan." Under CERCLA, there are two categories of cleanup actions, namely, removal actions and remedial actions. In general, removal actions are interim actions to clean up or remove contaminated materials when the materials are judged to pose an unacceptable immediate or short-term risk. In contrast, remedial actions are meant to be permanent actions to remediate a site to specified cleanup standards. These two types of CERCLA actions are briefly discussed in the following paragraphs.

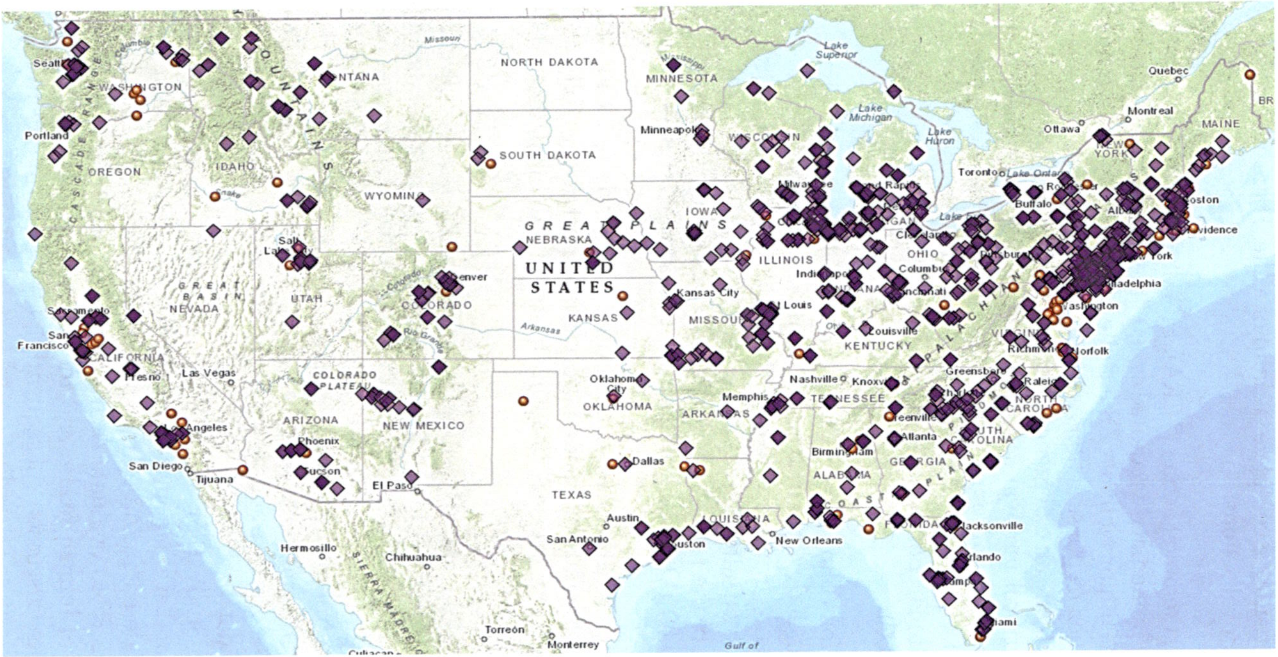

Figure 4-7. Locations of CERCLA sites around the United States.
Source: USEPA (2024r).
Note: The blue diamonds are sites being cleaned up by Responsible Parties under settlement agreements and orders with EPA. The red circles are federal facility sites being addressed through federal facility agreements between the facility owner and EPA. Abandoned sites for which responsible parties could not be found and addressed by EPA as orphan sites are not shown.

Removal actions are meant to be carried out more quickly than remedial actions to reduce or eliminate an immediate or short-term risk without necessarily effectuating a complete cleanup of a site. There are two broad categories of removal actions:

(1) Time-critical removal action (TCRA): EPA determines that response actions are immediately required when "there is an immediate risk to public health or welfare of the United States or the environment and continued response actions are immediately required to prevent, limit, or mitigate an emergency" [40 CFR §415(b)(5)(i)].
(2) Nontime critical removal action (NTCRA): EPA determines that a removal action is appropriate but that a planning and design period of 6 months or more is acceptable, based on exposure and risk considerations, before on-site cleanup activities must begin (USEPA 1993).

Regulations for conducting TCRAs and NTCRAs require that the removal action include a removal site evaluation and inspection (40 CFR §410), an engineering evaluation/cost analysis (EE/CA) that is a structured analysis or removal alternatives for a site [40 CFR §415(b)(4)], and implementation of the removal action using the selected alternative [40 CFR §415(d) and (e)]. Detailed guidance on conducting NTCRAs is found at USEPA (2024l).

CERCLA removal actions are carried out at relatively few CERCLA sites. In contrast, the remedial action process is almost always followed. The remedial action process has the following defined steps (40 CFR §430):

1. Remedial Investigation (RI): RI serves as the mechanism for collecting data to characterize site conditions, assess the nature and extent of contamination, including the affected media (structures, soil, sediment, groundwater, surface water), assess risk to human health and the environment (baseline human health and ecological risk assessments), and conduct treatability testing to evaluate the potential performance and cost of any treatment technologies that are being considered.
2. Feasibility Study (FS): FS is the mechanism for development, screening, and evaluation of the alternative remedial actions that are being considered, including the "no action" alternative. The criteria used in the FS to evaluate each remedial alternative include: (1) protection of human health and the environment; (2) compliance with ARARs (discussed subsequently); (3) long-term effectiveness and permanence; (4) reduction of toxicity, mobility, or volume

through treatment; (5) short-term effectiveness; (6) implementability; and (7) cost.

3. Proposed Plan (PP): PP announces to stakeholders, including the public, EPA's plan to select a specific remedy for the CERCLA site. The selected remedy is typically selected from one of the alternatives evaluated in the FS; however, the agency has the authority to modify the selected FS alternative if they deem it necessary. Once the PP is issued, stakeholders and the public are invited to comment on the plan.
4. Record of Decision (ROD): ROD is the EPA document where the agency presents its final decision on the selection of a specific remedy for implementation at the CERCLA site. As an appendix to the ROD, EPA must provide a Responsiveness Summary where they provide responses to the stakeholder and public questions received after issuance of the PP.
5. Remedial Design (RD): RD is the phase in the CERCLA site cleanup where the ROD-selected remedy is designed. The RD typically includes extensive engineering analysis, the development of drawings and specifications, and the development of a variety of work plans, including a construction quality assurance/control plan (CQACP).
6. Remedial Action (RA): RA follows the RD phase of the project. The RA involves the actual implementation and construction of the cleanup project.
7. Post Construction Completion (PCC): Once the RA is complete, the CERCLA site goes into a PCC phase. PCC activities, including long-term monitoring, are intended to ensure that the site remedy is maintained and remains protective of human health and the environment. PCC also includes post-RA operation and maintenance of the remedy and post-RA remedy optimization.
8. Five-Year Reviews: EPA conducts 5-year reviews of completed RAs when the level of hazardous substances left at the site remains above levels that permit unlimited use and unrestricted exposure. Five-year reviews provide an opportunity to evaluate the implementation and performance of a remedy to determine whether it remains protective of human health and the environment.

Guidance for conducting CERCLA RI/FSs is contained in USEPA (1988). Guidance for conducting CERCLA RD/RAs is contained in USEPA (1995). Guidance for how EPA prepares RODs is contained in USEPA (1999). A photograph of remedial construction at the Bailey Superfund Site is shown in Figure 4-8. The remedial design for this type of site is often led by geoenvironmental engineers.

Figure 4-8. Installation of cover system and leachate interceptor trenches, and contaminated sediment removal from marsh, Bailey Superfund Site, Orange, Texas.
Source: Courtesy of Geosyntec Consultants.

4.7.3 Comprehensive Environmental Response Compensation and Liability Act Applicable or Relevant and Appropriate Requirements

Unlike RCRA Subtitle C, the cleanup goals and requirements that must be achieved through the CERCLA RD/RA process are not spelled out but must be evaluated as part of the FS through the identification and evaluation of Applicable or Relevant and Appropriate Requirements (ARARs). Remedial actions under CERCLA are required to satisfy ARARs (unless a waiver is obtained from EPA), which are defined as any standard, requirement, criteria, or limitation under federal environmental law or more stringent promulgated standard, requirement, criteria or limitation under state environmental or facility siting law that is legally "applicable" to the hazardous substance (or pollutant or contaminant) concerned or is "relevant and appropriate" under the circumstances of the release. Additional information on ARARs can be found in USEPA (2024m). Definitions for "applicable" and "relevant and appropriate" are given in USEPA (1991). There are three types of ARARs under CERCLA. These are listed as follows and a few examples of ARARs that must commonly be considered for CERCLA sites are given:

- Location-specific ARARs restrict activities in certain specific areas. Examples of location-specific ARARs include the floodplain management and wetland protection portions of the Clean Water Act (CWA), the Endangered Species Act, the Rivers and Harbors Act of 1899, and the National Historic Preservation Act.

- Action-specific ARARs are usually technology- or activity-based requirements or limitations on actions or conditions involving specific substances. Examples of action-specific ARARs are the RCRA Subtitle C and D regulations applicable to hazardous waste and municipal waste, respectively. Another example is TSCA, which is applicable to wastes containing PCBs, as discussed subsequently.
- Chemical-specific ARARs are usually health- or riskbased numerical values or methodologies used to establish acceptable concentrations of chemicals that may be found in or discharged to the environment. An example of this type of ARAR is EPA drinking water maximum contaminant levels (MCLs) applied to groundwater, for example, MCLs that establish safe levels in drinking water. A second is soil cleanup standards (i.e., maximum allowable contaminant concentrations in soil) contained in the environmental regulations of many states.

4.7.4 Redevelopment of Remediated Comprehensive Environmental Response Compensation and Liability Act Sites

Geoenvironmental engineers often have a leading role in the design of redevelopment plans for remediated CERCLA sites. As appropriate to the contaminants remaining on-site after remediation is complete, EPA encourages the beneficial redevelopment of CERCLA sites. These sites often have many challenges that are best addressed by geoenvironmental engineers, including such things as (1) isolating residual site contaminants from redevelopment features, (2) addressing geotechnical issues on redevelopment sites that might contain closed landfills that have compressible wastes, (3) installing vapor barriers and gas collection systems at sites that generate residual vapors and gases, (4) designing foundations for structures that account for site limitations (e.g., the potential need for piles to penetrate cover systems, prohibitions on driving piles through waste, and others), and (5) designing long-term site monitoring features (e.g., groundwater and gas monitoring wells) in harmony with the site redevelopment features. EPA policy, guidance, and resources related to CERCLA site redevelopment can be found in USEPA (2002b, 2024n). An example of a successful CERCLA site redevelopment project in which geoenvironmental engineers had a major role is the McColl Superfund Site in Fullerton, California (Figure 4-9). The former industrial waste disposal facility had 12 unlined waste pits used for the disposal of petroleum sludges and other wastes. Remediation of the site involved installation of an RCRA Subtitle C-equivalent final cover system over the site, a vertical subsurface soil-bentonite cutoff wall around the site, and a subsurface gas collection and removal system. The site now contains several holes of an adjacent golf course.

Figure 4-9. Redeveloped McColl Superfund Site in Fullerton, California. Source: Courtesy of Geosyntec Consultants.

4.8 SPECIAL REQUIREMENTS FOR MANAGING AND DISPOSING POLYCHLORINATED BIPHENYL WASTES

The Toxic Substances Control Act (TSCA) of 1976 provides EPA with authority to require reporting, record-keeping, and testing requirements, and restrictions relating to chemical substances and/or mixtures. TSCA requires EPA to maintain and publish a list of each chemical substance that is manufactured or processed, including imports, in the United States. This "TSCA inventory" now includes more than 86,000 chemicals.

Of relevance to geoenvironmental engineers, TSCA addresses the production, use, and disposal of polychlorinated biphenyls (PCBs). PCBs are regulated by EPA through 40 CFR §761. Because of their nonflammability, chemical stability, high boiling point, and electrical insulating properties, PCBs were used in hundreds of industrial and commercial applications, including electrical, heat transfer, and hydraulic equipment, and plasticizers in paints, plastics, and rubber products. On the flip side, PCBs are toxic and pose serious health risks to fetuses, babies, and children, who may suffer developmental and neurological problems from prolonged or repeated exposure to even small amounts of PCBs. They are harmful to adults as well. Because of their widespread use and toxicity, PCBs are a relatively common remediation waste in the United States, and if present in the waste at high-enough concentrations, the waste must be managed as a TSCA waste and its disposal can only be in a TSCA-regulated landfill. EPA guidance and information on PCBs can be found in USEPA (2024o).

40 CFR §761.61 describes cleanup and disposal options for PCB remediation wastes in soils, sediments, dredged materials, and sewage

and industrial sludges (defined in 40 CFR §761.61 as bulk wastes). The PCB cleanup concentrations for high-occupancy areas are not more than 1 part per million (ppm) without further measures, with capping as an acceptable remedy for PCB concentrations between 1 and 10 ppm, and "cleaning, decontaminating, or removing" the waste for PCB concentrations above 10 ppm. The PCB cleanup concentrations for low occupancy areas are not more than 25 ppm without further measures, 25 to 50 ppm if the site is secured by a fence and has signage with markings required by the regulations, and 25 to 100 ppm with capping. The requirements for capping for both high- and low-occupancy areas allow for the use of concrete, asphalt (or similar materials), or soils to construct the cap. Concrete and asphalt must be at least 0.15 m thick. Soil caps must be at least 0.25 m thick. Whatever the construction materials, the cap must have enough strength to maintain its integrity for the design life. Additional information on PCB remediation waste cleanups can be found in USEPA (2024p).

Bulk PCB wastes (including remediation wastes) at concentrations of 50 ppm or greater can be land-disposed only at EPA-permitted TSCA chemical waste landfills (they may also be incinerated or managed using an EPA-approved alternative treatment method). Also, liquids with PCB concentrations of not more than 500 ppm may be disposed in a TSCA chemical waste landfill if the liquid does not have the ignitability characteristic and is first stabilized to a nonflowing consistency. The design and operating requirements for these landfills are contained in 40 CFR §761.75. In the United States, some RCRA hazardous waste landfills (discussed previously) include design elements that allow them to hold not only an RCRA permit but also a TSCA permit; the authors are not aware of any "TSCA only" landfills. The main additional element for a TSCA chemical waste landfill is that the compacted clay component of the liner system must be at least 1.22 m thick (in contrast to 0.9 m thick for RCRA hazardous waste landfills). The TSCA regulations also contain a variety of operating restrictions and requirements that are unique to PCB wastes. A list of approved TSCA commercial waste landfills is given in USEPA (2024q).

4.9 SPECIAL REQUIREMENTS FOR MIXED WASTES

A waste that has both hazardous waste components and a radioactive component is called a "mixed waste" and is regulated under both the RCRA and the Atomic Energy Act (AEA). Geoenvironmental engineers most commonly become involved in the remediation of sites containing mixed wastes, and the design of mixed waste landfills, at active or inactive Department of Energy (DOE) uranium and plutonium processing facilities, at permitted commercial mixed waste disposal facilities. The Department

Figure 4-10. Fernald on-site disposal facility in 2002. The northern-most cells are closed or closing, middle cells are active and being filled, and southern cells are under construction.
Source: Bonaparte et al. (2020).

of Energy facilities include, as examples, the Fernald Feed Materials Plant and Portsmouth Gaseous Diffusion Plant, both in Ohio, the Weldon Spring Feed Materials Plant in Ohio, the Oak Ridge Reservation in Tennessee, the Idaho National Laboratory in Idaho, and the Hanford Reservation in Washington. Examples of commercial mixed waste disposal facilities include the Energy Solutions Clive Disposal Facility in Utah, the US Ecology Facility in Washington, and the Waste Control Specialists Facility in Texas. Bonaparte et al. (2020) provide a detailed case study of the design, construction, closure, and performance of the DOE Fernald On-Site Disposal facility. Figure 4-10 shows the facility during construction, operation, and closure.

REFERENCES

Bonaparte, R. 1995. "Long-term performance of landfills." In *Proc., ASCE Specialty Conf. Geoenvironment 2000*, Geotechnical Special Publication 46, 515–553. Reston, VA: ASCE.

Bonaparte, R., J. P. Giroud, R. B. Wallace, C. Ah-Line, et al. 1987. *Background document: Proposed liner and leak detection rule.* EPA 530-SW-87-015. Washington, DC: USEPA.

Bonaparte, R., and B. A. Gross. 1990. "Field behavior of double-liner systems." In *Proc., Symp. on Waste Containment Systems*, Geotechnical Special Publication 26, 52–83. Reston, VA: ASCE.

Bonaparte, R., B. A. Gross, R. G. Gupta, J. F. Beech, et al. 2020. "Twenty-year performance of a mixed LLRW/RCRA waste disposal facility."

In *Proc., Geo-Congress 2020, University of Minnesota 68th Annual Geotechnical Engineering Conf,* Geotechnical Special Publication 321, 29–56. Reston, VA: ASCE.

Bonaparte, R., B. A. Gross, R. M. Koerner, D. E. Daniel, et al. 2004. *Draft technical guidance for RCRA/CERCLA final cover systems.* EPA 540-R-04-007. Washington, DC: USEPA, Office of Solid Waste and Emergency Response.

Bonaparte, R., R. M. Koerner, and D. E. Daniel. 2002. *Assessment and recommendations for improving the performance of waste containment systems.* EPA/600/R-02/099 (NTIS PB2003-103048). Washington, DC: USEPA, National Risk Management Research Laboratory.

Gross, B. A., R. Bonaparte, and J. P. Giroud. 1990. "Evaluation of flow from landfill leakage detection layers." In *Proc., 4th Int. Conf. on Geotextiles, Geomembranes and Related Products*, 481–486. Rotterdam, Netherlands: Balkema-Rotterdam.

NRC (National Research Council). 2007. *Assessment of the performance of engineered waste containment barriers.* Washington DC: National Academies Press.

USDOT (US Department of Transportation). 2024. "How to comply with federal hazardous material regulations." Accessed April 23, 2024. https://www.fmcsa.dot.gov/regulations/hazardous-materials/how-comply-federal-hazardous-materials-regulations.

USEPA (US Environmental Protection Agency). 1987. "Liners and leak detection for hazardous waste landfills." *Federal Register*, 52 (103): 20218–2031.

USEPA. 1988. *Guidance for conducting remedial investigations and feasibility studies under CERCLA.* EPA/540/G-89/004. Washington, DC: USEPA, Office of Emergency and Remedial Response.

USEPA. 1989. *Final covers on hazardous waste landfills and surface impoundments, technical guidance document.* EPA/530/SW-89/047. Washington, DC: USEPA, Office of Solid Waste and Emergency Response.

USEPA. 1991. *ARARs Q's & A's: General policy, RCRA, SWA, CWA, SDWA, post-rod information, and contingent waivers.* EPA 9234.2-01/FS-A. Washington, DC: USEPA, Office of Solid Waste and Emergency Response.

USEPA. 1992. "Method 1311 toxicity characteristic leaching procedure." Accessed February 26, 2024. https://www.epa.gov/sites/default/files/2015-12/documents/1311.pdf.

USEPA. 1993. *Conducting non-time-critical removal actions under CERCLA.* EPA/540/F-94/009. Washington, DC: USEPA, Office of Solid Waste and Emergency Response.

USEPA. 1995. *Remedial design/remedial action handbook.* EPA/540/R-95/059. Washington, DC: USEPA, Office of Emergency and Remedial Response.

USEPA. 1998. *Management of remediation waste under RCRA*. EPA/530-F-98-026. Washington, DC: USEPA, Office of Solid Waste and Emergency Response.

USEPA. 1999. *A guide to preparing superfund proposed plans, records of decision, and other remedy selection documents*. EPA/540-R-98-03. Washington DC: USEPA, Office of Solid Waste and Emergency Response.

USEPA. 2002a. *Guidance on demonstrating compliance with the land disposal restrictions (LDR) alternative treatment standards, final guidance*. EPA/530-R-02-002. Washington, DC: USEPA, Office of Solid Waste and Emergency Response.

USEPA. 2002b. *Reusing superfund sites: Commercial use where waste is left on site*. Washington, DC: USEPA, Office of Emergency and Remedial Response.

USEPA. 2004. "Method 1020B standard test methods for flash point by setaflash (small scale) closed-cup apparatus." Accessed February 26, 2024. https://www.epa.gov/hw-sw846/sw-846-test-method-1020c-standard-test-methods-flash-point-setaflash-small-scale-closed.

USEPA. 2005. *Hazardous waste identification. 40 CFR Part 261. EPA/530-K-05-012*. Washington, DC: USEPA, Office of Solid Waste and Emergency Response.

USEPA. 2014a. "Method 1030 ignitability of solids." Accessed February 26, 2024. https://www.epa.gov/sites/default/files/2015-12/documents/1030.pdf.

USEPA. 2014b. "The SW-846 compendium." Accessed February 26, 2024. https://www.epa.gov/hw-sw846/sw-846-compendium.

USEPA. 2017. *Post-closure performance of liner systems at RCRA Subtitle C landfills*. EPA/600/R-17/205. Washington, DC: USEPA, Office of Solid Waste and Emergency Response.

USEPA. 2024a. "Learn the basics of hazardous waste." Accessed February 26, 2024. https://www.epa.gov/hw/learn-basics-hazardous-waste.

USEPA 2024b. "Criteria for the definition of solid waste and solid and hazardous waste exclusions." Accessed April 23, 204. https://www.epa.gov/hw/criteria-definition-solid-waste-and-solid-and-hazardous-waste-exclusions.

USEPA. 2024c. "Hazardous waste generators." Accessed April 23, 2024. https://www.epa.gov/hwgenerators.

USEPA. 2024d. "Hazardous waste transportation." Accessed April 23, 2024. https://www.epa.gov/hw/hazardous-waste-transportation.

USEPA. 2024e. "Hazardous waste recycling." Accessed April 23, 2024. https://www.epa.gov/hw/hazardous-waste-recycling.

USEPA. 2024f. "Hazardous waste permitting." Accessed April 23, 2024. https://www.epa.gov/hwpermitting.

USEPA. 2024g. "Treatment standards for hazardous wastes subject to land disposal restrictions." Accessed April 23, 2024. https://www.epa.gov/

hw/treatment-standards-hazardous-wastes-subject-land-disposal-restrictions.

USEPA. 2024h. "Learn about corrective action." Accessed April 23, 2024. https://www.epa.gov/hw/learn-about-corrective-action.

USEPA. 2024i. "RCRA corrective action cleanup enforcement." Accessed April 23, 2024. https://www.epa.gov/enforcement/rcra-corrective-action-cleanup-enforcement.

USEPA. 2024j. "Love Canal, NY cleanup activities." Accessed April 23, 2024. https://cumulis.epa.gov/supercpad/SiteProfiles/index.cfm?fuseaction=second.Cleanup&id=0201290#bkground.

USEPA. 2024k. "A.L. Taylor (Valley of Drums) Brooks, KY cleanup activities." Accessed April 23, 2024. https://cumulis.epa.gov/supercpad/SiteProfiles/index.cfm?fuseaction=second.Cleanup&id=0402072#bkground.

USEPA. 2024l. "Non-time critical removal actions." Accessed April 23, 2024. https://www.epa.gov/superfund/non-time-critical-removal-actions.

USEPA. 2024m. "Applicable or Relevant and Appropriate Requirements (ARARs)." Accessed April 23, 2024. https://www.epa.gov/superfund/applicable-or-relevant-and-appropriate-requirements-arars.

USEPA. 2024n. "Superfund redevelopment policy, guidance and resources." Accessed April 23, 2024. https://www.epa.gov/superfund-redevelopment/superfund-redevelopment-policy-guidance-and-resources.

USEPA. 2024o. "Polychlorinated biphenyls (PCBs)." Accessed April 23, 2024. https://www.epa.gov/pcbs.

USEPA. 2024p. "Managing remediation waste from polychlorinated biphenyls (PCBs) cleanup." Accessed April 23, 2024. https://www.epa.gov/pcbs/managing-remediation-waste-polychlorinated-biphenyls-pcbs-cleanups.

USEPA. 2024q. "List of approved polychlorinated biphenyl (PCB) commercial storage and disposal facilities." Accessed April 23, 2024. https://www.epa.gov/pcbs/list-approved-polychlorinated-biphenyl-pcb-commercial-storage-and-disposal-facilities.

USEPA. 2024r. "Superfund site cleanup work through enforcement agreements and orders." Accessed April 23, 2024. https://www.epa.gov/enforcement/superfund-site-cleanup-work-through-enforcement-agreements-and-orders.

CHAPTER 5

MUNICIPAL SOLID WASTE: CHARACTERIZATION AND ENGINEERING PROPERTIES

Dimitrios Zekkos, Ph.D., P.E., M.ASCE; Xuede Qian, Ph.D., P.E.; Jeffrey Fassett, Ph.D., P.E.; and Christopher Bareither, Ph.D, P.E., M.ASCE

5.1 INTRODUCTION

Municipal solid waste (MSW) consists of discarded everyday items from residential, commercial, and institutional settings such as government, education, and medical facilities. MSW includes materials such as product packaging, yard trimmings, furniture, clothing, bottles and cans, food, paper, appliances, electronics, and batteries. The Environmental Protection Agency's (EPA) definition of MSW excludes industrial, hazardous, coal combustion residuals (CCR), and construction and demolition (C&D) wastes.

The amount of MSW generated in the United States and globally increases every year along with population growth. Studies have shown that waste production is connected to economic activity, as shown in Figure 5-1 for the United States. According to data collected by EPA, MSW generation in the United States has been increasing since the 1960s, as shown in Figure 5-2. In 2018, 292.4 million tons of MSW were generated, of which 24.9 million tons were composted, 69 million tons recycled, and 34.6 million tons combusted with energy recovery. The majority of MSW in the United States, approximately 145 million tons per year, is still disposed in landfills, and this trend is expected to remain until new waste management approaches or technologies are developed that are able handle large amounts of waste in a more cost-efficient and sustainable manner.

Design of MSW landfills requires an understanding of the engineering properties of MSW. This chapter aims to briefly summarize the current understanding of MSW engineering properties. In 2011, ASCE published "Geotechnical Characterization, Field Measurements, and Laboratory Testing of Municipal Solid Waste" (Zekkos 2011) that provides a review of

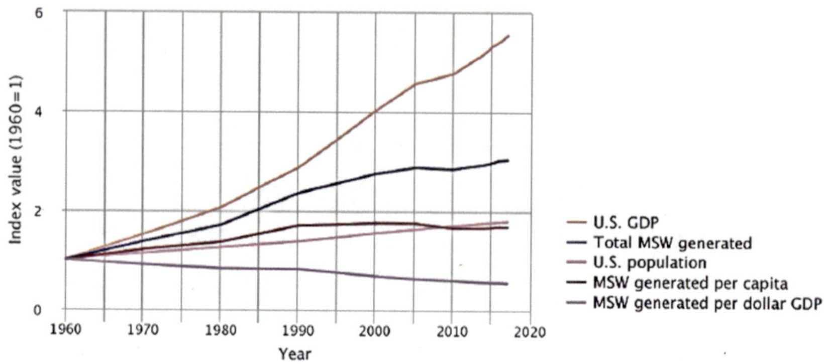

Figure 5-1. Evolution of municipal solid waste generation "intensity" and comparison with US GDP.
Source: USEPA (2021).

MSW properties. This chapter summarizes this information and incorporates additional advances made since then, with a focus on engineering practice. In selecting the engineering properties of MSW, one should consider a coordinated field characterization and field testing program, as described in Chapter 2, paired with targeted laboratory testing.

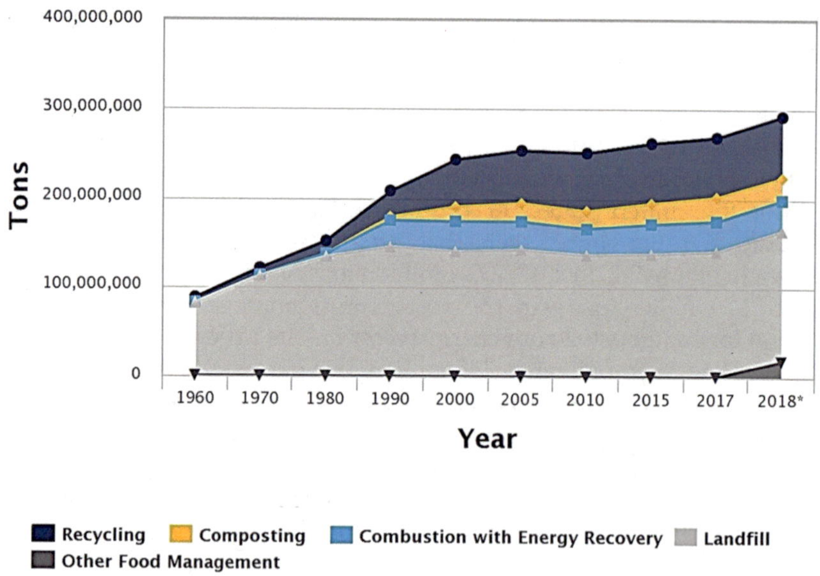

Figure 5-2. Municipal solid waste generated and managed in the United States from 1960 to 2018.
Source: USEPA (2023).

5.2 MUNICIPAL SOLID WASTE COMPOSITION

Municipal solid waste composition varies globally and also regionally within the United States. Composition is affected by many factors, including economic diversity and activity, social aspects, and consumer habits. MSW composition in landfills is also affected by waste diversion activities that cause a difference between the waste generated and the waste disposed in landfills. In addition, local business activities, seasons, and/or other operations may influence the solid waste disposed of at a particular MSW landfill. Natural disasters may also cause an abrupt influx of nonhazardous disaster-debris materials in MSW landfills.

In general, the majority of studies indicate that MSW consists primarily of paper, plastic, food waste, wood, and soil material. The soil material originates primarily from soil used as daily cover or from the disposal of nonhazardous contaminated soils. Experience has shown that soil is a significant portion (sometimes exceeding 50%) of the by-weight composition of waste in US landfills. The amount and type of daily soil cover used has also been observed to influence the engineering properties of MSW. The disposal of special wastes, such as sludges, in MSW landfills is also becoming more common. These wastes may have very high moisture contents and low shear strengths, which affect the engineering properties of codisposed MSW (Bareither et al. 2020, Bonaparte et al. 2020).

Waste compositional characterization is typically conducted by separating waste constituents to assess the type of waste. Such procedures have been developed by Dixon and Langer (2006) and Zekkos et al. (2010a), and variations of those are used in engineering practice, as described in Chapter 2.

5.3 SOLID WASTE STRUCTURE

Although MSW does not have a geologic history, because MSW is a human-placed material of young age, waste does have a pronounced structure (Tchobanoglous et al. 1993, Xu et al. 2013). Fibrous waste constituents tend to be oriented in a subhorizontal direction because of spreading and compaction of waste in layers and because of confining stress from overlying layers of waste placed subsequently (Zekkos 2013). In addition, daily soil cover contributes to the creation of a layered structure within the waste mass. As a result, MSW is found to have a highly anisotropic structure, which has implications in engineering practice owing to variations in the shear capacity, compressibility, and especially flow of liquids and gas through the waste mass.

5.4 BIOGAS GENERATION

Because of the presence of organics in MSW, and the lack of adequate oxygen, MSW degrades anaerobically generating primarily methane and carbon dioxide, although other constituents (such as sulfides) may also be present and cause odors (Barlaz et al. 1990). Overall, MSW has a significant biogas generation capacity and the rate of degradation is impacted by the presence of moisture (Tchobanoglous et al. 1993, Karimi and Bareither 2021). Thus, landfills in wetter environments tend to generate gas significantly faster than landfills in dry climates, and gas control systems are needed not only for the mitigation of methane emissions and energy generation, but also for the potential for high-temperature development and for addressing the issue of instability in the waste mass associated with high liquid/gas pressures.

5.5 UNIT WEIGHT

Total unit weight of MSW has been found to vary greatly among different landfills. It ranges from values as low as 5 kN/m^3 to as high as 18 kN/m^3, because of variations in waste composition, proportion of soil in the waste, climate, and landfill practices. Unit weight is found to be higher in modern regulated landfills that compact MSW, as well as landfills with higher amounts of soil material, as opposed to larger, bulkier, less dense waste constituents. MSW unit weight also varies with depth because of increasing confining stress (Zekkos et al. 2006).

5.6 MOISTURE CONTENT

The moisture content of typical MSW consists of water retained in macro pores that is freely draining and water "absorbed" into waste constituents such as paper, cardboard, food, and so on. In general, the moisture content of MSW within a landfill varies between the initial moisture content on collection and a moisture content representing fully saturated conditions. Moisture content may also decrease from its placed value because of various landfill processes such as water consumption in anaerobic reactions, compaction, and capillary drying, among other processes (Beaven et al. 2011).

Three main definitions of moisture content are used in landfill design. Most commonly, moisture content is defined as the percent by weight of water in the waste divided by the dry weight of waste (i.e., w_d). This is known as the "dry gravimetric moisture content definition" and is commonly used in geotechnical engineering analyses. In many cases in landfill practice, moisture content is defined on the basis of the wet weight of the waste. The wet gravimetric moisture content (i.e., w_w) is defined as the percent by weight of water in the waste on the basis of the total weight of waste. The third definition is the volumetric moisture content that is defined as the percent by volume of water in the waste divided by the total volume of waste (i.e., θ). This volumetric moisture content definition is widely used in hydrologic and environmental engineering analyses (Qian et al. 2002).

5.7 FIELD CAPACITY

Field capacity is defined as the amount of moisture content in a unit volume of waste or soil after downward gravity drainage has ceased following a thorough wetting (Bear 1979). Field capacity is the maximum amount of water that a medium such as waste can retain under the influence of gravity. Field capacity may be expressed as the ratio of the volume of water in the waste to the total volume of the waste [i.e., volumetric field capacity, $FC(\theta)$], the ratio of the weight of the water in the waste to the total weight of the waste [i.e., wet gravimetric field capacity, $FC(w_w)$], and the ratio of the weight of the water in the waste to the dry weight of the waste [i.e., dry gravimetric field capacity, $FC(w_d)$]. The concept of MSW field capacity is important in estimating leachate generation volumes in landfills because water in excess of the field capacity will be released as leachate. Field capacity is also an important parameter in recirculation and bioreactor landfill technology because field capacity influences the amount of moisture that can be absorbed into, and around, each waste particle. According to Qian et al. (2002) and Qian (2011), MSW near field capacity degrades faster than in drier conditions.

5.8 HYDRAULIC CONDUCTIVITY

Excluding high-moisture content special wastes (e.g., sludges), MSW is typically placed in an unsaturated state, usually at a moisture content below the waste's field capacity. After placement, the waste will typically remain in an unsaturated state, if covered with daily/intermediate cover in a timely manner, and if the landfill does not practice significant leachate recirculation or acceptance of high moisture content wastes. Given that MSW will frequently remain in a saturated state, the flow of liquids and gas in a landfill is affected by the waste's soil water retention curve. Such curves developed for MSW can be found in Breitmeyer et al. (2020) and in Stoltz et al. (2010). In addition, the saturated hydraulic conductivity of MSW varies greatly (Powrie and Beaven 1999). In general, the average hydraulic conductivity is shown to commonly vary from as high as 10^{-3} cm/s to as low as 10^{-8} cm/s depending on the waste composition, confining stress, and density of MSW (Breitmeyer et al. 2019). It is important to recognize that in the field, depending on variations in waste composition and the amount and type of daily soil cover used, hydraulic conductivity may vary spatially, and in some locations, wetter, or even perched saturated zones may occur, especially when finer grained daily soil cover is used.

As described previously, MSW is anisotropic in landfills because the waste material is placed in layers, covered by daily soil cover, and compacted (Townsend et al. 1995, Landva et al. 1998). This creates an anisotropic structure of MSW, and thus, hydraulic conductivity is higher in the subhorizontal direction (Xu et al. 2013). The anisotropy ratio is the ratio of horizontal to vertical hydraulic conductivity of landfilled MSW (i.e., k_h:k_v). Hudson et al. (2009) discussed the impact of anisotropy on the hydraulic conductivity of MSW. A review of the literature reveals a very wide variation in the anisotropy ratio of waste in landfills (Beaven et al. 2011).

5.9 SHEAR STRENGTH

The shear strength of MSW has been assessed through laboratory testing, field measurements, and back-calculation of stable and unstable slopes. Laboratory testing has the advantage that the waste composition is known and can be experimentally parameterized. However, specimens are typically reconstituted because an undisturbed sampling of MSW is typically impossible. Experience has shown that testing needs to be conducted on samples as large as possible to include many of the fibrous waste constituents, with 30 cm (12 in.) diameter or width being the minimum recommended large-size testing for unprocessed MSW (Athanasopoulos 2010). Field measurements, as described in Chapter 2 of this manual, have been conducted and have been used to assess the mechanical properties of MSW; however, careful interpretation of the observed data may be necessary as procedures used in soils may not always be appropriate. Back-calculations of slopes are also appealing and frequently used. Back-calculations of stable slopes may provide a lower-bound strength estimate as the actual factor of safety of the slopes remains unknown. Back-calculations of failed slopes are more meaningful but less common (Eid et al. 2000, Merry et al. 2005, Bonaparte et al. 2020).

Extensive laboratory testing on MSW has been conducted to measure shear strength. The most common test method has been the direct shear test, followed by triaxial compression tests. Triaxial extension tests and simple shear testing are less common. Studies have demonstrated that the shear strength of MSW is dependent on waste composition. Particularly when the fibrous waste constituents (paper, plastic, and wood) are present, MSW has a shear response that is similar to fiber-reinforced soils (Shi et al. 2009) that exhibit a shear resistance that is anisotropic and stress path–dependent. Direct shear tests are found to produce lower shear resistance compared to triaxial shear tests that shear across the fibrous constituents and yield greater shear resistance (Bray et al. 2009).

Recommended shear strength envelopes are generally based on direct shear data and back-calculation of slopes. Figure 5-3 illustrates several well-established shear strength envelopes presented in the literature (Kavazanjian et al. 1995, Manassero et al. 1996, Eid et al. 2000, Bray et al. 2009) along with direct shear experimental data. As illustrated in Figure 5-3, shear stresses at failure are highly variable at low normal stresses but show less variability in normal stress ranges typically present in modern MSW landfills; however, this may be the result of fewer available data at higher normal stresses. The strength envelopes generally involve an "apparent" cohesion of 5 to 30 kPa and friction angles ranging from 30 to 35 degrees. The strength resistance in direct shear is nonlinear, indicating lower friction angles at higher normal stresses. Many tests conducted on MSW were on materials prepared at below field capacity and tested under drained

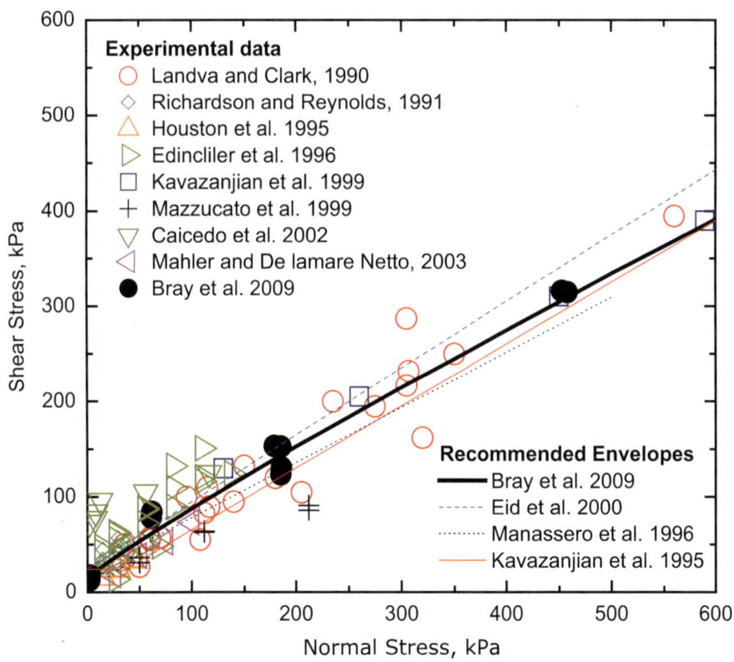

Figure 5-3. Shear strength envelopes for municipal solid waste.

conditions; thus, the influences of moisture content and drainage conditions are not necessarily captured. Recent studies have demonstrated that MSW at high moisture, saturated, or undrained conditions may exhibit lower shear resistances than unsaturated MSW tested under drained conditions (Zekkos and Fei 2017, Bareither et al. 2020).

The effect of biodegradation on waste shear strength is not commonly considered in engineering practice. Recent studies have shown that waste degradation affects the shear strength of MSW, but the impact remains largely inconclusive (i.e., in some cases, shear strength is found to increase), whereas in other cases, shear strength is found to decrease (Kavazanjian 2008, Bareither et al. 2012a). Zhan et al. (2007) investigated changes in the shear strength of MSW as a function of waste age by performing field and laboratory studies and found that as the age of the waste increased, the cohesion at a strain level of 10% decreased and the friction angle at the same strain level increased. Although the age of waste often is used as an indicator of the state of biodegradation, field conditions, and especially moisture environment, play a more critical role on waste biodegradation state compared to the age of waste.

5.10 COMPRESSIBILITY AND SETTLEMENT

MSW is a highly compressible material and landfill settlements resulting from this compressibility may affect not only the operation of a landfill (e.g., waste compaction, leachate collection, and gas collection), but also the landfill's closure and postclosure behaviors (Sharma and De 2007). When subjected to vertical stress, MSW may compress as much as 40%. The compressibility of MSW is found to be significantly affected by waste composition, with waste-rich MSW (i.e., MSW rich in paper, plastic, and food waste) being more compressible than soil-rich MSW (Bareither et al. 2012b, Zekkos et al. 2016).

In addition to its initial mechanical compressibility, MSW exhibits large secondary compressibility associated with its anaerobic biodegradation over years to decades. This secondary compression results in significant long-term settlements of MSW landfills. Biodegradation-induced secondary settlements are much more significant than the mechanical creep of MSW in generating long-term landfill settlements (Bareither et al. 2013). Geomechanical empirical models have been used to model the settlement of MSW, but these do not capture the true mechanism of biodegradation and can be used primarily to fit field measurements of settlement at a given landfill. More appropriate settlement models for full-scale landfill settlement incorporate physically significant model parameters that can be linked to MSW characteristics (Bareither and Kwak 2015, Datta et al. 2021). Changes in biodegradation conditions at a landfill, caused by changes in operational practices, construction of covers, or other interventions, may result in significant changes in biodegradation rates. The rate of biodegradation decreases over the long term as the organic fraction is consumed by microbes and becomes gradually depleted (Fei and Zekkos 2018).

5.11 DYNAMIC PROPERTIES

In seismic-impact zones, where MSW landfills are subjected to seismic excitations, the dynamic properties of MSW are of interest (Kavazanjian et al. 1995, Bray and Rathje 1998). These include primarily, the shear wave velocity or small-strain shear modulus, and the shear modulus reduction and material damping increase relationships with shear strain, which are described in detail in Zekkos et al. (2010b). Both analytical and experimental laboratory and field studies have been conducted to assess the dynamic properties of MSW. Similar to other engineering properties, dynamic properties are found to be significantly affected by waste composition and confining stress. Soil-rich and dense MSW is found to have higher shear wave velocities that generally increase with confining stress. Also, soil-rich MSW is found to exhibit a more pronounced reduction in shear modulus and also higher damping values than waste-rich MSW, which is attributed to the beneficial fiber-reinforcement effect of fibrous waste at larger strains (Zekkos et al. 2008, Yuan et al. 2011).

REFERENCES

Athanasopoulos, G. A. 2010. "Laboratory testing of municipal solid waste." In *Geotechnical characterization, field measurement, and laboratory testing of municipal solid waste*, Geotechnical Special Publication 209, edited by D. Zekkos, 195–205. Reston, VA: ASCE.

Bareither, C. A., C. H. Benson, and T. B. Edil. 2012a. "Effects of waste composition and decomposition on the shear strength of municipal solid waste." *J. Geotech. Geoenviron. Eng*. 138 (10): 1161–1174.

Bareither, C. A., C. H. Benson, and T. B. Edil. 2012b. "Compression behavior of municipal solid waste: Immediate compression." *J. Geotech. Geoenviron. Eng*. 138 (9): 1047–1062.

Bareither, C. A., C. H. Benson, and T. B. Edil. 2013. "Compression of municipal solid waste in bioreactor landfills: Mechanical creep and biocompression." *J. Geotech. Geoenviron. Eng*. 139 (7): 1007–1021.

Bareither, C. A., C. H. Benson, E. M. Rohlf, and J. Scalia IV. 2020. "Hydraulic and mechanical behavior of municipal solid waste and high-moisture waste mixtures." *Waste Manage*. 105: 540–549.

Bareither, C. A., and S. Kwak. 2015. "Assessment of municipal solid waste settlement models based on field-scale data analysis." *Waste Manage.* 42 (8): 101–117.

Barlaz, M. A., R. K. Ham, and D. M. Schaefer. 1990. "Methane production from municipal refuse: A review of enhancement techniques and microbial dynamics." *CRC Crit. Rev. Environ. Control* 19 (6): 541–557.

Bear, J. 1979. *Hydraulics of groundwater*. New York: McGraw-Hill.

Beaven, R. P., W. Powrie, and K. Zardava. 2011. "Hydraulic properties of MSW." In *Geotechnical characterization, field measurement, and laboratory testing of municipal solid waste*, Geotechnical Special Publication 209, edited by D. Zekkos, 1–43. Reston, VA: ASCE.

Bonaparte, R., R. C. Bachus, and B. A. Gross. 2020. "Geotechnical stability of waste fills: Lessons learned and continuing challenges." *J. Geotech. Geoenviron. Eng.* 146 (11): 05020010.

Bray, J. D., and E. R. Rathje. 1998. "Earthquake-induced displacements of solid-waste landfills." *J. Geotech. Geoenviron. Eng.* 124 (3): 242–253.

Bray, J. D., D. Zekkos, E. Kavazanjian Jr., G. A. Athanasopoulos, et al. 2009. "Shear strength of municipal solid waste." *J. Geotech. Geoenviron. Eng.* 135 (6): 709–722.

Breitmeyer, R. J., C. H. Benson, and T. B. Edil. 2019. "Effects of compression and decomposition on saturated hydraulic conductivity of municipal solid waste in bioreactor landfills." *J. Geotech. Geoenviron. Eng.* 145 (4): 04019011.

Breitmeyer, R. J., C. H. Benson, and T. B. Edil. 2020. "Effect of changing unit weight and decomposition on unsaturated hydraulics of municipal solid waste in bioreactor landfills." *J. Geotech. Geoenviron. Eng.* 146 (5): 04020021.

Datta, S., D. Zekkos, X. Fei, and J. McDougall. 2021. "Waste-composition-dependent 'HBM' model parameters based on degradation experiments." *Environ. Geotech.* 8 (2): 124–133.

Dixon, N., and U. Langer. 2006. "Development of a MSW classification system for the evaluation of mechanical properties." *Waste Manage.* 26 (3): 220–232.

Eid, H. T., T. D. Stark, W. D. Douglas, and P. E. Sherry. 2000. "Municipal solid waste slope failure. I: Waste and foundation soil properties." *J. Geotech. Geoenviron. Eng.* 126 (5): 397–407.

Fei, X., and D. Zekkos. 2018. "Coupled experimental assessment of physico-biochemical characteristics of municipal solid waste undergoing enhanced biodegradation." *Géotechnique* 68 (12): 1031–1043.

Hudson, A. P., R. P. Beaven, and W. Powrie. 2009. "Assessment of vertical and horizontal hydraulic conductivities of household waste in a large scale compression cell." In *Proc., 12th Waste Management and Landfill Symp.*, edited by R. Cossu, L. F. Diaz, and R. Stegman, 641–642. Sardinia, Italy: IWWG.

Karimi, S., and C. A. Bareither. 2021. "The influence of moisture enhancement on solid waste biodegradation." *Waste Manage.* 123 (15): 131–141.

Kavazanjian, E., Jr. 2008. "The impact of degradation on MSW shear strength." In *GeoCongress 2008: Geotechnics of Waste Management and Remediation*, 224–231. Reston, VA: ASCE.

Kavazanjian, E., Jr., N. Matasovic, R. Bonaparte, and G. R. Schmertmann. 1995. "Evaluation of MSW properties for seismic analysis." In *Geoenvironment 2000: Proc., Specialty Conf. on Geotechnical Practice in Waste Disposal, Part 1*, Geotechnical Special Publication 46, 1126–1141. Reston, VA: ASCE.

Landva, A. O., S. G. Pelkey, and A. J. Valsangkar. 1998. "Coefficient of permeability of municipal refuse." In *Proc., 3rd Int. Congress on Environmental Geotechnics*, 163–167. Netherlands: Balkema-Rotterdam.

Manassero, M., W. F. Van Impe, and A. Bouazza. 1996. "Waste disposal and containment." In *Vol. 3 of Proc., 2nd Int. Congress on Environmental Geotechnics*, 1425–1474. Rotterdam, Netherlands: A.A. Balkema.

Merry, S. M., E. Kavazanjian Jr., and W. U. Fritz. 2005. "Reconnaissance of the July 10, 2000, Payatas landfill failure." *J. Perform. Constr. Facil.* 19 (2): 100–107. Doi: 10.1061/(ASCE)0887-3828(2005)19:2(100)

Powrie, W., and R. P. Beaven. 1999. "Hydraulic properties of household waste and implications for landfills." *Proc. Inst. Civ. Eng. Geotech. Eng.* 137 (4): 235–237.

Qian, X. 2011. "Stability analyses and leachate head calculations for bioreactor landfills." In *Design and operation of bioreactor landfills*. ASCE Continuing Education. Reston, VA: ASCE.

Qian, X., R. M. Koerner, and D. H. Gray. 2002. *Geotechnical aspects of landfill design and construction*. Upper Saddle River, NJ: Prentice Hall.

Sharma, H. D., and A. De. 2007. "Municipal solid waste landfill settlement: Postclosure perspectives." *J. Geotech. Geoenviron. Eng.* 133 (6): 619–629.

Shi, J., X. Qian, J. Zhu, and Y. Li. 2009. "Application of shear strength of solid waste and multilayer liner in landfills." In *Advances in environmental geotechnics*, edited by Y. Chen, L. Zhan, and X. Tang, 286–294. Berlin, Heidelberg: Springer.

Stoltz, G., J. P. Gourc, and L. Oxarango. 2010. "Liquid and gas permeabilities of unsaturated municipal solid waste under compression." *J. Contam. Hydrol.* 118 (1–2): 27–42. Doi: 10.1016/j.jconhyd.2010.07.008

Tchobanoglous, G., H. Theisen, and S. Vigil. 1993. *Integrated solid waste management*. New York: McGraw-Hill.

Townsend, T. G., W. L. Miller, and J. F. K. Earle. 1995. "Leachate recycle infiltration ponds." *J. Environ. Eng.* 121 (6): 465–471.

USEPA (US Environmental Protection Agency). 2021. "Report on the environment." Accessed: March 12, 2021. https://cfpub.epa.gov/roe/indicator.cfm?i=53.

USEPA. 2023. "National overview: Facts and figures on materials, wastes and recycling." Accessed April 19, 2023. https://www.epa.gov/facts-and-figures-about-materials-waste-and-recycling/national-overview-facts-and-figures-materials.

Xu, Q., J. Powell, T. Tolaymat, and T. Townsend. 2013. "Seepage control strategies at bioreactor landfills." *J. Hazard. Toxic Radioact. Waste* 17 (4): 342–350.

Yuan, P., E. Kavazanjian Jr., W. Chen, and B. Seo. 2011. "Compositional effects on the dynamic properties of municipal solid waste." *Waste Manage.* 31 (12): 2380–2390.

Zekkos, D., ed. 2011. *Geotechnical characterization, field measurement, and laboratory testing of municipal solid waste.* Geotechnical Special Publication 209. Reston, VA: ASCE.

Zekkos, D. 2013. "Experimental evidence of anisotropy in municipal solid waste." In *Proc., Coupled Phenomena in Environmental Geotechnics*, 69–79. Boca Raton, FL: CRC Press.

Zekkos, D., and X. Fei. 2017. "Constant load and constant volume response of municipal solid waste in simple shear." *Waste Manage.* 63: 380–392.

Zekkos, D., J. D. Bray, and M. F. Riemer. 2008. "Shear modulus and material damping of municipal solid waste based on large-scale cyclic triaxial testing." *Can. Geotech. J.* 45 (1): 45–58.

Zekkos, D., J. D. Bray, E. Kavazanjian Jr., N. Matasovic, et al. 2006. "Unit weight of municipal solid waste." *J. Geotech.Geoenviron. Eng.* 132 (10): 1250–1261.

Zekkos, D., X. Fei, A. Grizi, and G. A. Athanasopoulos. 2016. "Response of municipal solid waste to mechanical compression." *J. Geotech. Geoenviron. Eng.* 143 (3): 04016101.

Zekkos, D., E. Kavazanjian Jr., J. D. Bray, N. Matasovic, et al. 2010a. "Physical characterization of municipal solid waste for geotechnical purposes." *J. Geotech. Geoenviron. Eng.* 136 (9): 1231–1241.

Zekkos, D., N. Matasovic, R. El-Sherbiny, A. Athanasopoulos-Zekkos, et al. 2010b. "Dynamic properties of municipal solid waste." In *Geotechnical characterization, field measurement, and laboratory testing of municipal solid waste*, edited by D. Zekkos, 112–134. Reston, VA: ASCE.

Zhan, L., Y. Chen, and W. Ling. 2007. "Shear strength characterization of municipal solid waste at the Suzhou landfill, China." *Eng. Geol.* 97 (2): 97–111.

CHAPTER 6
COAL COMBUSTION RESIDUALS

Prabir K. Kolay, Ph.D.; Bora Cetin, Ph.D.; and Lin Li, Ph.D.

6.1 BACKGROUND

Coal combustion residuals (CCRs) are one of the largest types of industrial waste generated in the United States. According to ACAA's Coal Combustion Product (CCP) Production and Use Survey Report (ACAA 2019), nearly 52 million tons of CCRs were generated in 2019. Between 2011 and 2019, the combined share of coal and natural gas had decreased from nearly 70% to a little more than 60% of total US electricity generation. During that time, the relative shares of output from these two fuels had shifted, with coal's share declining from 42.3% in 2011 to 23.4% in 2019 (EIA 2019). Over the same period, the natural gas share had increased from 24.7% to 38.4%. Coal units generated 1,733,430 GWh in 2011 and 965,000 GWh in 2019, a decrease of 44% compared to the year 2011 (EIA 2019). In 2019, the annual production of CCRs [i.e., fly ash, bottom ash, boiler slag, and flue gas desulfurization (FGD) waste were generated] was 79 million tons. According to the American Coal Ash Association (ACAA), approximately 50% of CCRs generated in 2019 were put to beneficial use (ACAA 2019).

6.2 ENVIRONMENTAL REGULATION

The US Environmental Protection Agency (USEPA 2015) has published a final rule in the Federal Register (FR) (FR 2015) to regulate the disposal of CCRs as solid waste under Subtitle D of the Resource Conservation and Recovery Act (RCRA) and 40 CFR §257.50 to §257.106. The final rule is the culmination of an extensive study on the effects of coal ash on the environment and public health. The rule established requirements for location restrictions, closure design and operating criteria, groundwater monitoring and corrective action, closure and postclosure care, and recordkeeping, notification, and posting requirements (USEPA 2015).

6.3 CHARACTERISTICS AND PROPERTIES OF COAL COMBUSTION RESIDUALS

6.3.1 Size and Shape

The coal fly ash particles are spherical, ranging from sand to fine silt in size. They consist primarily of silt-sized particles, which are generally spherical and typically ranging in size between 10 and 100 μm as shown in Figure 6-1. These small glass spheres improve the fluidity and workability of fresh concrete. Fineness is one of the important properties contributing to the pozzolanic reactivity of fly ash (FHWA 2003). EPRI (2016) reported that particles range from 0.001 to 0.1 mm in diameter with a median particle diameter of 0.02 to 0.025 mm.

Bottom ash particles are much coarser than fly ash. The grain size of bottom ash typically ranges from fine sand to fine gravel in size. The gradation of bottom ash can vary widely on the basis of the coal pulverization and burning processes in the power plant, the variety of coal burned, and the bottom ash handling equipment (Ramme and Tharaniyil 2013, Blanc et al. 2018, Balapour et al. 2020, Ongwandee et al. 2020).

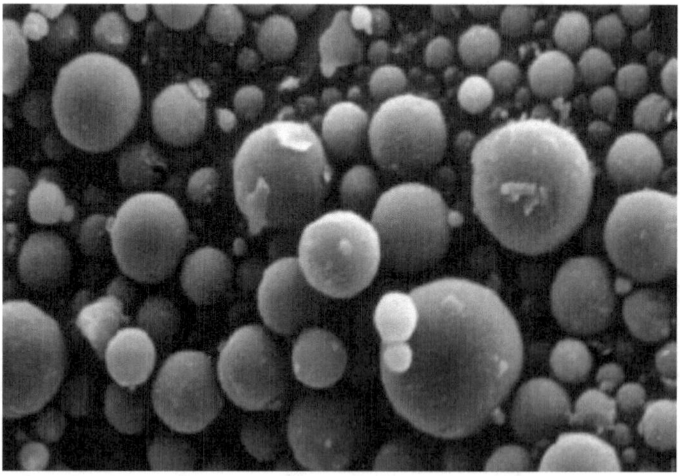

Figure 6-1. Fly ash particles at 2,000 times magnification.
Source: FHWA (2003).

Boiler slags are predominantly single-sized and within a range of 0.5 to 5.0 mm. In general, boiler slag particles have a smooth texture, but if gases are trapped in the slag as it is tapped from the furnace, the quenched slag will become somewhat vesicular or porous. Boiler slag from the burning of lignite or subbituminous coal tends to be more porous than that of bituminous coals (Ramme and Tharaniyil 2013, Blanc et al. 2018).

FGD scrubber material is initially generated as calcium sulfite. However, when power plants use wet FGD systems that utilize calcium-based sorbents and forced-air oxidation, calcium sulfite ($CaSO_3$) converts to calcium sulfate ($CaSO_4$) (FHWA 2011). Typical particle size for FGD material ranges from 1.3% to 16.5% sand-sized particles, 8.1% to 90.2% silt-sized particles, and 2.2% to 85% clay-sized particles (Ramme and Tharaniyil 2013).

6.3.2 Chemical Composition

Fly ash consists primarily of oxides of silicon, aluminum, iron, and calcium. Magnesium, potassium, sodium, titanium, and sulfur may also be present to a lesser degree. Fly ash elemental composition is variable and is associated with the source and pretreatment of the coal and furnace operating conditions. The major stable elements ($> 1\%$) that exist in fly ash particles are, in decreasing order, Si, Al, Fe, Ca, C, Mg, K, Na, S, Ti, P, and Mn. Also, trace elements in the fly ash, such as As, Cd, Cr, Cu, Pb, Ni, and Zn, may be released into aqueous environments depending on the geochemical conditions (Sivapullaiah and Moghal 2010; Komonweeraket et al. 2015a, b, c). The chemical composition of bottom ash is similar to that of fly ash, but bottom ash typically contains greater quantities of carbon. Bottom ash tends to be relatively more inert because the particles are larger and more fused than fly ash. Because these particles are highly fused, they tend to show less pozzolanic activity and are less suited as a binder constituent in concrete.

6.3.3 Classification of Fly Ashes

Fly ash is classified as either Class C or F ash on the basis of its chemical composition. American Association of State Highway Transportation Officials (AASHTO) M295 or ASTM International specification ASTM C618-15 (ASTM 2015) defines Classes C and F fly ashes depending on their chemical composition and coal types.

Class C ashes are generally derived from subbituminous coals and consist primarily of calcium alumino-sulfate glass, as well as quartz, tricalcium aluminate, and free lime (CaO). Class C ash is also referred to as high calcium fly ash because it typically contains more than 20% CaO and is rich in alkali and sulfate (SO_4). These are naturally pozzolanic, exhibiting self-cementing abilities. In the presence of water, these ashes harden and gain strength over time. For a Class C fly ash, the sum of $SiO_2 + Al_2O_3 + Fe_2O_3$ should be in the 50% to 70% range according to ASTM C618-15 (ASTM 2015).

Class F ashes are typically derived from bituminous and anthracite coals and consist primarily of an alumino-silicate glass, with quartz, mullite, and magnetite also present. Class F, or low calcium fly ash, has less than 10% CaO and is not naturally pozzolanic. However, when water and an activator (quicklime, hydrated lime, or ordinary portland cement) are added to Class F fly ash, the glassy silica and alumina react to produce cementitious compounds (Sivapullaiah et al. 1998). For a Class F fly ash, the sum of $Al_2O_3 + SiO_2 + Fe_2O_3$ should be greater than 70% according to ASTM C618-15 (ASTM 2015). Table 6-1 summarizes the criteria for fly ash

Table 6-1. ASTM C618 Chemical and Physical Criteria for Class C and Class F Fly Ash.

	ASTM requirements	
Chemical requirements	Class F	Class C
$SiO_2 + Al_2O_3 + Fe_2O_3$, min (%)	70	50–70
SO_3, max (%)	5	5
Moisture content (as received), max (%)	3	3
Loss on ignition, max (%)	6	6
	ASTM requirements	
Physical requirements	Class F	Class C
Fineness, max (%)	34	34
Strength activity at 7 days, min (%)	75	75
Strength activity at 28 days, min (%)	75	75
Water requirement, max (%)	115	115
Autoclave expansion, max (%)	0.8	0.8
Density variation, max (%)	5	5
Variation of % retained on 45 μm filter, max (%)	5	5

categorization in accordance with ASTM C618-15 (ASTM 2015). Any fly ashes that do not meet these criteria are referred to as off-specification fly ashes.

There have been studies of the characteristics and uses of off-specification fly ashes in construction material applications including cement, concrete, and soil. Conversely, it was found that these types of fly ashes could often not be used in cement and concrete applications because of their high unburned carbon content (LOI $> 6\%$). The unburned carbon reacts with air-entrainment agents and decreases the environmental durability of these materials. Conversely, previous studies showed that these fly ashes can be successfully used in geotechnical applications (Li et al. 2008; Cetin et al. 2010; Rosa et al. 2016, 2017).

6.3.4 Particle Size Distribution

As noted, fly ashes are predominantly silt-sized nonplastic materials with some sand size fraction. Leonards and Bailey (1982) reported the ranges of gradation for fly ashes and bottom ashes, which were classified as silty sands, or sandy silts according to the Unified Soil Classification Systems (USCS). A large percentage of fly ash particles (60% to 90%) can pass through a 75 μm screen (No. 200 sieve), whereas it also consists of 30% to 40% sand size particles. Fly ashes are generally classified as fine grained and usually fall within the silt range (2 to 74 μm) (Martin et al. 1990). Particle size distribution studies by Mehta and Monteiro (1997) show that the particles vary from less than 1 to 100 μm in diameter with more than 50% being under 20 μm. Sandy silt to silty sand fly ashes are usually poorly graded, with coefficient of curvature (C_c) values ranging from 0.61 to 3.70 and coefficient of uniformity (C_u) values ranging from 1.59 to 14.0 (Moghal 2017).

For bottom ashes, the particle size ranges from fine sand to fine gravel (Chan 1983).

6.3.5 Specific Gravity

The specific gravity of fly ash typically ranges from 2.05 to 2.68 (Ramme and Tharaniyil 2013). EPRI (2016) mentioned that the range of specific gravity for fly ash is 2.2 to 3.0. The range of specific gravities (G_s) observed in fly ash is the result of a combination of many factors such as gradation, particle shape, and chemical composition (Gray and Lin 1972). Pandian et al. (1998) observed that the low G_s is on account of some combination of the presence of hollow cenospheres from which entrapped air cannot be removed and the chemical composition (particularly, iron content). In general, G_s of coal ash lies around 2, but it can vary to a large extent. For example, McLaren and Digioia (1987) reported a G_s range of 1.6 to 3.1, whereas Pandian et al. (1998) reported a G_s range of 1.46 and 2.66. In most cases, fly ash is likely to have a higher G_s compared to bottom ashes. Because of lower G_s values compared to those for soils, fly ash fills possess lower dry densities. Therefore, the reduction in unit weight is advantageous when fly ash is used as a backfill material for retaining walls. A lower G_s is also favored for embankment materials built over compressible foundations. The specific gravity of bottom ash ranges from 2.28 to 2.33 depending on the coal used. The specific gravity of the boiler slag ranges from 2.3 to 2.9 and the specific gravity of FGD can vary from 2.36 to 2.57 (Ramme and Tharaniyil 2013).

6.3.6 Loss on Ignition

The presence of minerals can potentially affect the loss on ignition (LOI) tests. Minerals such as limestone, lime, and anhydrite can all appear in fly ashes collected from boilers. Besides mineral content, the LOI content of fly ashes is significantly higher at larger size particles. Although the carbon is concentrated in the larger size particles, large particles represent only a small portion of the total mass of the fly ash. For instance, DeBarr (1996) found that most carbon mass (50% to −70%) was in the particles that had diameters < 38 μm fraction.

6.3.7 Compaction

Usually fly ashes exhibit significant variations in compaction characteristics because of disparities in the nature of fly ash created at the same power plant over time (Yudhbir and Honjo 1991, Winter and Clarke 2002). The maximum dry density of fly ash varies from 60 to 110 lb/ft^3 (Ramme and Tharaniyil 2013). The variations in the dry density of fly ashes along with moisture contents are relatively less than those observed for a well-graded soil, possessing the same gradation in grain size (Moulton 1978). The fly ash is less sensitive to variations in moisture contents because of its higher air void ratio. Soils normally have air void content ranging between 1% and 5% at maximum dry density, whereas fly ash contains 5% to 15% air voids at this compaction level. The higher air void ratios tend to limit the buildup of pore pressures during compaction, thus allowing the fly ash to be compacted over a larger range of water content (Toth et al. 1988). Trivedi and Singh (2004) reported that fly ash being nonplastic in nature, the material could be compacted by vibration. However, because of a significant percentage of fines, it could be also compacted by using a standard Proctor hammer. There is a threshold moisture content at which a minimum value of dry unit weight is obtained. By further addition of moisture, dry unit weight continues to increase up to optimum moisture content.

The maximum dry densities for bottom ash are 13.2 and 10.5 kN/m^3 (84.2 and 67.1 lb/ft^3) with OMCs of 28.7% and 32.3%, respectively. For boiler slag, the maximum dry density ranges from 12.9 to 16 kN/m^3 (82 to 102 lb/ft^3), and for dewatered FGD gypsum, the maximum dry density is 12.8 kN/m^3 (81.5 lb/ft^3) at 35% OMC (CCP Handbook).

Furthermore, fly ash addition to fine-grained subgrade soils and coarse-grained base/subbase geomaterials typically decreases the maximum dry unit weight and increases the optimum moisture content compared to the values for the soil alone (Cetin et al. 2010; Rosa et al. 2016, 2017; Dayioglu et al. 2017; Coban and Cetin 2018; Mahedi et al. 2020b). In addition, fly ash inclusion in soils increases the workability of soils and other geomaterials such as recycled asphalt pavement (RAPs), recycled concrete aggregates (RCAs), and reclaimed pavement materials (RPMs).

6.3.8 Shear Strength

The strength of fly ashes assumes great importance when they are used in various geotechnical applications. Usually, self-cementing pozzolanic fly ashes originating from lignite and subbituminous sources are preferred. Class C fly ashes without an activator can exhibit unconfined compression strength values up to 1,000 kPa (Sivapullaiah et al. 1998), whereas Class F fly ashes do not exhibit unconfined compressive strength by themselves when compacted at their optimum moisture contents and cured between 7 and 28 days. The addition of activators such as lime (5% by weight) and gypsum (2.5% by weight) are found to increase the unconfined compressive strength of Class F fly ashes by 40 times (Sivapullaiah and Moghal 2011). One study, involving 51 different ashes, revealed the mean friction angle value to be approximately 34 degrees (McLaren and Digioia 1987).

Dayioglu et al. (2017) also showed that off-specification fly ash improved the unconfined compressive strength of expansive shale clays. The results of this study showed that the strength of shale clay increased from 1,200 to 5,000 kPa with the addition of fly ash at the range of 10% to 30% by weight (cured for 1, 7, and 28 days). In addition, fly ash was found to be more effective in increasing unconfined strength compared to lime addition at the range of 4% to 12% weight. Mahedi et al. (2020b) also found that Class C fly ash was efficient in improving the unconfined compressive strength of sulfate-rich soils.

6.4 BENEFICIAL USE OF COAL COMBUSTION RESIDUALS

The US Environmental Protection Agency has established four criteria for the beneficial use of CCRs (USEPA 2019):

1. Coal combustion residual must provide a functional benefit;
2. Coal combustion residual must substitute for the use of a virgin material;

3. Coal combustion residual must meet product specifications and/or design standards; and
4. When unencapsulated use of CCR involves placement on the land of 12,400 t or more in nonroadway applications, the user must demonstrate and provide documentation upon request, that environmental releases to groundwater, surface water, soil, and air are comparable to or lower than those from analogous products made without CCR, or that releases will be below relevant regulatory and health-based benchmarks for human and ecological receptors.

ACAA (2019) indicates that of the 79 million tons of CCR produced in 2019, 41 million tons (52%) were put to beneficial use.

6.4.1 Reuse of Coal Combustion Residual (Particularly Fly Ash)

6.4.1.1 Encapsulated Beneficial Uses. The two largest encapsulated uses reported by ACAA in 2014 are fly ash used in "concrete/concrete products/grout" (13.12 million tons) and flue gas desulfurization (FGD) material (i.e., gypsum) used in gypsum panel products (11.22 million tons), making up to 39% of the total amount of CCR put to beneficial use. In addition, 1,552,463 t of fly ash are currently being used in applications involving roadway base/subbase/subgrades, mining, waste solidification/stabilization, aggregate production, flowable fill, structural fills/embankments, soil modification/stabilization, and mineral filler in asphalt applications (ACAA 2019).

The US Environmental Protection Agency has evaluated the potential environmental impacts associated with using fly ash as a direct substitute for portland cement in concrete and FGD gypsum as a replacement for mined gypsum in wallboard. The US Environmental Protection Agency's evaluation concluded that the beneficial use of encapsulated CCR in concrete and wallboard is acceptable because environmental releases are comparable to or lower than those from analogous non-CCR products or are at or below relevant regulatory and health-based benchmarks (USEPA 2019).

6.4.1.2 Unencapsulated Beneficial Use. Unencapsulated uses of coal ash are those where the coal ash is used in a loose particulate, sludge, or other unbound form. In 2019, ACAA reported about 9% of generated CCRs (1.5 million tons) are beneficially used in unencapsulated applications (ACAA 2019). The largest unencapsulated use is for "structural fills/embankments" (0.5 million tons). EPA has developed an online resource to guide potential users through the evaluation of the beneficial use of CCRs (USEPA 2019).

Fly ash produced in recent decades from numerous coal utilities operating with low NO_x boilers contain a high percentage of unburned carbon, which is evaluated as loss-on-ignition (LOI). The LOI contents of the ash produced from these types of utilities can be as high as 12% to 25% (Aydilek and Cetin 2013). According to ASTM C618-15 (ASTM 2015), the upper limit for LOI of both Classes C and F fly ash is 6% maximum for use in concrete (Table 6-1). As previously described, these high-LOI fly ashes are usually referred to as off-specification fly ashes, and they cannot be used in cement and concrete production because of the potential for reactions between the unburned carbon and the air entrainment agents (Cetin et al. 2010).

The most beneficial use of CCRs occurs in the types of encapsulated applications described previously. CCRs are also used in unencapsulated applications such as stabilizing the road subgrade or base course, and these applications have significant potential to grow. Coal combustion residuals can increase strength and stiffness and reduce swelling in unencapsulated roadway applications, while decreasing or eliminating the amount of other stabilizing materials required. This improves the service life of roads, reduces GHG emissions, and reduces energy and water consumption, making CCR-stabilized soils a less-expensive and more-sustainable alternative than virgin aggregate or cement-stabilized soil (Bin-Shafique et al. 2004; Edil et al. 2006; Carpenter et al. 2007; Li et al. 2008, 2009; Lee et al. 2010, 2011, 2013; Tastan et al. 2011; Ebrahimi et al. 2012; Camargo et al. 2013; Soleimanbeigi and Edil 2014).

High calcium oxide content and self-cementing properties of subbituminous coal fly ash are suitable materials for the stabilization of soils, recycled pavement materials, and road surface gravel. Fly ash stabilization can result in improved properties, including increased stiffness, strength, and freeze–thaw durability; reduced hydraulic conductivity, plasticity, and swelling; and increased control of soil compressibility and moisture. Fly ash–stabilized materials may be used in roadway construction, such as working platforms during construction, and stabilized subgrade, subbase, and base layers. Fly ash stabilization can also be used in limiting the settlement of fills below buildings (ASTM D7762-11) (ASTM 2012).

In the fly ash stabilization of fine-grained soils, flocculation, agglomeration, and cementitious reactions may occur. Self-cementing fly ash has been demonstrated to be an effective stabilization agent for a range of fine-grained soils in increasing subgrade support capacity for pavements, in reducing the swelling potential of expansive soils, in increasing the shear strength of organic soils and fine-grained soils, and in reducing the compressibility of fills under foundations. A fly ash–stabilized subbase also provides a more stable working platform during pavement construction over very soft subgrades. Such stabilized working platforms can be treated as a subbase section in the pavement thickness design. In coarse-grained materials such as aggregate base, gravels, recycled pavement materials, and recycled road surface gravel, cementation through pozzolanic reactions and hydration with self-cementing fly ash can cause strength gain and enhance durability. The reaction rate depends on the self-cementing fly ash used (ASTM 2012).

Recycling or reclaiming existing flexible pavements with self-cementing fly ash as a stabilizing agent has been demonstrated to be an efficient and economical method of pavement rehabilitation for roadways. The process is accomplished by pulverizing the entire pavement section to the subgrade and stabilizing the pulverized mixture by adding self-cementing coal fly ash and water (as needed). The recycled section provides an enhanced base for a new hot mix asphalt (HMA)-wearing surface. Self-cementing fly ash–stabilized recycled sections have structural capacities (enhanced modulus and reduced plastic deformations), which are considerably better than a crushed-stone aggregate base and can be equivalent to an asphaltic concrete base section (ASTM 2012).

Recycling or reclaiming existing road surface gravel (RSG) with self-cementing fly ash stabilization is an economical method for converting gravel roads to paved roads. The process is accomplished by placing the appropriate amount of self-cementing fly ash over the entire road surface and then blending it into the RSG down to the required depth to form a base for the HMA layer. The self-cementing fly ash–stabilized-RSG has similar or better properties as a high-quality crushed aggregate base course in the stabilized section. The recycled and stabilized RSG section provides an enhanced base for a new HMA-wearing surface (ASTM 2012).

Recent data indicate that approximately 40% of CCRs being produced today are being placed in landfills, thereby consuming valuable land space at great cost (ACAA 2019). There have been efforts to reuse fly ash materials in construction to decrease the disposal rate. Cetin et al. (2010) tested the environmental impacts of several different off-specification (high carbon) fly ashes used in structural fill applications at the rate of 40% by weight. The results of this study showed that leached heavy metals from these fly ash–mixed structural fills were far below the EPA water drinking limits.

Bottom ash is mainly used in structural fills and embankments, soil modification, and for snow and ice control. Although bottom ash has an elemental composition that is similar to fly ash, the size of the material can range from fine sand to fine gravel, and it does not have any cementitious properties. Because bottom ash is not pozzolanic, it has more limited applications in the cement and concrete industry than fly ash (Joshi and Lohita 1997).

Bottom ash is typically used as a lightweight aggregate in precast concrete products, including concrete blocks and masonry units (Dienhart et al. 1998). The final product is lighter than when using conventional aggregates such as sand and gravel and has similar strength properties.

The demand for bottom ash over the next 20 years will be dependent on the end-use markets, especially US construction market activity, but it is expected to remain strong. As new technologies and uses emerge, increasing amounts of bottom ash will be used as an input for various construction materials. Bottom ash is also being used to replace fine aggregate in HMA, with research being conducted to evaluate the material's performance, stability, and moisture susceptibility (Nam 2014).

Bottom ash is also being studied as a replacement material in self-compaction concrete, which is a type of concrete "that will be leveled and compacted under its self-weight," with promising results (Kadir and Hassan 2014).

6.4.2 Environmental Concerns

Coal combustion residuals generated from coal power plants, if not properly managed, have the potential to cause water, air, and soil pollution. With respect to air emissions, wind-blown fugitive dust can create respiratory risks for people and settle on natural water bodies and flora and fauna (Moghal 2017). The wet ash disposal system employed by power stations leads to the discharge of ash particles into surface impoundments, which if not properly managed can overflow into nearby surface-water bodies or seep into groundwater. Fly ash can affect the health of humans who consume produce grown on CCR-impacted soils; contaminated groundwater used for drinking; inhalation of fine ash particles and by radiation emitted by fly ash–based building materials (Singh et al. 2020).

A number of researchers have investigated the environmental impact of the use of fly ashes in geotechnical engineering applications (Cetin et al. 2010; Komonweeraket et al. 2015a, b, c). The majority of their studies focused on evaluating the leaching of trace metals, sulfates, nitrates, organic/inorganic carbon along with the pH, electrical conductivity (EC), and oxidation–reduction potential (Eh) of the leachate collected from fly ash–stabilized/amended soils. These studies covered a wide range of soils and fly ashes that were mixed at different proportions. On the one hand, overall, it was found that an increase in fly ash content increased the pH, EC, and leaching concentrations of metals including arsenic (As), chromium (Cr), iron (Fe), calcium (Ca), copper (Cu), cobalt (Co), manganese (Mn), magnesium (Mg), and vanadium (V). It was also observed that fly ash addition increased sulfate concentrations in the leachate. On the other hand, some results showed that zinc (Zn) concentrations decreased with the addition of fly ash in soils (Cetin et al. 2010). Although fly ash addition increased the leached metal concentrations in general, it was observed that the increased level of metal concentrations were below the EPA maximum contaminant levels for drinking water.

REFERENCES

AASHTO M 295. 2021. "Standard Specification for Coal Fly Ash and Raw or Calcined Natural Pozzolan for Use in Concrete (ASTM Designation: C618-19)."

ACAA (American Coal Ash Association). 2019. "Coal combustion products production and use statistics." Accessed February 28, 2024. https://acaa-usa.org/wp-content/uploads/coal-combustion-products-use/2019-Survey-Results.pdf.

ASTM International. 2012. *Standard guide for laboratory mix design for stabilization of soil and soil-like materials with fly ash, annual book of standards*. ASTM D7762-11. West Conshohocken, PA: ASTM.

ASTM. 2015. *Standard specification for coal fly ash and raw or calcined natural pozzolan for use in concrete*. ASTM C618-15. West Conshohocken, PA: ASTM.

Aydilek, A., and B. Cetin. 2013. "Geoenvironmental impacts of using high carbon fly ash in structural fill applications." State Highway Administration Research Rep. SP909B4P. Accessed February 28, 2024. https://www.roads.maryland.gov/OPR_Research/MD-13-SP909B4P-Geoenvironmental-Impacts-Of-Using-HCFA-In-Structural-Fill-Applications.pdf.

Balapour, M., W. Zhao, E. J. Garboczi, N. Y. Oo, et al. 2020. "Potential use of lightweight aggregate (LWA) produced from bottom coal ash for internal curing of concrete systems." *Cem. Concr. Compos.* 105: 103428.

Bin-Shafique, S., T. B. Edil, C. Benson, and A. Senol. 2004. "Incorporating a fly-ash stabilised layer into pavement design." *Geotech. Eng.* 157 (4): 239–249.

Blanc, D., L. Gonzalez, M. Lupsea-Toader, and C. de Brauer. 2018. "Mineralogical evolution and leaching behaviour of a heap of bottom ash as a function of time: Influence on its valorization." *Waste Biomass Valorization* 9: 2517–2527.

Camargo, F., T. Edil, and C. Benson. 2013. "Strength and stiffness of recycled pavement materials stabilized with fly ash: A laboratory study." *Road Mater. Pavement Des.* 14 (2): 1–13.

Carpenter, A., K. Gardner, J. Fopiano, C. Benson, et al. 2007. "Life cycle-based risk assessment of recycled materials in roadway construction." *Waste Manage.* 27 (10): 1458–1464.

Cetin, B., A. H. Aydilek, and Y. Guney. 2010. "Stabilization of recycled base materials with high carbon fly ash." *Resour. Conserv. Recycl.* 54 (11): 879–892.

Chan, H. T. 1983. *Bottom ash: A potential aggregate for highway base and subbase courses. Tech. Rep. 83-222-K.* Toronto: Ontario Hydro Research Division.

Coban, H. S., and B. Cetin. 2018. "The use of lime sludge alone and with class C fly ash for subgrade soil stabilization." *Transp. Res. Board* 295: 495–504.

Dayioglu, M., B. Cetin, and S. Nam. 2017. "Stabilization of expansive Belle Fourche shale clay with different chemical additives." *Appl. Clay Sci.* 146: 56–69.

DeBarr, J. A. 1996. "Valuable products from utility fly ash." *Prepr. Pap. Am. Chem. Soc., Div. Fuel Chem.* 41 (2): 604–606.

Dienhart, G. J., B. R. Stewart, and S. S. Tyson. 1998. *Coal ash: Innovative applications of coal combustion products*. Alexandria, VA: American Coal Ash Association.

Ebrahimi, A., B. Kootstra, T. Edil, and C. Benson. 2012. "Practical approach for designing flexible pavements using recycled roadway materials as base course." *Road Mater. Pavement Des.* 13 (3): 1–18.

Edil, T. B., H. A. Acosta, and C. H. Benson. 2006. "Stabilizing soft fine-grained soils with fly ash." *J. Mater. Civ. Eng.* 18 (2): 283–294.

EIA (US Energy Information Administration). 2019. "Frequently asked questions (FAQs)." Accessed December 30, 2020. https://www.eia.gov/tools/faqs/faq.php?id=427&t=3.

EPRI (Electric Power Research Institute). 2016. *Corrective action technology profile: Practical feasibility of in situ stabilization/solidification as a source control for coal combustion residuals*. Palo Alto, CA: EPRI.

FHWA (Federal Highway Administration). 2003. "*Fly ash facts for highway engineers*." 4th ed. FHWA-IF-03-019. Washington, DC: FHWA.

FHWA. 2011. "User guidelines for waste and byproduct materials in pavement construction." FHWA-RD-97-148. Accessed February 28, 2024. https://www.fhwa.dot.gov/publications/research/infrastructure/structures/97148/toc.cfm.

FR (Federal Register). 2015. *Hazardous and solid waste management system: Disposal of coal combustion residuals from electric utilities*. 40 CFR Parts 257 and 261, 80(74). Washington, DC: USEPA.

Gray, D. H., and Y. K. Lin. 1972. "Engineering properties of compacted fly ash." *J. Soil Mech. Found. Eng.* 98 (4): 361–380.

Joshi, R. C., and R. P. Lohita. 1997. *Fly ash in concrete: Production, properties and uses*. Boca Raton, FL: CRC Press.

Kadir, A. B. A., and M. I. H. Hassan. 2014. "An overview of fly ash and bottom ash replacement in self-compaction concrete." *Key Eng. Mater.* 594: 465–470.

Komonweeraket, K., B. Cetin, C. H. Benson, A. H. Aydilek, et al. 2015a. "Geochemical analysis of leached elements from fly ash stabilized soils." *J. Geotech. Geoenviron. Eng.* 141 (5): 04015012.

Komonweeraket, K., B. Cetin, C. H. Benson, A. H. Aydilek, et al. 2015b. "Leaching characteristics of toxic constituents from coal fly ash mixed soils under the influence of pH." *Waste Manage.* 38: 174–184.

Komonweeraket, K., B. Cetin, C. H. Benson, A. H. Aydilek, et al. 2015c. "Effects of pH on the leaching mechanisms of elements from fly ash mixed soils." *Fuel* 140: 788–802.

Lee, J., T. B. Edil, C. H. Benson, and J. M. Tinjum. 2013. "Building environmentally and economically sustainable transportation infrastructure: Green highway rating system." *J. Constr. Eng. Manage.* 139 (12): A401300.

Lee, J. C., T. B. Edil, C. H. Benson, and J. M. Tinjum. 2011. "Evaluation of variables affecting sustainable highway design with BE2ST-in-highways system." *Transp. Res. Rec.* 2233 (1): 178–186.

Lee, J. C., T. B. Edil, J. M. Tinjum, and C. H. Benson. 2010. "Quantitative assessment of environmental and economic benefits of using recycled construction materials in highway construction." *Transp. Res. Rec.* 2158 (1): 1–11.

Leonards, G. A., and B. Bailey. 1982. "Pulverized coal ash as structural fill." *J. Geotech. Eng. Div.* 108 (4): 517–531.

Li, L., C. Benson, T. Edil, and B. Hatipoglu. 2008. "Sustainable construction case history: Fly ash stabilization of recycled asphalt pavement material." *Geotech. Geol. Eng.* 26 (2): 177–187.

Li, L., T. B. Edil, and C. H. Benson. 2009. "Mechanical performance of pavement geomaterials stabilized with fly ash in field applications." *Coal Combust. Gasific. Prod.* 1 (1): 43–49.

Mahedi, M., B. Cetin, and D. J. White. 2020b. "Cement, lime, and fly ashes in stabilizing expansive soils: Performance evaluation and comparison." *J. Mater. Civ. Eng.* 32 (7): 04020177.

Martin, J., R. Collins, J. Browning, and F. Beihl. 1990. "Properties and use of fly ashes for embankments." *J. Energy Eng.* 116 (2): 71–86.

McLaren, R. J., and A. M. Digioia. 1987. "The typical engineering properties of fly ash." In *Proc., Conf. on Geotechnical Practice for Waste Disposal*, ASCE Geotechnical Special Publication 13, edited by E. Wood, New York, 683–697.

Mehta, P. K., and P. J. M. Monteiro. 1997. *Concrete: Microstructure, properties and material*. Chennai, India: Indian Concrete Institute.

Moghal, A. A. B. 2017. "State-of-the-art review on the role of fly ashes in geotechnical and geoenvironmental applications." *J. Mater. Civ. Eng.* 29 (8): 04017072.

Moulton, K. L. 1978. *Technology and utilization of power plant ash in structural fills and embankments*. Morgantown, WV: West Virginia University.

Nam, B. H. 2014. "Evaluating the use of waste-to-energy bottom ash as road construction materials." *Ph.D. thesis*, University of Central Florida, Dept. of Civil, Environmental, and Construction Engineering.

Natusch, D. F. S., Wallace, J. R., and Evans, C. A., Jr. (1974). "Toxic trace elements: Preferential concentration in respirable particles." *Science* 183 (4121): 202–204.

Ongwandee, M., Namepol, K., Yongprapat, K., Homwuttiwong, S., et al. 2020. "Coal bottom ash use in traditional ceramic production: Evaluation of engineering properties and indoor air pollution removal ability." *J. Mater. Cycles Waste Manage.* 22 (6): 2218–2129.

Pandian, N. S., C. Rajasekhar, and A. Sridharan. 1998. "Studies on the specific gravity of some Indian coal ashes." *J. Test. Eval.* 26 (3): 177–186.

Ramme, B. W., and M. P. Tharaniyil. 2013. *We Energies' coal combustion products utilization handbook.* 3rd ed. Milwaukee: Wisconsin Electric Power, https://www.we-energies.com/environment/pdf/ccp_handbook.pdf

Rosa, M. G., B. Cetin, T. B. Edil, and C. H. Benson. 2016. "Development of a test procedure for freeze–thaw durability of geomaterials stabilized with fly ash." *Geotech. Test. J.* 39 (6): 938–953.

Rosa, M. G., B. Cetin, T. B. Edil, and C. H. Benson. 2017. "Freeze–thaw performance of fly ash stabilized materials and recycled pavement materials." *J. Mater. Civ. Eng.* 29 (6): 04017015.

Singh, A. K., R. E. Masto, B. Hazra, J. Esterle, et al. 2020. *Ash from coal and biomass combustion.* Cham, Switzerland: Springer.

Sivapullaiah, P. V., and A. A. B. Moghal. 2010. "Leachability of trace elements from two stabilized low lime Indian fly ashes." *Environ. Earth Sci.* 61 (8): 1735–1744.

Sivapullaiah, P. V., and A. A. B. Moghal. 2011. "Role of gypsum in the strength development of fly ashes with lime." *J. Mater. Civ. Eng.* 23 (2): 197–206.

Sivapullaiah, P. V., J. P. Prashanth, A. Sridharan, and B. V. Narayana. 1998. "Reactive silica and strength of fly ashes." *J. Geotech. Geol. Eng.* 16 (3): 239–250.

Soleimanbeigi, A., and T. B. Edil. 2014. "Settlement of highway embankments constructed with recycled materials." In *Mid-Continent Transportation Symp.* Madison, WI: Madison Concourse Hotel.

Tastan, E. O., T. B. Edil, C. H. Benson, and A. H. Aydilek. 2011. "Stabilization of organic soils with fly ash." *J. Geotech. Geoenviron. Eng.* 137 (9): 819–833.

Toth, P. S., H. T. Chan, and C. B. Cragg. 1988. "Coal ash as structural fill with special reference to Ontario experience." *Can. Geotech. J.* 25 (4): 694–704.

Trivedi, A., and S. Singh. 2004. "Geotechnical and geoenvironmental properties of power plant ash." *J. Inst. Eng.* 85: 93–99.

USEPA (US Environmental Protection Agency). 2019. "Coal ash reuse." Accessed February 28, 2024. https://www.epa.gov/coalash/coal-ash-reuse.

Winter, M. G., and B. G. Clarke. 2002. "Improved use of pulverised fuel ash as general fill." *Proc. Inst. Civ. Eng. Geotech. Eng.* 155 (2): 133–141.

Yudhbir, H. Y., and Y. Honjo. 1991. "Application of geotechnical engineering to environmental control." In *Proc., 9th Asian Regional Conf. on Soil Mechanics and Foundation Engineering*, 431–469. Bangkok, Thailand.

CHAPTER 7

WOODY BIOMASS FLY ASH: PROPERTIES AND ENGINEERING APPLICATIONS

Susan E. Burns, F.ASCE; N. N. Nortey Yeboah, M.ASCE; Xenia Wirth, M.ASCE; and Shaivan Shivaprakash

7.1 INTRODUCTION

Biomass energy, which refers to the fuel source derived from plant or plant-based materials, has gained popularity as a fuel in recent years because of its inherent short-term carbon cycling. Primarily, biomass energy is generated from the combustion of wood and other biomass sources such as grassy or woody plants, wood waste such as sawdust, bark, or pallets, or some organic components of municipal solid waste, which can be combusted to produce energy. Although biomass has been used as a source of energy throughout human history, more recent efforts have focused on the development of industrial-scale higher-efficiency power plants that are fueled entirely by biomass sources. Biomass is viewed as a more carbon neutral fuel when compared to other fossil fuels because the CO_2 capture by biomass during photosynthesis and its subsequent release during combustion is on roughly the same timescale, as opposed to a fossil fuel that formed millions of years ago. However, the process is not carbon neutral because energy is used in harvesting, transportation, and preparation of biomass for combustion.

Current renewable energy standards require utilities to obtain significant percentages of their fuel sources from renewable inputs such as solar, wind, or biomass. According to the Solar Energy Industries Association (SEIA 2024), roughly 75% of states in the United States currently have some type of renewable energy standards in place. These standards exist on a state-by-state basis only, as there is no federal standard in the United States that governs renewable energy. In addition, the current ASTM standard that governs the beneficial use of fly ash in concrete (ASTM C618) requires that fly ash used in concrete must be generated using only coal as the fuel source and must also have an organic content determined by loss on ignition (LOI) of 6% or less (ASTM 2023). The current ASTM standard does not allow the beneficial use of biomass fly ash in concrete, even in the case where the properties of the biomass ash may meet acceptable performance requirements.

Similar to coal fly ash, discussed in Chapter 6, biomass fly ash is produced through the combustion of biomass in a biomass-fired or a coal-fired boiler that has been converted for the combustion of biomass. The resulting ash is a particulate material that is carried by flue gas through the combustion process and is collected at the end of the treatment train by electrostatic precipitators, baghouses, or scrubbers (ACAA 2023b). The properties of biomass fly ash produced in a commercial boiler will be a strong function of the type of biomass and combustion conditions, and the resulting properties, such as unit weight, specific gravity, composition, and pozzolanic or cementitious properties, are critical to the ability to beneficially use biomass fly ash in applications such as concrete. More recent requirements to meet air pollution control standards have resulted in changed ambient gas concentrations within a boiler, along with decreased temperature within the boiler, which may lead to changes in the grain size distribution and residual carbon content of the resulting ash.

7.2 BIOMASS ASH

Most research on biomass fly ash has focused on either investigating power plant operations, burner efficiency, ash formation, and emissions with biomass addition (Belosevic 2010; Blevins and Cauley 2005; Demirbas 2005; Hansson et al. 2009; Higgins et al. 2009; NREL 2000; Robinson et al. 2002, 2003; van Loo and Koppejan 2007; Wang et al. 2011; Wei et al. 2002) or investigating productive use of the resulting biomass derived ash, mainly in concrete and cement applications (Fouad et al. 1998; Johnson et al. 2010; Wang and Baxter 2007; Wang et al. 2008a, b). Biomass is a mixture of organic and inorganic material and solid and fluid phases, which are intimately associated with each other, along with minerals from both natural and anthropogenic sources (Demirbas and Demirbas 2009, Vassilev et al. 2010), and it is well established that compared to coal, biomass chemistry is highly complex and extremely variable (Vassilev et al. 2013a, b; Voshell et al. 2018).

Woody and herbaceous biomasses are the most common forms of biomass fuels that are directly combusted or co-fired with coal for energy. Vassilev et al. (2010) compared elemental C, H, O, N, and S content (ultimate analysis), as well as volatile matter and ash content of both woody and herbaceous biomass fuel types to typical coal values. Dry weight (%) of volatile matter and ash produced by biomass is typically higher than that produced by coal combustion (Belosevic 2010, Chmielniak and Sciazko 2003, Demirbas 2005, Demirbas and Demirbas 2009, Robinson et al. 2003, Wei et al. 2002). The data show that despite the inter- and intraspecies biomass variability, when compared to coal, there are some consistent characteristic differences. Biomass is enriched in volatile matter, oxygen, and to a lesser extent, hydrogen, when compared to coal. In contrast, coal is enriched in carbon, ash content, and to a lesser extent, nitrogen and sulfur.

During the combustion of biomass, both natural and anthropogenic materials are burned, and the resulting ash is composed of some of the original mineral phases in the biomass, in addition to new phases that are generated during combustion. Chemical composition and inorganic content of biomass ash depend not only on the biomass source (e.g., woody or herbaceous) but also on the growth and production conditions (Wang et al. 2008b). The most common elemental constituents in biomass ash are calcium and potassium with lower parts magnesium, sodium, manganese, and iron. Silicates, oxalates, and sulfates are common anions in biomass ash (Demirbas and Demirbas 2009). Compared to coal ash, biomass ash has lower concentrations of silica, aluminum, iron, and sulfur (Yeboah et al. 2014). However, it should be noted that there are some biomass species that can have significant silica and sulfur content.

Compared to coal, biomass and biomass ash has been shown to be more likely to be enriched with more reactive functional groups (e.g., $-COOH$, $-OCH_3$, $-OH$), chelates, light hydrocarbons, carbohydrates, oxyhydroxides, carbonates, chlorides, and phosphates, but fewer aromatic carbon species that are predominant in coal (Netzel 1991, Vassilev et al. 2010, Yeboah et al. 2014). In addition, biomass and its ash (herbaceous ash in particular) typically contain more alkali (K, Na, Mg, Ca) and chloride content than coal, which can contribute to operational difficulties within a power plant (AGS 2007, Robinson et al. 2002, van Loo and Koppejan 2007, Wei et al. 2002). In addition to the high organic content that is often found in biomass ash, water-soluble alkalis can react with air-entraining agents in concrete, which reduces its potential as a partial replacement for portland cement (van Loo and Koppejan 2007). There is no indication, however, that biomass ash cannot be used in other reuse applications such as structural fills, soil stabilization, and waste solidification/stabilization applications.

In addition to fly ash produced from the combustion of coal or the combustion of biomass, some power plants have produced ash from the co-combustion of coal and biomass. The resulting properties of the ash generated from co-combustion of coal and biomass will be a function of the mass concentration of each fuel source used in combustion. This chapter focuses on the properties of woody biomass fly ash that results from combustion of 100% biomass as the fuel source. A comprehensive review of the properties of biomass fly ash can be found in Belviso (2018) and Voshell et al. (2018).

7.3 CHARACTERIZATION OF PHYSICAL AND CHEMICAL PROPERTIES

7.3.1 Particle Morphology and Particle Size

Similar to coal ash, biomass fly ash contains some concentration of glassy aluminosilicate spheres, including cenospheres, which are hollow aluminosilicate spheres, and plerospheres, which are spheres that enclose a number of smaller particles or spheres, as well as residual fibrous, woody particles (Yeboah et al. 2014) (Figure 7-1). Although coal fly ash typically classifies as silt-sized particles, some studies have demonstrated that biomass fly ash has larger particle sizes that would classify as sand-sized. Yeboah et al. (2014) reported a median particle size for three samples of

(a)

(b)

Figure 7-1. Scanning electron microscope images of (a) coal fly ash with aluminosilicate cenospheres, (b) biomass fly ash with a residual wood structure in a biomass sample.

biomass ash that ranged between 168 and 1440 μm. However, Wirth et al. (2017) reported that another biomass fly ash produced from woody biomass in a full-scale biomass power plant had a particle morphology and particle size similar to coal fly ash, some small cenospheres, and a median particle size equal to 26 μm.

7.3.2 Specific Gravity

Although the specific gravity of biomass is close to 1, the specific gravity of biomass fly ash is measurably higher and can range from approximately 1.25 to 2.80 (Yeboah et al. 2014) (Table 7-1). It is important to note that the ash formed from biomass will have a residual organic content (ROC) that varies as a function of combustion conditions. Higher values of ROC in a biomass fly ash will result in lower values of specific gravity.

7.3.3 Chemical Composition and Classification of Biomass Fly Ash

The major elemental composition of ash is measured as the oxide content using X-ray fluorescence (XRF) spectrometry, which, as described in Chapter 3, is an elemental analysis technique that identifies oxides commonly found in fly ash (e.g., SiO_2, Al_2O_3, Fe_2O_3, CaO, MgO, SO_3, Na_2O, K_2O, P_2O_5, TiO_2, Mn_2O_3, Cr_2O_3, ZnO) on the basis of energy emissions after

Table 7-1. Ranges of Woody Biomass Ash Physical and Chemical Properties.

Physical	
Specific gravity, G_s	1.25–2.80
Median particle size (μm)	26–1440
Common morphologies	Unburnt carbon fragments, ash, aluminosilicate spheres
Chemical	
SiO_2 (%)	1.7–70
Al_2O_3 (%)	0.75–4.7
CaO (%)	1.2–15.4
LOI (%)	0.2–95.9
TOC (%)	0.53–80.5

excitation of the particles with X-rays. Mineralogy of the ash samples is identified with X-ray diffraction (XRD).

Similar to coal fly ash, biomass ash can be classified on the basis of its primary oxide content (POC = $SiO_2 + Al_2O_3 + Fe_2O_3$). Class C fly ash is ash that demonstrates both pozzolanic and cementitious properties, with a minimum of 50% POC as well as CaO content above 20% (typically). As described in Chapter 6, Class F fly ash typically has CaO content much lower than 10% and thus demonstrates mainly pozzolanic properties with a POC of 70% or higher, according to ASTM C 618. The standard for use in concrete also requires LOI < 6%, moisture content < 3%, and sulfur trioxide < 5%.

The oxides that are present in biomass will vary as a function of the combusted material but most commonly will include SiO_2, Al_2O_3, CaO, MgO, K_2O, P_2O_5, and SO_3 (Belviso 2018). Although the primary oxide content of biomass fly ash is low compared to coal ash (approximately < 20% for biomass ash versus 50% to 99% for coal ash), and the CaO content is variable, SO_3 and MgO contents are similar for coal and biomass fly ash (Yeboah et al. 2014). Because of the low POC values measured for biomass fly ash, Yeboah et al. (2014) concluded that it could not be classified as either Class C or F ash.

Biomass fly ash properties reported by Wirth et al. (2017) demonstrated that the POC may be misleading for classifying the pozzolanicity of biomass fly ash when silica sand is used as a bed material in the boiler. The biomass fly ash tested by Wirth et al. (2017) had an average POC of 77% because of high percentages of silica, and the aluminum and iron oxides were low, less than 6% (Table 7-1). The POC content of this fly ash would classify this material as a Class F fly ash. However, silica sand is used as a bed material in the biomass-fired fluidized bed boiler, and high silica values may have been a result of contamination of the fly ash material with bed sand. Because this silica is not reactive silica, but inert quartz, it does not contribute to pozzolanicity (Ohenoja et al. 2016). The CaO, SO_3, and MgO contents were similar to those reported by Yeboah et al. (2014).

As with coal and virtually all other ash types, biomass ash contains trace elements such as arsenic, cadmium, chromium, and selenium. These constituents are naturally occurring and remain in the solid phase after combustion. Hansen et al. (2001) reported on the mobility of cadmium from biomass ash derived from straw and wood.

7.3.4 Organic Content: Loss on Ignition versus Residual Organic Carbon

The concrete industry uses mass loss on ignition (i.e., heating of sample in an oxygen-containing atmosphere) as a surrogate for organic content in combustion residuals (ASTM D7348) (ASTM 2013). Organic carbon content in fly ash is an issue because the carbon acts as a sorbent for air-entraining agents and other additives, limiting their effectiveness in concrete or requiring an increase in the added concentrations. LOI is measured by heating combustion residuals under controlled conditions of temperature and time, while measuring changes in mass. For applications in the concrete industry, samples are normally heated to 950 °C, although for some material applications, heating to 750 °C is sufficient. Although mass LOI is indicative of material change, it is not always correlated to mass loss of organic carbon because other materials, such as sulfur, will also exhibit mass loss on heating.

As an alternate method, the total organic carbon (TOC) content of a sample can be measured using a total organic carbon analyzer, as described in Chapter 3. In this method, the total carbon and the inorganic carbon are measured using oxidation and acidification, and the remaining organic carbon is determined as the difference between the total and the inorganic carbon. The advantage to TOC analysis is that the measurement of carbon is not combined with volatilization from other molecular constituents.

7.4 ENGINEERING PROPERTIES AND APPLICATIONS

7.4.1 Biomass Fly Ash in Concrete

Because of its pozzolanic activity, coal fly ash is beneficially used as a cement substitute in roughly half of the concrete produced in the United States (ACAA 2023a). Although ASTM C618 does not specifically address the use of biomass fly ash in concrete, there are indications that biomass fly ash could be utilized as a cement replacement if its properties conform with the requirements outlined in ASTM C618 (ASTM 2017). Consequently, research on the beneficial use of biomass fly ash in concrete and cement has been an area of active interest. Rajamma et al. (2009) studied the use of biomass fly ash produced from the combustion of forest residue in a power plant in Portugal and demonstrated that biomass fly ash could be used as a cement replacement up to 20% without degrading concrete strength, as long as the carbon, chloride, and sulfate contents are within acceptable limits. Lessard et al. (2017) demonstrated acceptable flexural strength in roller compacted concrete made with 20% or less biomass fly ash produced from the by-products of paper manufacturing, when the biomass fly ash was used as a cement replacement. In contrast, Shearer and Kurtis (2015) tested biomass fly ash produced from the combustion of forest, sawmill, and urban wood waste for use as a supplementary cementitious material and measured reduced strength in mortars made with biomass fly ash. They noted that these results were obtained for only one biomass ash sample, which had a very high loss on ignition (approximately 44%).

7.4.2 Biomass Fly Ash for Soil Improvement

Biomass ash is often used as a soil amendment in agricultural applications (van Eljk et al. 2012, Schiemenz and Eichler-Löbermann 2010, Vassilev et al. 2013a), as adding alkaline ash balances the acidic effects of introducing nitrogen fertilizers while simultaneously providing necessary nutrients for plant growth (van Eljk et al. 2012, Niu et al. 2016). However, biomass ash derived from industrial wood waste may be contaminated with trace elements (As, Cr Hg, Se, among others), and its use should be evaluated on an individual basis (Vassilev et al. 2013a, Tosti et al. 2019).

The high lime content of woody biomass ash has been shown to be beneficial for geotechnical site improvement applications such as soil mixing. Vichan and Rachan (2013) found that combinations of calcium carbide residue and biomass ash (42% rice husk, 24% bark, 23% eucalyptus chips, and 6% board) increased the unconfined compressive strength of a high-plasticity clay up to approximately 350%. In addition, Barišić et al. (2019) found that agricultural biomass ash can be used as a binder to stabilize a low-plasticity clay for pavement design, although chemical composition has a significant effect on strength gains.

7.4.3 Biomass Fly Ash as Mine Backfill

A bulk reuse option for biomass fly ash is as a solidifying slurry mixture for backfilling mines. Biomass fly ash is mixed with water as a dense slurry and pumped into an open mine to provide ground support and stabilize used tunnels. Fuller et al. (2015) identified mine backfill as a likely beneficial use alternative for biomass fly ash combusted from wood dust, as the resultant ash satisfied the maximum metal requirements for mine backfills in Germany. Stefaniak et al. (2018) found that biomass fly ash–stabilized mixtures developed high uniaxial compressive strength, higher than many coal fly ash mixtures. The high-stabilized compressive strength was attributed to high lime content in the ash.

7.4.4 Biomass Fly Ash for Waste Solidification/Stabilization

An additional potential beneficial use alternative for woody biomass ash includes its use as a solidifier in waste solidification/stabilization applications. Biomass fly ashes with high reactive lime, silicon, and aluminum content have self-hardening properties, which makes them successful binders for waste solidification (Illikainen et al. 2014, Ohenoja et al. 2016). Ohenoja et al. (2016) compared self-hardening properties of four biomass fly ashes of varying composition. Free lime content increased pH, dissolving reactive silica and aluminum, thus making them available for self-hardening reactions (Ohenoja et al. 2016, Illikainen et al. 2014). In addition, a higher-reactive silica content correlated to higher-strength gains, but total silica content was not an appropriate metric for predicting self-hardening potential (Ohenoja et al. 2016).

7.4.5 Biomass Fly Ash as Sorptive Agents

Yeboah et al. (2014) presented results of a study examining adsorption capacity of three pure biomass ash samples for Pb[II], As[V], and Se[VI] from aqueous solution. Pure coal ash, ash from co-firing coal with biomass, as well as coal- and wood-based–activated carbon samples (for reference) were included in the study for comparison. Of the three heavy metal contaminants investigated, Pb[II] was by far the most easily adsorbed by all the various ash and activated carbon samples included in the study. As[V] and Se[VI] showed little affinity for any of the potential sorbents used in the study.

The pure biomass ash samples showed a higher capacity for lead (Pb) adsorption than the other ash samples, as well as the coal- and wood based–activated carbon reference samples. Notably, all the biomass ash samples had a lower surface area than either of the activated carbon reference samples. These results suggested that functional groups might be present on the biomass ash surfaces that are not present on the activated carbon surfaces. In particular, one of the biomass ash samples with high calcite ($CaCO_3$) content showed the highest sorption capacity for Pb. It is possible that along with adsorption to the carbon surfaces, a coprecipitation of lead carbonate was also occurring.

Overall, the results from the Yeboah et al. (2014) study showed promise for the potential utility of biomass ash as low-cost adsorptive media for heavy metals. In particular, biomass ashes may be used to remove heavy metals like Pb, which are commonly present in industrial wastewaters.

REFERENCES

ACAA (American Coal Ash Association). 2023a. "Fly ash." Accessed February 28, 2024. https://acaa-usa.org/about-coal-ash/what-are-ccps/#flyash.

ACAA. 2023b. "What are coal combustion products?" Accessed February 28, 2024. https://acaa-usa.org/about-coal-ash/what-are-ccps/.

AGS (Alliance for Global Sustainability). 2007. "Co-combustion, a summary of technology." Accessed November 18, 2011. http://www.energy-pathways.org/pdf/R5_co-combustion.pdf.

ASTM International. 2013. *Standard test methods for loss on ignition (LOI) of solid combustion residues*, ASTM D7348, 1–7. West Conshohocken, PA: ASTM.

ASTM. 2023. *Standard specification for coal fly ash and raw or calcined natural pozzolan for use in concrete.* ASTM C618, 3–6. West Conshohocken, PA: ASTM.

Barišić, I., N. I. Grubeša, T. Dokšanović, H. Draganić, et al. 2019. "Agricultural biomass ash in civil engineering – from energy to sustainable development." In *Sustainable industrial processing summit SIPS2019 Vol. 11: New and advanced materials, technologies, and manufacturing*, 47–60. Montreal: FLOGEN Star Outreach.

Belosevic, S. 2010. "Modeling approaches to predict biomass co-firing with pulverized coal." *Open Thermodynam. J.* 4: 50–70.

Belviso, C. 2018. "State-of-the-art applications of fly ash from coal and biomass: A focus on zeolite synthesis processes and issues." *Prog. Energ. Combust. Sci.* 65: 109–135.

Blevins, L. G., and T. H. Cauley. 2005. "Fine particulate formation during switchgrass/coal cofiring." *J. Eng. Gas Turbines Power* 127 (3): 457–463.

Chmielniak, T., and M. Sciazko. 2003. "Co-gasification of biomass and coal for methanol synthesis." *Appl. Energ.* 74 (3–4): 393–403.

Demirbas, A. 2005. "Potential applications of renewable energy sources, biomass combustion problems in boiler power systems and combustion related environmental issues." *Prog. Energ. Combust. Sci.* 31 (2): 171–192.

Demirbas, T., and C. Demirbas. 2009. "Fuel properties of wood species." *Energ. Sourc. Part A* 31 (16): 1464–1472.

Fouad, F. H., C. A. Copham, and J. M. Donovan. 1998. *Evaluation of concrete containing fly ash with high carbon content and/or small amounts of wood. Southern Company Rep. TR-110633.* Palo Alto, CA: EPRI.

Fuller, A., M. Carbo, P. Savat, J. Kalivodova, et al. 2015. "Results of fly ash quality for disposal options from high thermal shares up to pure biomass combustion in a pilot-scale and large scale pulverized fuel power plants." *Renew. Energ.* 75: 899–910.

Hansen, H. K., A. J. Pedersen, L. M. Ottosen, and A. Villumsen. 2001. "Speciation and mobility of cadmium in straw and wood combustion fly ash." *Chemosphere* 45 (1): 123–128.

Hansson, J., G. Berndes, F. Johnsson, and J. Kjärstad. 2009. "Co-firing biomass with coal for electricity generation—An assessment of the potential in EU27." *Energ. Policy* 37 (4): 1444–1455.

Higgins, B., L. Yan, H. Gadalla, J. Meier, et al. 2009. "Biomass co-firing retrofit with ROFA for NO_x reduction at EdF-Wrocław Kogeneracja." In *Proc., Energy Efficiency and Air Pollutant Control Conf.*, 1-27. Wroclaw, Poland.

Illikainen, M., P. Tanskanen, P. Kinnunen, M. Körkkö, et al. 2014. "Reactivity and self-hardening of fly ash from the fluidized bed combustion of wood and peat." *Fuel* 135: 69–75.

Johnson, A., L. J. J. Catalan, and S. D. Kinrade. 2010. "Characterization and evaluation of fly-ash from co-combustion of lignite and wood pellets for use as cement admixture." *Fuel* 89 (10): 3042–3050.

Lessard, J.-M., A. Omran, A. Tagnit-Hamou, and R. Gagne. 2017. "Production of RCC using biomass fly and bottom ashes: From laboratory to fieldwork." *J. Mater. Civ. Eng.* 29 (11): 4017225.

Netzel, D. A. 1991. "Multinuclear NMR approach to coal fly ash characterization." Accessed November 28, 2011. http://www.osti.gov/bridge/servlets/purl/10137808-jIjjQ0/10137808.pdf.

Niu, Y., H. Tan, and S. Hui. 2016. "Ash-related issues during biomass combustion: Alkali-induced slagging, silicate melt-induced slagging (ash fusion), agglomeration, corrosion, ash utilization, and related countermeasures." *Prog. Energy Combust. Sci.* 52: 1–61.

NREL (National Renewable Energy Laboratory). 2000. "Biomass cofiring: A renewable alternative for utilities." Accessed December 5, 2011. http://www.nrel.gov/docs/fy00osti/28009.pdf.

Ohenoja, K., P. Tanskanen, V. Wigren, P. Kinnunen, et al. 2016. "Self-hardening of fly ashes from a bubbling fluidized bed combustion of peat, forest industry residuals, and wastes." *Fuel* 165: 440–446.

Rajamma, R., J. Ball, L. A. C. Tarelho, G. C. Allen, et al. 2009. "Characterisation and use of biomass fly ash in cement-based materials." *J. Hazard. Mater.* 172 (2–3): 1049–1060.

Robinson, A. L., H. Junker, and L. L. Baxter. 2002. "Pilot-scale investigation of the influence of coal-biomass cofiring on ash deposition." *Energ. Fuels* 16 (2): 343–355.

Robinson, A. L., J. S. Rhodes, and D. W. Keith. 2003. "Assessment of potential carbon dioxide reductions due to biomass-coal cofiring in the United States." *Environ. Sci. Technol.* 37 (22): 5081–5089.

Schiemenz, K., and B. Eichler-Löbermann. 2010. "Biomass ashes and their phosphorus fertilizing effect on different crops." *Nutr. Cycl. Agroecosys.* 87: 471-482.

SEIA (Solar Energy Industries Association). 2024. "Building a strong solar industry to power America." Accessed February 29, 2024. https://www.seia.org/initiatives/renewable-energy-standards

Shearer, C. R., and K. E. Kurtis. 2015. "Use of biomass and co-fired fly ash in concrete." *ACI Mater. J.* 112 (2): 209–218.

Stefaniak, S., E. Kmiecik, E. Miszczak, F. Plewa, et al. 2018. "Geotechnical properties of products of alternative fuel combustion and co-firing with hard coal in the context of their use as solidifying dense mixtures." *Waste Manage. Res.* 36 (12): 1127–1136.

Tosti, L., A. van Zomeren, J. R. Pels, J. J. Dijkstra, et al. 2019. "Assessment of biomass ash applications in soil and cement mortars." *Chemosphere* 223: 425–437.

van Eljk, R. J., I. Obernberger, and K. Supancic. 2012. *Options for increased utilization of ash from biomass combustion and co-firing*, 1–39. IEA Bioenergy Task 32 Report. Arnhem, Netherlands: IEA.

van Loo, S., and J. Koppejan. 2007. *The handbook of biomass combustion and co-firing*. London: Earthscan.

Vassilev, S. V., D. Baxter, L. K. Andersen, and C. G. Vassileva. 2010. "An overview of the chemical composition of biomass." *Fuel* 89 (5): 913–933.

Vassilev, S. V., D. Baxter, L. K. Andersen, and C. G. Vassileva. 2013a. "An overview of the composition and application of biomass ash. Part 1. Phase–mineral and chemical composition and classification." *Fuel* 105: 40–76.

Vassilev, S. V., D. Baxter, L. K. Andersen, and C. G. Vassileva. 2013b. "An overview of the composition and application of biomass ash: Part 2. Potential utilisation, technological and ecological advantages and challenges." *Fuel* 105: 19–39.

Vichan, S., and R. Rachan. 2013. "Chemical stabilization of soft Bangkok clay using the blend of calcium carbide residue and biomass ash." *Soils Found.* 53(2): 272-281.

Voshell, S., M. Mäkelä, and O. Dahl. 2018. "A review of biomass ash properties towards treatment and recycling." *Renew. Sustain. Energ. Rev.* 96: 479–486.

Wang, S. Z., and L. Baxter. 2007. "Comprehensive study of biomass fly ash in concrete: Strength, microscopy, kinetics and durability." *Fuel Process. Technol.* 88 (11–12): 1165–1170.

Wang, S. Z., E. Llamazos, L. Baxter, and F. Fonseca. 2008a. "Durability of biomass fly ash concrete: Freezing and thawing and rapid chloride permeability tests." *Fuel* 87 (3): 359–364.

Wang, S. Z., A. Miller, E. Llamazos, F. Fonseca, et al. 2008b. "Biomass fly ash in concrete: Mixture proportioning and mechanical properties." *Fuel* 87 (3): 365–371.

Wang, X., H. Tan, Y. Niu, M. Pourkashanian, et al. 2011. "Experimental investigation on biomass co-firing in a 300 MW pulverized coal-fired utility furnace in China." In *Proc. Combust. Inst.* 33 (2): 2725–2733.

Wei, X. L., C. Lopez, T. von Puttkamer, U. Schnell, et al. 2002. "Assessment of chlorine–alkali–mineral interactions during co-combustion of coal and straw." *Energ. Fuels* 16 (5): 1095–1108.

Wirth, X., N. N. N. Yeboah, and S. E. Burns. 2017. "The effects of wood source and combustion conditions on properties of biomass fly ash." In *Proc., World of Coal Ash Conf.*, 1–9. Lexington, Kentucky.

Yeboah, N. N. N., C. R. Shearer, S. E. Burns, and K. E. Kurtis. 2014. "Characterization of biomass and high carbon content coal ash for productive reuse applications." *Fuel* 116: 438–447.

PART 4
CONTAINMENT SYSTEMS

CHAPTER 8

DESIGN, ANALYSIS, AND CONSTRUCTION OF LINER SYSTEMS FOR CONTAINMENT FACILITIES

John S. McCartney, Ph.D. P.E., F.ASCE and Keaton Botelho, P.E.

8.1 INTRODUCTION

The main goal of a liner system in solid or liquid waste containment facilities is to minimize transport and/or transport rate of contaminants from the contained waste into the environment, while remaining functional and stable during facility construction, operation, closure, and postclosure care. Many details of landfill liner systems are prescribed by US federal and state regulations, including the minimum required components and their configurations, hydraulic boundary conditions, and material hydraulic and mechanical properties. Therefore, the first part of this chapter briefly summarizes the major regulatory requirements in the United States. Although the minimum design requirements are set by regulation, the regulations also allow for the use of alternate designs if they either satisfy performance criteria spelled out in the regulations or are demonstrated to provide equivalent or superior performance to a liner system constructed to the regulatory minimum design requirements. Further, there are still several challenges with the design of liner systems that must be considered by engineers when selecting materials, developing construction specifications, and analyzing site-specific hydromechanical performance issues. This is an established topic that has been discussed in various textbooks (e.g., Daniel 1993, Rowe et al. 2004, Koerner 2005), synthesis papers (e.g., Rowe 1998) and widely cited reports to the Environmental Protection Agency (USEPA) (Bonaparte et al. 2002), and the National Research Council (NRC 2007). Accordingly, this chapter discusses relevant material properties and analysis regarding the hydraulic performance, mechanical performance, and construction of different components of liner systems. Although it is critical to understand the behavior and role of the individual components of a liner system, the overall hydromechanical performance of the liner system depends on the redundancy and interaction among various components that are expected to work together to minimize the likelihood and effects of contaminant release.

8.2 REGULATORY REQUIREMENTS FOR LINERS

8.2.1 Background

Although regulatory requirements vary throughout the world, this section focuses on requirements in the United States. The Resource Conservation and Recovery Act (RCRA) describes the waste management program mandated by the US Congress and enforced by USEPA. RCRA includes a framework for the management of both hazardous and nonhazardous solid waste. Hazardous waste is regulated under RCRA Subtitle C (in 40 CFR §264 and §265), which requires, at a minimum, a double composite liner system, a leachate collection and removal system, a leak detection system, controls for runon and runoff, a construction quality assurance (CQA) program, and other features. Nonhazardous solid waste is regulated under RCRA Subtitle D, and specifically, municipal solid waste (MSW) is regulated under 40 CFR §258 of Subtitle D. Under these federal regulations, MSW landfill liner systems must, at a minimum, include a composite liner system, a leachate collection and removal system, controls for runon and runoff, a CQA program, and other features. In addition to covering minimum liner system design criteria for landfills relevant to the content of this section, RCRA also includes location restrictions for landfills (i.e., requirements to avoid areas near active faults, airports, and the 100-year flood plain), closure criteria for landfill final cover systems, closure and postclosure care requirements, and groundwater monitoring requirements during the operation, closure, and postclosure care periods, including definition of a point of compliance to monitor groundwater down-gradient of the landfill. The regulations are intended to provide minimum national requirements, whereas the individual states may impose more restrictive regulations. Some state regulations require double-liner systems for MSW landfills. Regulatory agencies may also impose in the permit they issue for a specific landfill more restrictive design or operating requirements owing to the site-specific hydrogeological conditions or the nature of the facility. Alternate liner system designs are permissible through consultation with USEPA and/or state regulatory agencies based on a combination of analytical evaluations, numerical simulations, and field trials with performance criteria mutually defined by stakeholders and regulators.

The main philosophy of RCRA regulations in Subtitles C and D is to encapsulate the solid waste using a liner system and minimize infiltration of rainwater with a final cover system to avoid or reduce the rate of contaminant leaching into groundwater. Water entering the landfill is referred to as leachate and is considered contaminated. Without proper drainage, leachate may increase the hydraulic head on the liner system leading to increased leakage through the liner system and into the environment. To effectuate the drainage of leachate that migrates through the waste to the bottom of the landfill, the liner system is required to have a minimum slope (typically 2%, but sometimes steeper) that drains leachate through the leachate collection and removal system to a sump, where the leachate can be removed, either by gravity drainage or by a pumping system, for storage and treatment. A typical configuration of a hazardous waste landfill that meets the minimum design requirements of RCRA Subtitle C is shown in Figure 8-1.

8.2.2 Prescriptive Designs

Examples of the prescriptive minimum design requirements set by RCRA Subtitles C and D are shown in Figure 8-2a, b, respectively. The prescriptive designs in both cases rely on hydraulic barriers in the form of a composite liner, which is a combination of low-permeability soil–based liner in intimate contact with an overlying geomembrane. The soil-based liner is typically a compacted clay liner (CCL), a geosynthetic clay liner

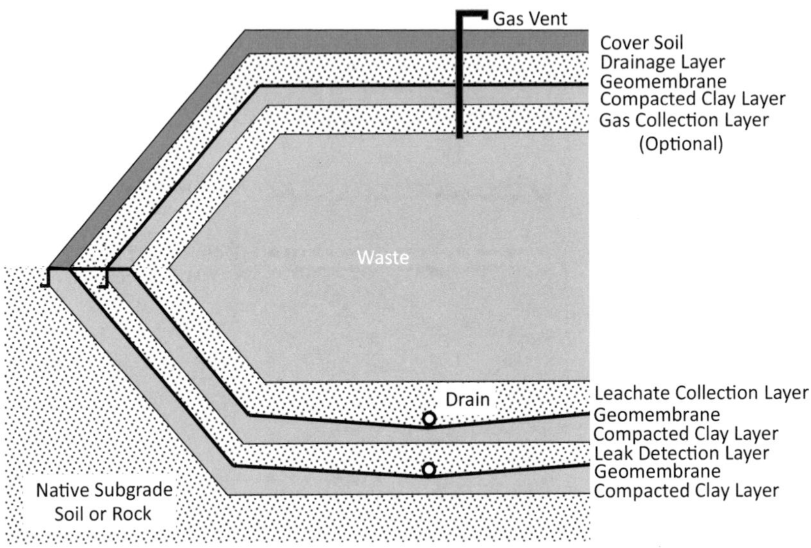

Figure 8-1. Typical configuration for a hazardous waste landfill.

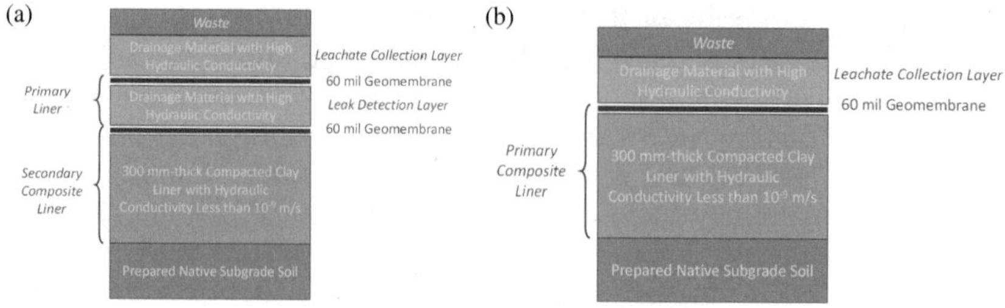

Figure 8-2. Minimum liner requirements prescribed by RCRA: (a) Hazardous solid waste landfills, (b) Nonhazardous (municipal) solid waste landfills.

(GCL), or a GCL/CCL combination. The double-liner system in Figure 8-3 is an example that goes beyond the minimum design requirements of RCRA Subtitle C as it includes two composite liners, whereas RCRA Subtitle C only requires a single composite liner (i.e., for the bottom liner component of the double-liner system). Composite liners perform much better than either a geomembrane alone or a CCL or GCL alone because a composite liner takes advantage of the complementary properties of the materials. Geomembranes are nonporous, but they are subject to chemical diffusion and leakage may occur through holes or faulty seams and they do not respond well to organic contaminants like chlorinated solvents. Compacted clay liners have the advantage of a relatively large thickness (a minimum of 900 mm in the prescribed Subtitle C design) and a maximum hydraulic conductivity of 10^{-9} m/s (10^{-7} cm/s), which serve to retard contaminant transport by both advection and diffusion. Compacted clay liners in intimate contact with an overlying geomembrane tend to impede

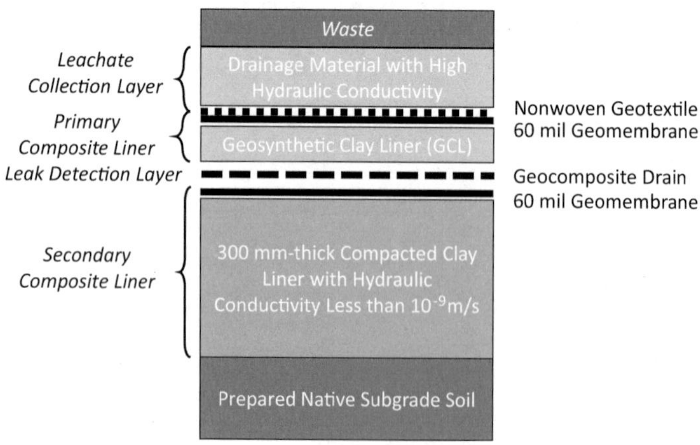

Figure 8-3. Example of a liner system for a hazardous solid waste landfill incorporating geosynthetic drainage and barrier materials.

leakage through geomembrane holes or faulty seams (they in a sense serve to plug the holes). Geosynthetic clay liners incorporate granular or powdered bentonite clay, which has a lower hydraulic conductivity than the clay typically used in CCLs, encapsulated between carrier geotextiles. GCLs are delivered to a landfill as rolls and thus have faster construction times than CCLs. However, their smaller thickness leads to higher gradients and shorter breakthrough times when calculating contaminant transport. Also, thin clay-based liners are more susceptible to damage during construction. Also, clay-based liners are not effective against hydrocarbon-based organic contaminants as these contaminants are not attracted to the negatively charged clay particles. In this case, an organoclay blanket may be more appropriate in a liner system.

As evident in Figures 8-1 through 8-3, the hydraulic barrier liner is only one component of a liner system used to achieve the overall objective of reducing transport of contaminants through the liner system. In both hazardous and nonhazardous waste landfills, drainage systems are a second key component used to ensure adequate liner performance by collecting leachate draining from the waste to limit the maximum hydraulic head above the liner to 300 mm. In this application, the drainage system above the top-most liner is referred to as the "primary leachate collection and system." In the case of double-liner systems, drainage systems are also used to collect potential leakage through the primary liner system. In this application, the drainage system is referred to as a leak detection system (LDS). Mineral-based drainage systems consisting of sand or gravel are typically required by regulation to have a minimum hydraulic conductivity of 10^{-4} m/s (10^{-2} cm/s). Geosynthetic-based drainage systems can be used as an alternative to mineral-based systems. Geosynthetic-based drainage systems typically have the advantage of incorporating high-transmissivity geonets between nonwoven geotextiles that serve as filters on one or both sides of the geonet.

8.3 GEOMEMBRANE LINERS

8.3.1 Overview

The most common types of geomembranes used in landfills in the United States are nonporous, thermoplastic polymeric materials. The polymer used in a given landfill liner system depends on the solid or liquid waste being contained, constructability, stability, and physical/mechanical considerations. Common polymeric materials include high-density polyethylene (HDPE), very low-density polyethylene (VLDPE), linear low-density polyethylene (LLDPE), polypropylene (PP), and polyvinyl chloride (PVC). Chlorosulfonated polyethylene (CSPE) and/or ethylene propylene diene terpolymer (EPDM) are examples of thermoset polymers that are only occasionally used in landfills but find broader use in other applications. The most common geomembrane polymer used in landfills is HDPE, as it has excellent durability, chemical compatibility, and reasonable flexibility. Seaming methods for HDPE geomembranes are well developed and reliable. Geomembranes made from different polymers have different weights, flexibility, tensile stiffnesses, ultimate tensile strengths, chemical compatibility, seaming approach, and means of delivery (folded sheets or rolls). Additives may be included in geomembrane resin to modify these properties on the basis of the expected environmental conditions. Selection of a geomembrane polymer for a given landfill is typically made by considering the different physical, durability, mechanical, chemical, and thermal properties needed for the application and comparing those to the properties provided by the manufacturer. The other variable that can be selected during design is the geomembrane thickness and the surface treatment. The geomembranes used in landfill liner systems typically have a thickness of at least 1.5 mm (60 mils) and may have a smooth or textured surface, with the latter used to increase interface shear strength. Although texturing is essential for the containment facility side slopes, it may also be necessary for the geomembrane on the base of the containment facility. The potential sliding surfaces along the base, identified in slope stability analyses, may not have adequate factors of safety unless textured geomembranes are used. In some landfills, a geomembrane with texturing on only one side is used to create a preferential slip plane to protect sensitive interfaces. Studies such as Peggs and Thiel (1998) provide design frameworks for geomembrane selection considering different placement environments. In these frameworks, geomembrane selection involves the use of manufacturer-provided material properties. However, samples of as-received and as-installed geomembrane used on a project are typically evaluated in a laboratory as part of construction quality control (CQC) and CQA protocols. Databases of geomembrane chemical compatibility characteristics are typically used to select geomembranes for design, but laboratory testing can be used to evaluate the chemical compatibility of a geomembrane with the site-specific leachate, as well as the hydraulic and mechanical performance of seams.

The in-service conditions that geomembrane liners will be exposed to in a landfill will depend on whether the geomembrane is used as a solitary liner or as part of a composite liner, or whether it is a component of the primary or secondary liner system. During placement, geomembranes will be exposed to ultraviolet (UV) radiation for different periods of exposure. The potential negative effect of UV radiation on the mechanical and hydraulic properties of geomembranes is mitigated using additives during manufacturing that are selected during specification of the geomembrane.

UV resistance is a conformance test requirement for geomembranes according to GRI GM13 (GRI 2021) and is typically performed in accordance with ASTM D7238 (ASTM 2017a), ASTM D8117 (ASTM 2021), and ASTM D5885/D5885M (ASTM 2017b). An issue with geomembrane construction in liner systems considered in design is the thermal expansion and contraction after placement, which may lead to the formation of wrinkles in the geomembrane. Wrinkles may change the contact conditions with an underlying CCL or GCL after waste placement and induce stress concentrations that can lead to stress cracking of the geomembrane and reduction of its service life (Gudina and Brachman 2006, 2011; Take et al. 2007). Specifications for placement of geomembranes and overlying materials along with a good CQA program minimize the wrinkling potential during installation.

8.3.2 Hydraulic Performance

This section summarizes leakage of liquids through solitary geomembranes, as they may be used in some situations as a single liner. A discussion of the combined leakage through a composite liner consisting of a geomembrane and a CCL or GCL is presented later in the chapter. Giroud and Bonaparte (1989a) established the mechanisms of leakage through geomembranes, whereas Giroud and Bonaparte (1989b) established the mechanisms of leakage through composite liners. Giroud et al. (1989, 1992), Bonaparte et al. (1989), and Bonaparte et al. (2002) provided additional data and analyses on this topic. These studies found that, for an intact geomembrane, the main mechanism for liquid flow is via diffusion. Giroud and Bonaparte (1989a) noted that the rate of diffusion varies with the hydraulic head difference, and therefore, water flow through geomembranes does not follow the laminar flow mechanism that occurs in soils. This is one of the reasons that the pressure-dependent diffusion coefficient is used to represent the hydraulic behavior of intact geomembranes.

The leakage rate per unit surface area owing to diffusion through a geomembrane is extremely low. Although the assumption of a low rate of leakage via diffusion through geomembranes may not be valid for free-phase organic liquids or for volatile organic compounds that can readily diffuse through intact polymers, it is rare to find these liquids in a landfill, and therefore, this assumption is commonly applied to landfill leachates in liquid containment facilities. The main mechanism for liquid flow and advective contaminant transport through geomembranes in landfill liners is typically assumed to be flow through holes, tears, or faulty seams. Even in the case of good construction quality control, these defects in geomembranes are likely to occur at a given frequency that depends on the area of the liner (Giroud and Bonaparte 1989a). Giroud and Bonaparte

(1989a) defined holes as openings that are at least as large as the thickness of a geomembrane and noted that they may be caused by defective seams, puncturing by stones during loading by construction equipment or the weight of contained material, tensile stresses caused by anchorage on a slope or thermal expansion and contraction, and so on. Giroud and Bonaparte (1989a) used Bernoulli's equation for flow through an orifice to estimate the volumetric flow rate through a hole in a geomembrane between two infinitely pervious media. Giroud and Bonaparte (1989a) evaluated the frequency of occurrence of holes in different landfill liners and recommended assuming one defect per 4,000 m^2 of liner area. This frequency of occurrence assumed that a construction quality assurance program is performed. With regard to the hole size to use in design, a small hole size of 1 cm^2 is recommended for sizing liner components, and a large hole size of 3.1 cm^2 is recommended when analyzing the expected performance of the leachate collection and removal system. Although the Hydraulic Evaluation of Landfill Performance (HELP) model developed by USEPA is typically used in practice to estimate leakage rates through landfill liners (Tolaymat and Krause 2020), a common simplified assumption is that the leakage rates assuming one defect every 4,000 m^2 are 1,000 L/hectare/day (lphd) for a small hole and 30,000 lphd for a large hole. Lower leakage rates may be used to size the liner components when an electric leak location survey is used to minimize the number of holes and tears in landfills (e.g., Gilson 2021).

8.3.3 Mechanical Performance

Geomembranes may experience complex stress states in a landfill liner system, and many scenarios need to be considered in design. The first issue is related to the mobilization of tensile stresses in anchored geomembranes caused by self-weight on a slope, downdrag owing to the settlement of overlying waste, thermal expansion and contraction, penetration by gravel or other mineral materials used in the leachate collection and removal system and rocks protruding from an underlying subgrade or CCL, and wind uplift (Koerner 2005; Giroud et al. 1995a, b). Analyses have been developed to estimate the magnitude of tensile stresses in each of these scenarios so that a geomembrane with appropriate tensile strength can be selected in design, in addition to identifying the appropriate anchorage depth at the crest of the side slopes of a landfill, and whether a textured or smooth geomembrane should be specified as a measure to reduce downdrag stresses. Although the occurrence of high tensile forces in geomembranes may lead to rupture, it is also possible that geomembranes may experience creep deformations under lower but sustained tensile forces, and this creep can lead to brittle stress fracture and a reduction in the service life of the geomembrane. The performance of welded seams must also be considered during tensile loading. Seam strength requirements for common thermally bonded geomembranes are provided in GRI-GM19a and GM19b to address the performance of welded seams, and this topic has been studied by Haxo and Kamp (1990) and Peggs (1990). Another issue that should be considered for geomembranes is puncture resistance (e.g., Connor et al. 2008). The vertical stresses on landfill liners may be relatively high because of the weight of the overlying waste, and therefore, puncture from rocks in the drainage system or in the underlying subgrade may occur depending on the flexibility of the geomembrane. In addition to selecting an appropriate geomembrane for a given stress state, geomembranes may be protected using heavy-weight nonwoven geotextiles or by preparing the subgrade before placing the geomembrane (e.g., removal of rocks, smooth rolling).

Another critical issue that should be considered in design is the interface shear strength of geomembranes with CCLs, GCLs, and other geosynthetics (e.g., Triplett and Fox 2001, Chiu and Fox 2004, McCartney et al. 2009, Fox and Stark 2015). The laboratory measurement of the shear strength of geosynthetic interfaces has been standardized according to ASTM D5321 (ASTM 2012) and ASTM D6243/D6243M (ASTM 2016). The stability of landfill liners is typically assessed using an infinite slope analysis, although more advanced veneer stability analyses considering the finite length of a slope have been developed by Giroud et al. (1995a, b) and Koerner (2005). As the interface shear strength of geomembranes with GCLs may be highly variable, reliability-based design approaches have been developed to consider the stability of landfill liner slopes with geosynthetic interfaces (McCartney et al. 2004, Dixon et al. 2006).

8.4 COMPACTED CLAY LINERS

8.4.1 Overview

Although RCRA Subtitles C and D prescribe the minimum allowable thickness of a compacted clay liner (CCL), approximately 900 and 600 mm, respectively, and the maximum allowable hydraulic conductivity of the clay (10^{-9} m/s), there are still challenges in selecting the appropriate soil and defining construction specifications to attain hydraulic conductivity values of less than 10^{-9} m/s in the field. Although it was well known that clay mineralogy has a major effect on the hydraulic conductivity (e.g., Mesri and Olson 1971), the placement conditions for clays also play a key role. The criteria used for selecting clays in landfill liners builds upon material property characterization experience from earthen dam construction. For example, Mitchell et al. (1965) evaluated the impact of compaction conditions (i.e., the compaction water content and the dry unit weight) on the hydraulic conductivity, undrained shear strength, and shrinkage potential of clays. Mitchell et al. (1965) found that compaction dry of optimum led to a flocculated soil structure, whereas compaction wet of optimum led to a dispersed soil structure. A flocculated soil structure leads to greater undrained shear strength and lower shrinkage potential but also leads to higher hydraulic conductivity. This finding implies that the hydraulic behavior of a given clay used to construct a CCL may change depending on how it is placed and compacted during construction. The hydraulic conductivity experiments of Mitchell et al. (1965) were performed using rigid wall permeameters, which were later found to overestimate the hydraulic conductivity of compacted clays because of issues like side-wall leakage and poor control over the stress state. Daniel et al. (1985) found that these issues could be addressed using a flexible-wall permeameter to measure the hydraulic conductivity of clays. Daniel and Benson (1990) used the results from flexible-wall permeameter and shear strength tests to identify the range of gravimetric water content and dry density that would lead to appropriate hydraulic conductivity and shear strength for use of soils in CCLs, which leads to different compaction criteria than that used in dams or other earthen structures. A wide range of soils may meet the hydraulic conductivity requirements of CCLs in landfill liners. For example, Benson et al. (1994) performed an analysis of a database of hydraulic conductivity measurements on soils used in CCLs from 67 landfills and found that the geometric mean of the hydraulic conductivity is less than 10^{-9} m/s (10^{-7} cm/s) and this can be achieved if the soil used in the CCL has a liquid limit greater than 20, a plasticity index greater than 7, a percent fines greater than 30%, and a percent clay content greater than 15%. At the same time, Benson et al. (1999) performed an analysis of a database of hydraulic conductivity measurements on soils used in 85 full-scale compacted clay liners and field test pads and found that only 74% of the liners had a hydraulic conductivity less than 10^{-9} m/s (10^{-7} cm/s), and that success of the liner was correlated primarily with the compaction water content relative to the line of optimums and the number of lifts or the thickness of the liner. Daniel et al. (1985) found that effective stress plays a critical role in the hydraulic conductivity of compacted clays used in liner systems. Specifically, desiccation cracks that may form during construction will close during the application of elevated effective stresses, minimizing their effect on seepage through the CCL. A benefit of using CCLs is that the use of multiple compaction lifts helps balance out the variations in hydraulic conductivity with minor variations in clay compaction. Further, a CCL with multiple lifts may minimize the risk that desiccation cracks pass from one lift into another.

Contaminants in contact with CCLs in a liner system may potentially influence the clay's hydraulic conductivity. For example, Anderson (1982) used a rigid-wall permeameter to investigate the flow of pure-phase contaminants like acetic acid, acetone, methanol, and xylene through a clay layer and found a sharp increase in hydraulic conductivity with increasing pore volumes of flow for each contaminant. However, the experiments of Anderson (1982) were performed under negligible effective stress, and therefore, the soil structure may have changed during flow and side-wall leakage probably occurred. Further, it is rare for contaminants in landfill leachate to be in pure phase, but these are instead typically diluted or present in miscible (aqueous or dissolved) form. However, some containment facilities may have a greater concentration of hydrocarbons, in which case a CCL with compacted clay alone may not be sufficient in restricting the movement of these contaminants. In this case, organoclay amendments that can sorb organic compounds may be needed. A detailed discussion on the effects of chemical compatibility of CCLs can be found in NRC (2007).

8.4.2 Hydraulic Performance

Contaminant transport through CCLs may occur by both advection and diffusion. Accordingly, this section first describes the hydraulic performance of a solitary CCL resulting from advective water flow and then introduces contaminant transport caused by combined advection and diffusion. The hydraulic performance of CCLs within a composite liner system is discussed subsequently. In the case of contaminant transport through water-saturated CCLs via steady-state advection, this calculation assumes a datum at the base of the CCL and the water pressure head at the bottom of the CCL is zero. Based on regulatory requirements for Subtitle C facilities, the maximum allowable value of saturated hydraulic conductivity for the CCL is 10^{-9} m/s, the maximum allowable value of the height of leachate ponding is 305 mm, and the minimum allowable value of the CCL thickness is 905 mm. The seepage rate for these regulatory-prescribed values is 1,155 lphd, which is greater than the leakage rate corresponding to diffusion through a geomembrane but less than the leakage rate through a large hole in a solitary geomembrane. The other mechanisms of contaminant transport through CCLs are diffusion, mechanical dispersion, and retardation (i.e., sorption of chemical species onto clay particles, which is discussed in more detail in Chapter 1). Rowe et al. (2004) derived the differential equation for one-dimensional contaminant transport for a single, nondegrading reactive solute through a saturated porous medium such as a CCL or GCL. Unlike hydraulic conductivity, the effective diffusion coefficient does not change significantly with clay type. Retardation is an important component of contaminant transport in CCLs, which may delay the transport of contaminants across a barrier. Batch testing is typically performed to evaluate the retardation factors to be used for a given leachate and clay used in the CCL.

8.4.3 Mechanical Performance

The long-term slope stability of landfill liners is typically governed by drained conditions. In this case, the potential for slope failure should be assessed using an infinite slope analysis that uses the drained friction angle of the clay and negligible drained cohesion. Issues of importance to constructed CCLs are differential settlement, freeze–thaw, desiccation, and interface and internal strength, and sometimes consolidation properties if it is necessary to maintain drained conditions in the CCL for stability purposes. The construction techniques used to form a CCL may influence the mechanical performance of a CCL. For example, the compaction procedures for the CCL on the side slopes may affect the shear strength depending on whether lifts are placed parallel to the slope or in horizontal layers.

8.4.4 Construction Issues and Specifications

Construction specifications for CCLs should consider both the mechanical and the hydraulic performance of the clay. Daniel and Benson (1990) and Daniel and Wu (1993) developed procedures that could be used to develop the compaction specifications for clays involving testing of specimens compacted at a range of water contents and compaction efforts to identify the optimal zone of performance, with the goals of having low hydraulic conductivity, high undrained shear strength, and low shrinkage potential. Benson et al. (1999) evaluated the performance of 85 CCLs constructed in the field and found that the main reason for a clay not having successful hydraulic performance was that the clay would not have been compacted above the line of optimums. A summary on the field behavior of CCLs is provided in the USEPA report by Bonaparte et al. (2002).

A summary of compaction issues that may affect the performance of CCLs is provided by Mitchell and Jaber (1990). Clays that have sufficiently low hydraulic conductivity often have high plasticity and have the tendency to form clods during the hydration process. Compaction of CCLs without breaking up clods may have a significant negative impact on seepage through a CCL (Benson and Daniel 1990). When the clay is dry of optimum, clods should be broken up using a roller, whereas when the clay is wet of optimum, clods may be sufficiently workable to compact normally. The use of a sheepsfoot compactor helps minimize the likelihood of preferential pathways through the CCL. Another condition that may occur during construction is rapid temperature changes leading to freezing of the CCL. Freezing will have a negative effect on the hydraulic conductivity of clays, with an accumulated effect during multiple freeze–thaw cycles (Zimmie and La Plante 1990, Othman and Benson 1993, Benson and

Othman 1993). The impact of freeze–thaw cycles can be mitigated by placing a drainage layer or waste atop the CCL as soon as possible, leading to some insulation of the CCL and increased effective stress. This may be difficult in practice as CCLs must undergo certification and approval before they can be approved, and therefore, temporary insulation layers like tire-derived aggregate (TDA), encapsulated fiberglass, or extruded polystyrene may be useful (Benson et al. 1996).

Although the relative impacts of construction conditions can be assessed using field infiltrometer tests on as-constructed CCLs (Daniel 1984, 1989; Benson et al. 1999; Bonaparte et al. 2002) or ponding tests on field-scale test pads (e.g., Day and Daniel 1985, Elsbury et al. 1990), the results of these tests may be affected by the low effective stresses associated with the lack of overlying waste. Accordingly, the effects of desiccation cracks occurring shortly after construction or preferential pathways because of clods may be overemphasized in these tests. However, they are useful for construction quality control and to compare the relative performance of different CCL construction techniques. Bonaparte et al. (2002) analyzed results from large-scale field hydraulic conductivity tests on 89 natural CCLs and 12 soil–bentonite CCLs and found that it was critical to compact the soil in the zone of gravimetric water content and dry density that will yield low hydraulic conductivity. They also found that there was no clear correlation between field hydraulic conductivity and parameters like plasticity index and percentage of clay, indicating that CCLs can be successfully constructed with a relatively broad range of soil materials. They also found that hydraulic conductivity decreased with increasing CCL thickness, up to a thickness of 1 m.

8.5 GEOSYNTHETIC CLAY LINERS

8.5.1 Overview

In some situations, the CCL in a landfill liner can be replaced or complemented with a GCL. GCLs have the advantage of incorporating sodium bentonite clay bonded to one or more carrier geosynthetics, facilitating rapid installation in the field as they are deployed in rolls. The hydraulic conductivity to typical potable water of sodium bentonite used in GCLs (in powdered or granular form) is lower than 10^{-10} m/s, which is at least one order of magnitude lower than the maximum hydraulic conductivity allowed for CCLs (i.e., 10^{-9} m/s). This lower hydraulic conductivity permits GCLs, which have a thickness of approximately 10 to 15 mm, to potentially have a similar hydraulic performance to thicker CCLs, assuming that the hydraulic conductivity of the GCL to the waste leachate is similar to the hydraulic conductivity to water. In addition,

GCLs can exhibit membrane behavior that gives rise to hyperfiltration, chemico-osmotic flow, and reduced diffusion (Malusis et al. 2003). GCLs incorporating polymer-modified bentonites may maintain low hydraulic conductivity when subjected to permeant liquids containing appreciable concentrations of higher-valence cations (e.g., Ca^{2+}), which are known to cause increases in hydraulic conductivity of natural sodium bentonite (e.g., Shackelford et al. 2000, Scalia et al. 2011).

8.5.2 Mechanical Performance

It is widely recognized that the sodium bentonite making up the core of GCLs has very low shear strength and could play a role in liner stability (Daniel et al. 1998). Mesri and Olson (1970) and Olson (1974) found that the drained friction angle of sodium bentonite can be as low as 8 degrees. To address this issue, many GCLs incorporate strategies for increasing the shear strength, including needle-punching of fibers through the bentonite layer and possible subsequent heating bonding of the needle punched fibers. Although needle-punching generally leads to an increase in shear strength, swelling during hydration under low or high vertical stresses may lead to complex behavior that results in a wide variation in shear strength (Gilbert et al. 1996; Eid and Stark 1997; Triplett and Fox 2001; Zornberg et al. 2005; Fox and Stark 2004, 2015). As a result, interface direct shear testing is commonly performed as part of the design of a landfill to establish project-specific GCL internal and interface shear parameters.

8.5.3 Construction Issues and Specifications

Although the placement of GCLs is relatively straightforward compared to constructing CCLs, there are still issues that must be considered. It is typically specified that a geomembrane is placed over the top of the GCL on the same day that it is installed to minimize hydration from above. It is also recommended to place a confining stress atop a GCL as soon as possible after placement, as hydration of the GCL from the underlying subgrade may occur rapidly after placement. In some cases, the underlying subgrade is compacted to a water content greater than the optimum water content to facilitate hydration and osmotic swell of the GCL bentonite (e.g., Scalia and Benson 2011), which also emphasizes the importance of applying a confining stress quickly. Third, GCL rolls are not connected using seams in the way that geomembranes are connected and are typically simply overlapped with powdered bentonite between the seams. Some GCLs can be heat-bonded to reduce overlap lengths. It is important to have an adequate overlap of GCL rolls to ensure that shrinkage or necking caused by tensile stresses does not lead to a reduction in the overlap or potential panel separation.

8.6 PERFORMANCE OF COMPOSITE LINERS

8.6.1 Hydraulic Performance

Although contaminant transport for diffusion should always be considered for an intact geomembrane, as this may play an important role in some situations, most of the liquid migration through a composite liner occurs through holes in the geomembrane or defects in the seams. However, the low hydraulic conductivity of the underlying CCL or GCL will reduce the flow rate through a geomembrane hole. Rowe (1998) noted that the simplest equation to estimate leakage through a composite liner is Forchheimer's equation, which assumes that there is no leakage along the interface between the geomembrane and the underlying low-permeability layer. However, Giroud and Bonaparte (1989b) noted that no contact is perfect and that once water flows through a defect in the geomembrane liner, it will flow laterally some distance between the geomembrane and the CCL or GCL. Flow through the interface between the geomembrane and the CCL or GCL is referred to as interface flow, and the area around the hole through which water flows is referred to as the wetted area. This interface flow also means that the hydraulic head applied to the CCL or GCL is greatest under the location of the hole in the geomembrane and decreases with radial distance away from the hole. The quality of contact between the geomembrane and the CCL or GCL is a key factor leading to leakage through the composite liner system. Good contact conditions correspond to a geomembrane installed with minimal wrinkles on top of a GCL or CCL that has been adequately compacted and has a smooth surface, whereas poor contact conditions correspond to a geomembrane with wrinkles placed atop a CCL that does not have a smooth surface (or a GCL atop a variable subgrade) (Bonaparte et al. 1989). The weight of the overlying waste is typically assumed to keep the geomembrane in contact with the GCL or CCL in the case that they are initially in good contact, but this overburden stress may not be sufficient to close the interface between a geomembrane and a GCL or CCL with poor contact.

Giroud and Bonaparte (1989b) used multiple-variable regression to define an empirical equation to estimate the leakage through a composite liner on the basis of the results from flow cell tests under low effective stress performed by Brown et al. (1987). The empirical equation of Giroud and Bonaparte (1989b) was revised in subsequent studies (Giroud et al. 1989, 1992). Rowe (1998) presented a general analytical solution to leakage through composite liners that is based on the differential equation for the head distribution on the top of the clay liner. This solution requires a calculation of Bessel functions, which can be readily programmed into a spreadsheet program. Rowe (1998) tabulated leakage rates through composite liners under a variety of conditions and found that leakage rates through composite liners are lower than the leakage through a geomembrane hole and leakage through a CCL or GCL.

Rowe (1998) also developed an equation to evaluate leakage through a composite liner for the case that the hole in the geomembrane was on a wrinkle. Touze-Foltz et al. (2000) and Foose et al. (2001) revisited the problem of leakage through composite liners using numerical simulations and provided correction factors that could be applied to the analytical solution of Rowe (1998) to consider the effects of different contact conditions.

Contaminant transport through composite liners is also affected by advection–diffusion processes. Lake and Rowe (2005) evaluated the performance of a composite liner consisting of a 2.9 m-thick CCL overlain by a geomembrane, 14 years after construction. They found that leachate was able to migrate through the geomembrane into the CCL, with chloride and sodium detected approximately halfway through the CCL and potassium detected a quarter of the way through the CCL (probably because of a greater sorptive uptake of the potassium by the clay). The study showed the importance of the CCL as a diffusion barrier in a composite liner.

8.6.2 Temperature and Desiccation Effects

An issue that has been identified in the performance of base liner systems is the potential for elevated temperatures in the waste resulting in desiccation of the clay in the CCL or GCL (Southen and Rowe 2005). Desiccation may occur in the case that the geomembrane is intact, and leachate is not flowing into the clay. In this case, the temperature gradient across the liner system will result in a thermally induced flow of water in unsaturated soils. Specifically, water will tend to evaporate from the soil nearest the high-temperature source, migrate by vapor diffusion away from the temperature source, and condense further away (Philip and De Vries 1957). Southen and Rowe (2005) performed column tests on a composite liner consisting of a GCL overlain by a geomembrane and found that a loss in water content of the clay in the GCL may lead to desiccation cracking. However, their experiments were performed under a vertical stress of 80 kPa. Desiccation cracking may be dependent on effective stress, and therefore, further study is needed on this topic. A strategy that could be explored to minimize thermal gradients across liner systems is to incorporate a geothermal heat extraction system atop the liner system to control the temperature at this location (Coccia et al. 2013, Nocko et al. 2020). This may be a useful approach to maintain liner integrity while extracting heat for use in buildings.

8.7 DRAINAGE SYSTEMS

8.7.1 Overview

Drainage layers in landfill liner systems are needed to collect and remove leachate from the landfill to maintain the hydraulic head on the primary liner system to less than the regulatory-prescribed maximum allowable value of 300 mm. In double-liner systems, drainage layers are used to create an LDS between the primary and the secondary liners. Drainage layers should be designed to have both adequate capacity for flow (i.e., high hydraulic conductivity) and low likelihood of clogging (i.e., good filtration characteristics). Drainage layers may be constructed of either granular or geosynthetic materials. A dendritic system of pipes is typically included within the drainage layer to help transmit large volumes of flow collected from the drainage layer. The most common type of a geosynthetic drainage layer is usually described as a "geocomposite drainage layer" and consists of a geonet core and a nonwoven geotextile filter. In some cases, geotextile filters are included on both sides of the core depending on slope stability requirements. Granular layers typically have a large thickness and relatively high transmissivity (hydraulic conductivity multiplied by the thickness of the drain) but can be challenging to construct on slopes. Geosynthetic drainage layers are easier to construct and have higher transmissivity and better filtration than granular layers, but their small thickness may be an issue, especially if clogging occurs and they do not provide the level of protection of the underlying liners provided by a 0.3 m-thick, or thicker, granular drainage layer (Rowe et al. 1997). In design, it is not uncommon to have the drainage layer on the base of a landfill be constructed of granular material, whereas the drainage layer on the side slopes is a geocomposite drainage layer.

8.7.2 Hydraulic Performance

Many filter criteria have been developed for granular and geosynthetic drainage systems. Sherard et al. (1984) used a variety of experimental techniques to define filter criteria for granular drainage layers. The gradient ratio test is the main test that has been used to evaluate soil interaction with geosynthetic drainage layers to ensure proper filtration performance (Fannin et al. 1994, Palmeira and Gardoni 2002, Palmeira et al. 2005, McCartney and Berends 2010). Palmeira et al. (2005) observed the importance of considering the impact of the vertical stress on the filtration response of nonwoven geotextiles. Another risk is that the lateral transmissivity of drains will decrease if filtration is not adequate. To investigate this, McCartney and Berends (2010) developed a flexible wall permeameter that incorporated needles to measure the gradient in a soil layer overlying a geocomposite drainage layer, as well as ports that could measure changes in the transmissivity of the geosynthetic drainage layer after a stable cross-plane flow is established. Biological activity in the leachate may also lead to a time-dependent clogging of the drainage layer (Rowe et al. 1997), and therefore, reduction factors on the permissible transmissivity to use in design may be used as part of the hydraulic performance analysis. A summary of the state of the art in filtration criteria for geotextiles is found in Giroud (2010), including the use of different geotextile characteristics to compare with soil characteristics. The hydraulic performance of drainage systems is typically evaluated using analytical solutions for the shape of the phreatic surface above a sloped liner such as that of McEnroe (1993) and Giroud et al. (2000). The main objective of these analyses is to select the proper drainage layer transmissivity, drain spacing, and liner slope to ensure that the leachate head will be less than 300 mm given a certain impinging flow rate from the overlying waste. The impinging flow rate is likely to change throughout the lifetime of a landfill, with greater rates expected before the final cover is installed. In practice, the HELP model is commonly used to design drainage layers for site-specific impinging flow rates (Tolaymat and Krause 2020).

8.8 FINAL COMMENTS

Landfill liners are complex systems consisting of mineral-based and polymer-based materials in the form of both hydraulic barriers and drainage systems, with the overall objective of minimizing transport of contaminants from contained waste into the surrounding subsurface. Although the design of landfill liners is prescribed by regulations, it is critical to be able to consider transport through hydraulic barriers via advection and diffusion analyses considering interactions between the geosynthetic and the clay-based layers in composite liners. These analyses are useful in confirming the adequate performance of landfill liners with different waste constituents and in different settings, and in verifying the adequacy of alternate liner designs.

REFERENCES

Anderson, D. 1982. "Does landfill leachate make landfill liners more permeable?" *Civ. Eng.* 52: 66–69.

ASTM International. 2012. *Standard test method for determining the shear strength of soil–geosynthetic and geosynthetic–geosynthetic interfaces by direct shear*. ASTM D5321. West Conshohocken, PA: ASTM.

ASTM. 2016. *Standard test method for determining the internal and interface shear strength of geosynthetic clay liner by the direct shear method.* ASTM D6243/D6243M. West Conshohocken, PA: ASTM.

ASTM. 2017a. *Standard test method for effect of exposure of unreinforced polyolefin geomembrane using fluorescent UV condensation apparatus.* ASTM D7238. West Conshohocken, PA: ASTM.

ASTM. 2017b. *Standard test method for oxidative induction time of polyolefin geosynthetics by high-pressure differential scanning calorimetry.* ASTM D5885/D5885M. West Conshohocken, PA: ASTM.

ASTM. 2021. *Standard test method for oxidative induction time of polyolefin geosynthetics by differential scanning calorimetry.* ASTM D8117. West Conshohocken, PA: ASTM.

Benson, C. H., and D. E. Daniel. 1990. "Influence of clods on the hydraulic conductivity of compacted clay." *J. Geotech. Eng.* 116 (8): 1231–1248.

Benson, C. H., D. E. Daniel, and G. P. Boutwell. 1999. "Field performance of compacted clay liners." *J. Geotech. Geoenviron. Eng.* 125 (5): 390–403.

Benson, C. H., M. A. Olson, and W. R. Bergstrom. 1996. "Temperatures of insulated landfill liner." *J. Transp. Res. Board* 1534 (1): 24–31.

Benson, C. H., and M. A. Othman. 1993. "Hydraulic conductivity of compacted clay frozen and thawed in situ." *J. Geotech. Eng.* 119 (2): 276–294.

Benson, C. H., H. Zhai, and X. Wang. 1994. "Estimating hydraulic conductivity of compacted clay liners." *J. Geotech. Eng.* 120 (2): 366–387.

Bonaparte, R., J. P. Giroud, and B. A. Gross. 1989. "Rates of leakage through landfill liners." In Vol. 1 of *Proc., Geosynthetics '89*, 18–29. Roseville, MN: Industrial Fabrics Association International.

Bonaparte, R., R. M. Koerner, and D. E. Daniel. 2002. *Assessment and recommendations for improving the performance of waste containment systems.* EPA/600/R-02/099. Washington, DC: US Environmental Protection Agency, National Risk Management Research Laboratory.

Brown, K. W., J. C. Thomas, R. L. Lytton, P. Jayawickrama, et al. 1987. *Quantification of leakage rates through holes in landfill liners.* EPA/CR810940. Cincinnati: US Environmental Protection Agency.

Chiu, P., and P. J. Fox. 2004. "Internal and interface shear strengths of unreinforced and needle-punched geosynthetic clay liners." *Geosynth. Int.* 11 (3): 176–199.

Coccia, C. J., R. Gupta, J. Morris, and J. S. McCartney. 2013. "Municipal solid waste landfills as geothermal heat sources." *Renewable Sustainable Energy Rev* 19: 463–474.

Connor, C. J., T. D. Stark, and T. R. Boerman. 2008. "Puncture resistance of PVC geomembranes using the truncated cone test." *Geosynth. Int.* 15 (6): 480–486.

Daniel, D. E. 1984. "Predicting hydraulic conductivity of clay liners." *J. Geotech. Eng.* 110 (2): 285–300.

Daniel, D. E. 1989. "In situ hydraulic conductivity tests for compacted clay." *J. Geotech. Eng.* 115 (9): 1205–1226.

Daniel, D. E. 1993. *Geotechnical practice for waste disposal.* London: Chapman and Hall.

Daniel, D. E., D. C. Anderson, and S. S. Boynton. 1985. "Fixed-wall versus flexible-wall permeameters." In *Hydraulic Barriers Soil Rock,* 107–126. ASTM STP 874. West Conshohocken, PA: ASTM.

Daniel, D. E., and C. H. Benson. 1990. "Water content-density criteria for compacted soil liners." *J. Geotech. Eng.* 116 (12): 1811–1830.

Daniel, D. E., R. M. Koerner, R. Bonaparte, R. E. Landreth, et al. 1998. "Slope stability of geosynthetic clay liner test plots." *J. Geotech. Geoenviron. Eng.* 124 (7): 628–637.

Daniel, D. E., and Y.-K. Wu. 1993. "Compacted clay liners and covers for arid sites." *J. Geotech. Eng.* 119 (2): 223–237.

Day, S. R., and D. E. Daniel. 1985. "Hydraulic conductivity of two prototype clay liners." *J. Geotech. Eng.* 111 (8): 957–970.

Dixon, N., D. R. V. Jones, and G. J. Fowmes. 2006. "Interface shear strength variability and its use in reliability-based landfill stability analysis." *Geosynth. Int.* 13 (1): 1–14.

Eid, H. T., and T. D. Stark. 1997. "Shear behavior of an unreinforced geosynthetic clay liner." *Geosynth. Int.* 4 (6): 645–659.

Elsbury, B. R., D. E. Daniel, G. A. Sraders, and D. C. Anderson. 1990. "Lessons learned from compacted clay liner." *J. Geotech. Eng.* 116 (11): 1641–1660.

Fannin, R. J., Y. P. Vaid, and Y. C. Shi. 1994. "Filtration behaviour of nonwoven geotextiles." *Can. Geotech. J.* 31 (4): 555–563.

Foose, G. J., C. H. Benson, and T. B. Edil. 2001. "Predicting leakage through composite landfill liners." *J. Geotech. Geoenviron. Eng.* 127 (6): 510–520.

Fox, P. J., and T. D. Stark. 2004. "State-of-the-art report: GCL shear strength and its measurement." *Geosynth. Int.* 11 (3): 141–175.

Fox, P. J., and T. D. Stark. 2015. "State-of-the-art report: GCL shear strength and its measurement–ten-year update." *Geosynth. Int.* 22 (1): 3–47. 10.1680/gein.14.00030

Gilbert, R. B., F. F. Fernandez, and D. Horsfield. 1996. "Shear strength of a reinforced geosynthetic clay liner." *J. Geotech. Eng.* 122 (4): 259–266.

Gilson, A. 2021. "Advancements in field testing for locating geomembrane installation damage." *Int. J. Geosynth. Ground Eng.* 7 (3): 72.

Giroud, J. P. 2010. "Development of criteria for geotextiles and granular filters." In *Proc., 9th Int. Conf. on Geosynthetics,* 45–64. Guaruja, Brazil: International Geosynthetics Society.

Giroud, J. P., R. C. Bachus, and R. Bonaparte. 1995a. "Influence of water flow on the stability of geosynthetic-soil layered systems on slopes." *Geosynth. Int.* 2 (6): 1149–1180.

Giroud, J. P., K. Badu-Tweneboah, and R. Bonaparte. 1992. "Rate of leakage through a composite liner due to geomembrane defects." *Geotext. Geomembr.* 11 (1): 1–28.

Giroud, J. P., and R. Bonaparte. 1989a. "Leakage through liners constructed with geomembranes. Part I. Geomembrane liners." *Geotext. Geomembr.* 8 (1): 27–67.

Giroud, J. P., and R. Bonaparte. 1989b. "Leakage through liners constructed with geomembranes. Part II. Composite liners." *Geotext. Geomembr.* 8 (2): 71–111.

Giroud, J. P., A. Khatami, and K. Badu-Tweneboah. 1989. "Evaluation of the rate of leakage through composite liners." *Geotext. Geomembr.* 8 (4): 337–340.

Giroud, J. P., T. Pelte, and R. J. Bathurst. 1995b. "Uplift of geomembranes by wind." *Geosynth. Int.* 2 (6): 897–952.

Giroud, J. P., J. G. Zornberg, and A. Zhao. 2000. "Hydraulic design of geosynthetic and granular liquid collection layers." *Geosynth. Int.* 7 (4-6): 285–380.

GRI (Geosynthetics Research Institute). 2021. *Test methods, test properties and testing frequency for high density polyethylene (HDPE) smooth and textured geomembranes.* GM13. Folsom, PA: Geosynthetics Institute.

Gudina, S., and R. W. I. Brachman. 2006. "Physical response of geomembrane wrinkles overlying compacted clay." *J. Geotech. Geoenviron. Eng.* 132 (10): 1346–1353.

Gudina, S., and R. W. I. Brachman. 2011. "Geomembrane strains from wrinkle deformations." *Geotext. Geomembr.* 29 (2): 181–189.

Haxo, H. E., and L. C. Kamp. 1990. "Destructive testing of geomembrane seams: Shear and peel testing of seam strength." *Geotext. Geomembr.* 9 (4–6): 369–395.

Koerner, R. M. 2005. *Designing with geosynthetics.* 5th ed. Upper Saddle River, NJ: Pearson College Division.

Lake, C. B., and R. K. Rowe. 2005. "The 14-year performance of a compacted clay liner used as part of a composite liner system for a leachate lagoon." *Geotech. Geol. Eng.* 23 (6): 657–678.

Malusis, M., C. Shackelford, and H. Olsen. 2003. "Flow and transport through clay membrane barriers." *Eng. Geol.* 70 (3–4): 235–248.

McCartney, J. S., and R. Berends. 2010. "Measurement of filtration effects on the transmissivity of geocomposite drains for phosphogypsum." *Geotext. Geomembr.* 28 (2): 226–235.

McCartney, J. S., J. G. Zornberg, and R. H. Swan. 2009. "Analysis of a large database of GCL–geomembrane interface shear strength results." *J. Geotech. Geoenviron. Eng.* 135 (2): 209–223.

McCartney, J. S., J. G. Zornberg, R. H. Swan Jr., and R. B. Gilbert. 2004. "Reliability-based stability analysis considering GCL shear strength variability." *Geosynth. Int.* 11 (3): 212–232.

McEnroe, B. M. 1993. "Maximum saturated depth over landfill liner." *J. Environ. Eng.* 114 (5): 1052–1062.

Mesri, G., and R. E. Olson. 1970. "Shear strength of montmorillonite." *Géotechnique* 20 (3): 261–270.

Mesri, G., and R. E. Olson. 1971. "Mechanisms controlling the permeability of clays." *Clays Clay Min.* 19 (3): 151–158.

Mitchell, J. K., D. R. Hooper, and R. G. Campanella. 1965. "Permeability of compacted clay." *J. Soil Mech. Found. Div.* 91 (SM4): 41–65.

Mitchell, J. K., and M. Jaber. 1990. "Factors controlling the long-term properties of clay liners." In *Waste containment systems: Construction, regulation and performance*, Geotechnical Special Publication 26, edited by R. Bonaparte, 84–105. Reston, VA: ASCE.

Nocko, L., K. Botelho, J. Morris, R. Gupta, et al. 2020. "Evaluation of thermal conductivity of municipal solid waste using in-situ heat extraction tests." *J. Geotech. Geoenviron. Eng.* 146 (9): 04020077.

NRC (National Research Council). 2007. *Assessment of the performance of waste containment barriers*. Washington, DC: National Academies Press.

Olson, R.E. 1974. "Shearing strengths of kaolinite, illite, and montmorillonite." *J. Geotech. Eng. Div.* 100 (11): 1215–1229.

Othman, M. A., and C. H. Benson. 1993. "Effect of freeze–thaw on the hydraulic conductivity and morphology of compacted clay." *Can. Geotech. J.* 30 (2): 236–246.

Palmeira, E. M., and M. G. Gardoni. 2002. "Drainage and filtration properties of nonwoven geotextiles under confinement using different experimental techniques." *Geotext. Geomembr.* 20 (2): 97–115.

Palmeira, E. M., M. G. Gardoni, and D. W. Bessa da Luz. 2005. "Soil–geotextile filter interaction under high stress levels in the gradient ratio test." *Geosynth. Int.* 12 (4): 162–175.

Peggs, I. D. 1990. "Destructive testing of polyethylene geomembrane seams: Various methods to evaluate seam strength." *Geotext. Geomembr.* 9 (4–6): 405–414.

Peggs, I. D., and R. Thiel. 1998. "Selecting a geomembrane material." In *Proc., 6th Int. Conf. on Geosynthetics*, 381–388. Roseville, MN: Industrial Fabrics Association International.

Philip, J. R., and D. A. de Vries. 1957. "Moisture movement in porous materials under temperature gradients." *Eos Trans. Am. Geophys. Union* 38 (2): 222–232.

Rowe, R. K. 1998. "Geosynthetics and the minimization of contaminant migration through barrier systems beneath solid waste." In *Proc., 6th Int. Conf. on Geosynthetics*, 27–102. Roseville, MN: Industrial Fabrics Association International.

Rowe, R. K., I. R. Fleming, M. D. Armstrong, and A. J. Cooke. 1997. "Recent advances in understanding the clogging of leachate collection systems." In *Proc., 6th Int. Landfill Symp.*, Padova: CISA, 383–390.

Rowe, R. K., R. M. Quigley, R. W. I. Brachman, and J. R. Booker. 2004. *Barrier systems for waste disposal facilities*. London: E & FN Spon.

Scalia, J., and C. H. Benson. 2011. "Hydraulic conductivity of geosynthetic clay liners exhumed from landfill final covers with composite barriers." *J. Geotech. Geoenviron. Eng*. 137 (1): 1–13.

Scalia, J., C. H. Benson, T. B. Edil, G. L. Bohnhoff, et al. 2011. "Geosynthetic clay liners containing bentonite polymer nanocomposites." In *Proc., Geo-Frontiers 2011*, 2001–2009. Reston, VA: ASCE.

Shackelford, C. D., C. H. Benson, T. Katsumi, T. B. Edil, and L. Lin. 2000. "Evaluating the hydraulic conductivity of GCLs permeated with non-standard liquids." *Geotext. Geomembr.* 18 (2–4): 133–161.

Sherard, J. R., L. P. Dunnigan, and J. R. Talbot. 1984. "Basic properties of sand and gravel filters." *J. Geotech. Eng*. 110 (6): 684–700.

Southen, J. M., and R. J. Rowe. 2005. "Laboratory investigation of geosynthetic clay liner desiccation in a composite liner subjected to thermal gradients." *J. Geotech. Geoenviron. Eng*. 131 (7): 925–935.

Take, W. A., M. J. Chappel, R. W. I. Brachman, and R. K. Rowe. 2007. "Quantifying geomembrane wrinkles using aerial photography and digital image processing." *Geosynth. Int*. 14 (4): 219–227.

Tolaymat, T., and M. Krause. 2020. *Hydrologic evaluation of landfill performance: HELP 4.0 user manual*. EPA/600/B-20/219. Cincinnati: US Environmental Protection Agency.

Touze-Foltz, N., R. K. Rowe, and C. Duquennoi. 2000. "Liquid flow through composite liners due to geomembrane defects: Analytical solutions for axi-symmetric and two-dimensional problems." *Geosynth. Int*. 6 (6): 455–479.

Triplett, E. J., and P. J. Fox. 2001. "Shear strength of HDPE geomembrane/geosynthetic clay liner interfaces." *J. Geotech. Geoenviron. Eng*. 127 (6): 543–552.

Zimmie, T. F., and C. La Plante. 1990. "The effect of freeze–thaw cycles on the permeability of a fine-grained soil." In *Proc., 22nd Mid-Atlantic Industrial Waste Conf.*, 580–593.

Zornberg, J. G., J. S. McCartney, and R. H. Swan. 2005. "Internal shear strength of geosynthetic clay liners." *J. Geotech. Geoenviron. Eng*. 131 (3): 1–14. 10.1061/(ASCE)1090-0241(2005)131:3(367)

CHAPTER 9

FINAL COVERS FOR LANDFILLS AND WASTE PILES

Richard Thiel, P.E., M.ASCE; Joe Scalia, Ph.D., A.M.ASCE; and N. Yesiller, Ph.D., A.M.ASCE

9.1 INTRODUCTION

Final covers on landfills and waste piles serve three main functions: (1) provide physical isolation of the waste from surficial exposure for sanitary and safety reasons, (2) limit the amount of meteoric fluids (usually precipitation, but could also be air) entering the waste mass that can generate additional leachate, contact-water runoff, and/or cause or exacerbate bio/chemical processes within the waste, and (3) limit the amount of landfill gas (e.g., methane and radon) emissions to the atmosphere and facilitate collection of the landfill gas. (Landfills and waste piles can include municipal solid waste, industrial waste, agricultural waste, residual radioactive material repositories, or mining waste that consist of waste rock piles or closed tailings storage facilities. For the remainder of this chapter, the term *landfill* is used to describe these types of waste piles and waste masses.) Other functions of a landfill cover include providing an aesthetically pleasing surface and serving as a final surface or foundation upon which end use may occur. Technical design considerations typically associated with final covers include minimization of meteoric fluids, stormwater and erosion control, gas control, odor control, leachate seep control, slope stability, and various aspects of access, maintenance, and monitoring related to the aforementioned functions and other technical considerations. Often, there are regulatory requirements that need to be addressed that are associated with these functions and technical considerations. Several different types of final covers exist that are used to meet the required functions and technical needs of a project. The final selection of the appropriate cover is typically based on considerations related to the ability of the engineered cover to adequately meet the defined functions and satisfy social, regulatory, cost, durability, and reliability criteria.

9.2 GENERAL REQUIREMENTS FOR FINAL COVERS

9.2.1 Definition of a Final Cover

A final cover is a multilayer-engineered system that is designed and constructed to provide the multiple required functions listed subsequently. The degree of importance of these various functions will be specific for a given containment facility.

Of note, the finality of covered facilities is usually indefinite and is almost always multigenerational. Although final covers are accessible for maintenance, owners and institutions typically plan for and fund postclosure maintenance for a limited amount of time, such as the US minimum postclosure period of 30 years for municipal solid waste (MSW) landfills. Legacy sites that require maintenance after their postclosure funds are exhausted may become a societal liability. Acknowledgment of this position is important when considering final cover alternatives and making design decisions related to aging, durability, and reliability.

The term *closure* is used herein to refer to the construction of a final cover, although the concept of closure is functionally broader and includes other institutional elements and activities such as monitoring and security.

9.2.2 Physical Isolation of Waste

One of the main functions of a final cover is physical isolation of the waste for health and safety reasons. Considering that the final cover is normally expected to be multigenerational, there are numerous possible scenarios where the physical integrity of the final cover can be challenged over such an extended duration. These situations include loadings from equipment, people, animals, extreme weather, climate impacts, plant intrusion, aging degradation, and seismic movements.

9.2.3 Fluid Barrier

The second main function of final covers is to provide fluid barriers. Beyond physical isolation of the waste, final covers serve to minimize the amount of liquids (and in some cases, gases) entering the waste mass and gases emitted from the waste mass. Liquids, typically precipitation, can add significantly to the amount of leachate generated in a landfill that may potentially contaminate underlying soils and groundwater. Limiting this leachate production pathway is a critical function of final covers. In some cases, such as for waste rock piles, ingress of air/oxygen into the waste mass (together with water) can contribute to oxidation of the wastes and generation of acid rock drainage, and limiting this oxidation pathway via minimization of oxygen ingress is a critical function of final covers. Final covers also facilitate collection and management of gases generated within waste masses because of decomposition of organic waste constituents, chemical and biochemical transformation of waste masses, and direct volatilization of inorganic chemicals. In MSW landfills, the primary landfill gas constituents are methane and carbon dioxide (both greenhouse gases). Methane is also a potential fire and explosion hazard. Trace components in landfill gas include various inorganic constituents and nonmethane volatile organic compounds (VOCs) that are greenhouse gases and ozone-depleting substances. These compounds can have strong odors, such as hydrogen sulfide, and can have direct and indirect climate impacts as well as contain chemicals that are harmful to human health. For residual radioactive material repositories, final covers play an important role in the control of the release of radon gas.

9.2.4 Regulatory Requirements

Final covers must meet the regulatory requirements of the governing jurisdiction(s). Although there are differences in regulations for different jurisdictions, regulations for final covers commonly address some or all the following:

- Timing for performing final closure;
- Minimum-allowable soil thickness;
- Prescriptive layers, or performance-based goals, to achieve levels of low permeability to meteoric percolation or the uncontrolled escape of gases;
- Resistance to erosion to specified levels of storm events;
- Slope stability criteria;
- Maximum-allowable and minimum-required slopes;
- Landfill gas collection from below the covers, possibly combined with enhanced lateral gas migration monitoring and control (in the case of MSW or industrial waste landfills that might generate landfill gas);
- Maintenance, inspection, and reporting requirements;
- Longevity requirements; and
- Site security and access control requirements.

Other closure requirements may result from additional stakeholders, including neighbors, and be enforceable through land-use codes and permits, environmental impact approvals, or negotiations with affected parties. These could include issues such as odor control (typically related to MSW or industrial landfill gas management), vector control, noise (also typically related to MSW landfill gas management because of the sound generated by flares or co-gen facilities), aesthetics of the final cover, habitat caused or affected by the closure of the facility, end use, and concerns related to the real estate occupied by, or adjacent to, the closed landfill.

9.3 TYPES OF LANDFILL FINAL COVERS

There are three general categories of final covers:

1. Soil covers for which the purpose is primarily to provide physical separation of the waste from the environment and to avoid surface runoff of water that has contacted the waste. These are typically single-layer final covers.
2. Water balance covers, also known as *evapotranspiration* (ET) covers, which use natural hydrologic processes to meet cover performance requirements. The two common configurations for water balance covers are monolithic (single layer of soil) and capillary break (a two- or more-layer system with finer soil placed above one or more layers of coarser soil or a geosynthetic with an open pore structure). Water balance covers are designed to store meteoric water entering the cover system during wet weather and then evaporate or transpire (through vegetation that is established in the cover soil) the water during periods where ET exceeds percolation, or to promote the water to flow out through a drainage layer before percolating into the underlying waste mass.
3. Resistive covers that incorporate low-permeability barrier layers to resist fluid movement. Resistive covers are generally constructed from compacted clayey soils, geosynthetic clay liners (GCLs), and/or geomembranes. These covers are sometimes referred to as *conventional* covers because they were conventionally the prescriptive design in early environmental regulations.

Selection of the appropriate general cover category for a given project is usually determined by a combination of regulatory compliance, stewardship goals, end use, climate, pollution potential of the underlying waste, standard accepted local practice, aesthetics, and cost.

9.3.1 Soil Covers

Soil covers have historically been prescribed for nonputrescible inert waste where leachate does not present a significant potential to degrade groundwater, and for which atmospheric ingress or generated gas egress are not a concern, but surficial exposure is not desirable from a habitat, safety, and stormwater runoff standpoint. Types of landfills that have historically been closed using soil covers include mine waste rock and tailings piles that do not have the potential for acid rock drainage or metal leaching, construction and demolition debris fills, and some forest-product waste fills. Old waste fills, where leachate and gas generation potential are no longer an environmental concern, might also be candidates for this type of cover. Design goals for soil covers relate primarily to providing a minimum cover thickness, generally 0.6 m or more, establishing a grading and drainage plan that will not cause ponding water to accumulate, while accounting for estimated future settlement, control of stormwater runoff to minimize erosion, and revegetating the surface of the final cover.

As an example, a unique soil-only 3 m-thick soil cover for the Puente Hills Landfill near Los Angeles, California, is shown in Figure 9-1. This very thick soil cover supports trees and diverse habitat and appears to provide good control of landfill gas emissions, even though the soil cover was used for an MSW landfill.

9.3.2 Water Balance Covers

Water balance covers rely on the natural processes of soil water storage and ET to minimize percolation. Infiltrating water is temporarily stored in near-surface soil pores and subsequently released via evaporation and transpiration to limit percolation to underlying waste to acceptable levels (Benson and Bareither 2012). Water balance covers can be constructed as monolithic layers above the top of the waste mass or as capillary-break barriers that are composed of layers of finer-grained soils placed above layers of coarser-grained soils. Geotextiles, geonets, and geocomposite

Figure 9-1. A 3 m-thick soil-only cover at Puente Hills landfill near Los Angeles.

drainage layers can also be used to create a capillary barrier effect in lieu of, or in addition to, coarser-grained soils. The surfaces of the covers are vegetated and this surficial vegetation component is critical to remove moisture as well as to provide an erosion-control function. Thus, the vegetation needs to be carefully selected on the basis of local climatic conditions and appropriate plant types.

The development of water balance covers as an engineered cover technology in the United States began in the mid-1990s through the Alternative Landfill Cover Demonstration Project under the leadership of Sandia National Laboratories (Dwyer 1995). A comprehensive summary of the principles of and practices for water balance covers is provided by Albright et al. (2010), and an instructive summary of the state-of-the-practice is described using a case history example by Benson and Bareither (2012). Water balance covers have been shown to be effective in climate zones where potential evapotranspiration, *PET*, dominates precipitation, such as in arid and semiarid climates in the Western United States. Water balance covers may not be effective in climates where substantial groundwater recharge occurs in natural landscapes (an indication of periods of precipitation exceeding soil storage capacity and *PET*).

The principle of a water balance cover can be described by the following equation (Ogorzalek et al. 2008):

$$P_r = P - R - D - ET - \Delta S \tag{9-1}$$

where

P_r = Percolation below the final cover,
P = Meteoric precipitation,
R = Runoff from the surface,
D = Lateral drainage within the cover system whereby water is removed,
ET = Evapotranspiration back to the atmosphere, and
ΔS = Change in water storage in the soil cover over the period of time under consideration.

The metric units of these parameters are often depth of water (e.g., m) or total volume of water per unit area (e.g., m^3/m^2). The design goal is to provide adequate storage capacity, S, which depends on soil type and thickness, to minimize P_r given the site-specific values of P, R, D, and ET. The acceptable value of P_r may be driven by regulations, negotiations, or risk management. Often, the term D is conservatively assumed to be zero, except for steep slopes and/or where there is an underlying drainage/capillary layer below the fine-grained cover soil.

Several numerical models, including UNSATH (Fayer 2000), HYDRUS (Šimůnek et al. 1999), LEACHM (Hutson and Wagenet 1992), and

VADOSE/W (Krahn 2004), have been used for modeling the water balance for site-specific soil and climate conditions. These models simulate unsaturated flow of water through soils using a modified form of Richard's equation. Historically, the Hydrologic Evaluation of Landfill Performance (HELP) model (Schroeder et al. 1994) has also been used. However, because the HELP model uses a more simplistic and limited water-routing approach, the model is now considered outdated for final cover water balance design.

Albright et al. (2010) provided a simplified semiempirical method for estimating the minimum required thickness of monolithic and capillary-break covers. This semiempirical method (the Alternative Cover Assessment Program, or ACAP, method) is based on a database provided by the ACAP program (28 final cover test sections in 11 states that were monitored for 4 to 8 years), site-specific soil properties, and site-specific P and PET. If local meteorological data are not available, surrogate data can be used to predict hydrology on the basis of the methods described in Bareither et al. (2016).

Experience has shown that the most desirable water balance cover soils are classified as "loam" (Albright et al. 2010), an agricultural term that references the proportions of sand ($\leq$ 52%), silt ($\geq$ 28%, $<$ 50%), and clay ($\geq$ 7%, $<$ 27%) as defined using the soil texture triangle by the US Department of Agriculture (USDA 2017). Loam soils provide a good balance of water storage, erosion resistance, percolation resistance, and resistance to shrinkage and swelling (and subsequent formation of preferential flow pathways through the cover) and are favorable for vegetative growth. For many arid and semiarid sites, typical designs result in a requirement for a loam soil with a minimum thickness of 1 to 2 m. Although calculations may sometimes indicate that thinner covers could achieve the water balance function, there are often concerns as to the viability of long-term native plant survival with less than 1 m of cover soil.

The second water balance cover configuration incorporates a capillary barrier consisting of a fine-textured soil layer in which vegetation grows overlying a coarse-textured layer. The capillary barrier concept uses the low matric suction of a dry coarse-grained soil to limit the downward flow of water under gravity and increases the storage capacity of the overlying fine-grained soil. Compared to monolithic covers, capillary barriers are effective for extending the range of climatic conditions within which a water balance cover might function (e.g., temperate climates), for providing greater reliability, or allowing for a thinner cover. Capillary barriers only increase storage in the overlying fine-grained soil but do not provide a resistive barrier to water ingress if the storage capacity of the overlying fine-grained soil is exceeded. The principles and design of capillary barrier layers are described in detail in Albright et al. (2010). Flow in sloped capillary barriers, evaluated by Ross (1990) largely for capping radioactive waste repositories, has been extended to the design of double-capillary barriers for landfill covers even in temperate climates in Germany, Canada, and the United States (Dwyer 1995, Kampf et al. 1999, Abdolahzadeh et al. 2008).

In a double-capillary barrier, there are two layers of progressively coarser-grained soils below the fine-grained cover soil (Figure 9-2). Both the interfaces between the finer vegetative soil layer and the sand, and the sand and gravel form capillary barriers limiting downward percolation.

Both monolithic and capillary barrier covers are typically placed atop foundation layers described in the following section for resistive covers.

Critical to water balance covers is the availability of a local soil that meets the cover design criteria (e.g., loam or similar). Typically, soils are characterized on the basis of grain size distribution (e.g., sand, silt, and clay), Atterberg limits, and hydrologic characteristics such as the saturated hydraulic conductivity and the soil-water characteristic curve. Placement of water balance cover soils is usually performed in lifts up to 0.5 m thick by dozer spreading, with minimal compaction. Overcompaction is typically a larger problem for water balance covers than undercompaction because these covers rely on large soil pores to store water near the cover surface and to encourage plant growth. A typical goal is to achieve 85% relative density in accordance with Standard Proctor compaction (ASTM D698), with a range of 80% to 90% typically being acceptable. Haul roads over the cover that become heavily compacted during construction are often ripped and refinished with a light dozer at the end of a project.

Figure 9-2. The construction of a water balance cover with a double-capillary barrier near McPherson, Kansas, showing the spread of vegetative cover soil (V.C.) over a fine capillary (sand) break layer, which is on top of a coarse capillary (gravel) break layer, and which is on top of the clay foundation layer.
Source: Jack Chapelle.

For most water balance projects, long-term effectiveness depends on successful establishment of a healthy, functioning plant cover. For this reason, a minimum cover soil thickness of 1 m, or more, is usually required. Plants that are native to the project area are often preferred over nonnative seed mixes. Deeper-rooted plants are more effective in performing the ET function than shallow-rooted plants. However, plant rooting characteristics should be evaluated to ensure that plant roots do not pass through the cover layer into the waste or, if they do, such penetration should be nondetrimental to plant survivability and the functionality of the cover and not cause an upward migration of contained contaminants of concern.

9.3.3 Resistive Covers

Resistive covers are conventional designs that have been used for final covers ever since the need for such systems was first recognized.

Resistive covers incorporate relatively impermeable barriers such as compacted clays, GCLs, and geomembranes to resist advective flux. Most resistive covers include one or more layers of soil, although there is a special class of exposed geomembrane covers (EGCs) that does not necessitate the use of soil as a barrier layer. In addition to low-permeability layer(s), resistive covers also typically include surface soil layers and a water drainage layer above the resistive barrier component. Other layers incorporated in a resistive final cover often include a foundation soil layer over the waste and a gas collection layer below the barrier layer system (for systems that include gas collection, such as MSW or industrial waste landfill covers). Common resistive cover components are shown in Figure 9-3 and labeled by the primary function of each layer.

A typical example of a resistive final cover system, which represents minimum technology guidance for a USEPA Resource Conservation and Recovery Act (RCRA) Subtitle C hazardous waste landfill, is shown in Figure 9-4 (USEPA 2004). The components of the final cover, from bottom to top, include 0.6 m-thick compacted low-permeability soil having a maximum permeability of 1×10^{-9} m/s, a geomembrane, a drainage layer, and 0.6 m-thick layer of an erosion-resistant soil. A filter layer should normally be used above the drainage layer to prevent fines migration and fouling. A similar design would meet the USEPA Subtitle D prescriptive minimum criteria for a system with a geomembrane/compacted clay liner (CCL) composite bottom liner, except that the low-permeability soil layer below the geomembrane would have a minimum required thickness of 0.45 m with a maximum permeability of 1×10^{-7} m/s, the cover soil layer above the geomembrane would have a required minimum thickness of 0.15 m, and there would be no mandatory requirement for a drainage layer. However, a drainage layer, although not required for USEPA Subtitle D, is commonly included in the design to satisfy slope stability requirements.

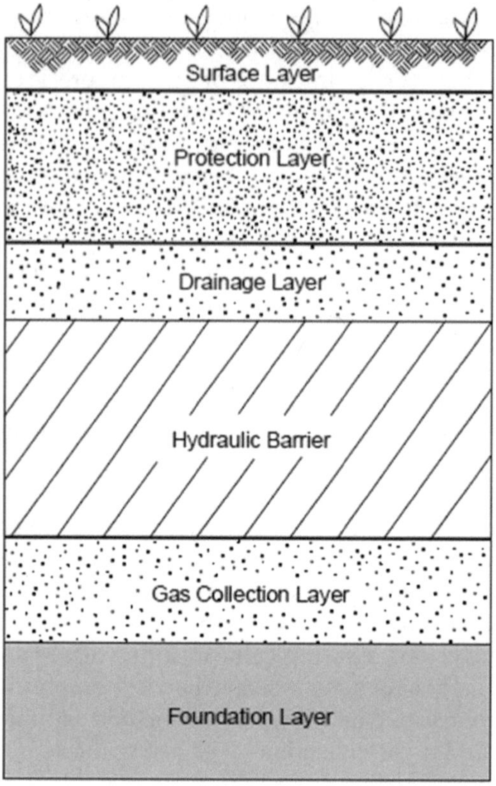

Figure 9-3. Typical layers in a landfill-resistive cover system.
Source: USEPA (2004).

The layers in a resistive final cover are described in the following paragraphs. The descriptions provided are ordered starting from the lower-most layer above the waste and moving to the top.

9.3.3.1 Foundation Layer. The foundation layer typically consists of the interim cover (usually soil) placed over the waste between the time of waste disposal and the final cover construction. The foundation layer serves as a clean transitional layer and structural foundation for construction of the overlying final cover components. The thickness of the foundation layer is typically a minimum of 0.3 to 0.6 m, depending on the regulatory requirements for interim cover soil thickness and on project-specific requirements to provide a structural foundation for compaction/placement of overlying layers.

9.3.3.2 Gas Collection Layer. This layer, if needed or used, can be constructed from either coarse-grained soils or geosynthetics. Gas collection

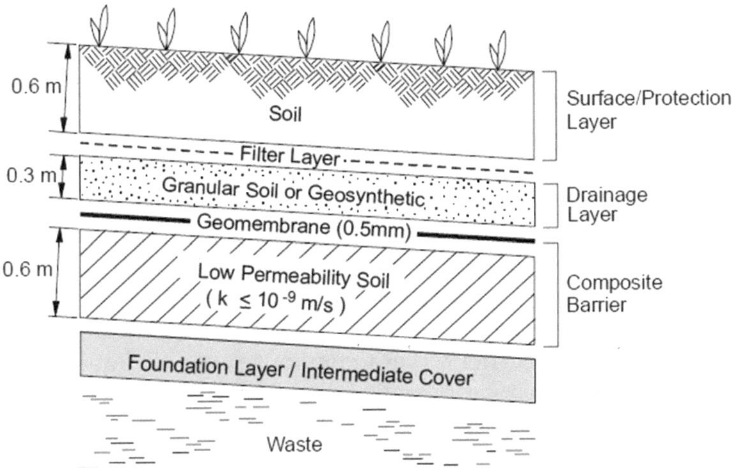

Figure 9-4. A resistive final cover in accordance with USEPA Subtitle C requirement for hazardous waste landfills.
Source: USEPA (2004).

layers can be important for several reasons including: (1) the need to control gas migration by directing gas flux up and out of the landfill in a controlled manner, rather than forcing the gas to move downward or laterally in an uncontrolled manner; (2) to reduce and control pressures that may potentially build up along the bottom of the final cover and generate destabilizing forces; (3) the need to collect gases to reduce emissions of greenhouse gases, ozone-depleting substances, and toxic constituents; (4) the need to collect the gases for odor control; (5) the need to collect gases that may adversely impact the cover vegetation by displacing oxygen; (6) the need to remove near-surface potential fire and explosive hazards; and (7) to collect gas (specifically methane) for energy production. The gas collection layer also commonly serves a secondary function of collecting liquid seeps that are released out of the waste mass toward outer slopes. While the seeps flow downslope by gravity, the gas can migrate through the collection layer in any direction because the overall gas density is close to that of air. Provisions should be included in design for the toe of the slope to collect liquid seeps and direct these as appropriate.

9.3.3.2.1 Hydraulic Barrier Layer. The acceptable configuration and specification of hydraulic barrier layers is often a result of regulatory requirements (e.g., the requirements of USEPA Subtitle C or D, or other local requirements), a negotiated alternative, or risk management decisions made by the facility owner. Estimates of potential advective leakage rates, analogous to the percolation term P_r in Equation (9-1), through barrier layers in resistive covers can be based on the water balance models discussed previously. When geomembrane barriers are modeled, it is necessary to input a frequency and size of defects (holes) in the geomembrane. In this regard, the HELP model (Schroeder et al. 1994) was often used in the past because it contains calculations to account for geomembrane defects, despite the limitation imposed by the simplistic water-routing approach. For representative leakage modeling, the other models listed previously for water balance covers can be used to determine the water flux through the soil and/or geosynthetic layers (Figure 9-3), which would result in head buildup above the barrier layer. Descriptions of the various barrier layer materials and requirements are provided in detail in the next section. Although the barrier serves to minimize not only liquid transport but also gas emissions, the overall design of the barrier component is largely based on hydraulic considerations and not gas transport principles. Significantly more data and analyses are available for hydraulic considerations in comparison with gas transport issues. Guidance is also provided for gas barrier function in a subsequent section of this chapter.

9.3.3.2.2 Drainage Layer. A drainage layer is commonly placed above a hydraulic barrier layer on slopes to improve slope stability, especially when the hydraulic barrier layer is a geosynthetic. Numerous small to relatively large veneer stability failures on final covers have occurred because of the lack of incorporation of an adequate drainage layer above a low-permeability barrier layer. (The term *veneer* is commonly used in the United States to describe a relatively thin layer of soil that is placed on a geosynthetic liner system, such as for a final cover system.) The drainage layer can consist of a clean coarse sand or gravel (as shown in Figure 9-5),

Figure 9-5. A drainage layer being placed on a single geomembrane (white-surfaced HDPE) cover at a forest product landfill in Washington State.

or a geosynthetic drainage material (example shown in Figure 9-6). Design guidance for this layer is provided in a subsequent section of this chapter.

9.3.3.2.3 Protection Layer. Depending on the configuration of the hydraulic barrier and drainage layers, and the type of vegetation or erosion control provided on the surface, there may be a need for maintaining a minimum thickness of protective cover soil. USEPA (2004) provides guidance for selecting the thickness of the protection layer on the basis of the

- Need to protect underlying layers from damage caused by wet–dry and freeze–thaw cycles;
- Need to prevent accidental human intrusion, penetration by burrowing animals, or root penetration into underlying materials;
- Need to support vegetative growth by accommodating plant roots and providing adequate water storage to sustain vegetation (this is the most common design basis for this layer);
- Need to temporarily store water in the protection layer to attenuate rainfall percolation into the underlying layers; and
- Need to provide other types of protection unique to a particular waste (e.g., attenuate potential radon emissions from underlying radon-bearing waste).

Commonly, when suitable materials are available, the same soil borrow source is used for the protective soil and topsoil layers, with both placed at the same time, with the final surface amended as needed with compost, mulch, pH-adjusters, microbes, and fertilizers to establish the initial vegetation.

Figure 9-6. The placement of a final cover soil on a composite cover for a Subtitle C hazardous waste landfill in the Western United States. The drainage layer being covered, shown in the lower right of the photo, is a geocomposite material, which is overlaying a 1.5 mm HDPE geomembrane and a GCL on a 3:1 slope.

9.3.3.2.4 Surface Layer. The uppermost surface is often regulated to be erosion resistant. For the purposes of practicality and aesthetics, grassy vegetation mixed with various legumes, herbaceous vegetation, flowers, and other nonwoody plants are commonly used. Thus, the surface layer is often comprised of topsoil to support vegetative growth. Woody vegetation such as bushes and trees are often discouraged and controlled, except where thick topsoil or vegetative growth layers are included, because of potential root intrusion that may drainage and barrier layers and penetrate the waste mass. Other types of nonvegetative surface layers can be used to control both water and wind erosion including gravel/cobbles/riprap, tire shreds disked into the soil, asphalt or other pavements, concrete articulated mats in localized areas, erosion control mats, and geocells filled with either planted soil, gravel, or cemented soil. Steep slopes may also incorporate various types of retaining-wall or reinforced-earth facings.

9.4 RESISTIVE LAYERS

Materials used as resistive or barrier layers include single CCLs, single GCLs, single geomembranes, a composite system consisting of a geomembrane over a CCL, or a composite system consisting of a geomembrane over a GCL.

9.4.1 Single Compacted Clay Liners

Compacted clay barrier layers are the earliest form of engineered barrier used in the construction of landfill final covers. Although CCLs have proven to be effective in bottom liners, these barriers have been reported to degrade over time in cover systems because of desiccation cracking, freeze–thaw susceptibility, vegetation penetration, and cracking caused by settlement deformation (Daniel 1995, USEPA 2004). Field studies reported by NRC (2011) indicate that compacted saturated low-hydraulic-conductivity soil materials that are not covered with a geomembrane and used in final covers do not retain as built properties but transform to soils with properties typical of surrounding soils within 5 to 10 years after installation, with larger changes observed for soils with lower as-built saturated hydraulic conductivity. The report suggests that saturated hydraulic conductivity of fine-textured earthen storage and barrier layers not covered by geomembranes range between 1×10^{-7} m/s and 5×10^{-6} m/s regardless of the as-built condition, with a saturated hydraulic conductivity of 5×10^{-7} m/s considered typical. A follow-up study by NRC (2022) reported that these conclusions were determined to be applicable along a cover profile up to 2.1 m depth. Therefore, the use of compacted fine-grained soils without an overlying geomembrane is not expected to be effective as a long-term low-permeability barrier. If such barriers are proposed for use in a final cover to limit percolation, they should be evaluated in terms of their performance in a water balance model considering both their saturated and unsaturated material properties. For example, as discussed subsequently, Hanson et al. (2020) determined, on the basis of extensive field measurements, that single CCLs can be more effective than water balance covers in the control of surface emissions of landfill gases in California.

9.4.2 Single Geosynthetic Clay Liners

Single GCLs have been used as single-layer barriers for resistive landfill covers with varying degrees of performance. Studies presented by Meer and Benson (2007), NRC (2011), and Rowe et al. (2016) indicated that GCLs placed against soils undergo cation exchange over time in the bentonite layer that reduces the effectiveness of the GCL as a hydraulic barrier layer to varying degrees depending on several factors. GCLs are likely to interact with water moving through the overlying soil layers and at a given time may be undergoing hydration (wetting) or dehydration (drying) along hysteretic cycles (Risken et al. 2015, Yesiller et al. 2019). Other factors noted in these studies that affect the long-term performance of single GCLs as fluid barriers in covers include the granularity of the bentonite, the hydraulic gradient, presence of plant roots, width of seam overlaps (which can affect root penetration and maintenance of GCL areal coverage of landfill surface after potential shrinkage or necking), depth of soil cover, size and intensity of needle-punched fiber bundles, the rate and completeness of initial hydration to successfully create a gel structure, chemistry of soil porewater, and extent of drying after cation exchange has occurred.

Single GCLs are likely to perform well if placed in covers such that the bentonite is hydrated relatively rapidly from adjacent soil porewater and does not lose the initial gel structure of the hydrated bentonite (i.e., is not allowed to desiccate or undergo wet/dry cycles) after installation. By maintaining the GCL in a hydrated state after initial hydration, a gel structure is maintained, and the GCL can potentially maintain relatively low permeability even with the significant cation exchange that inevitably occurs under most field conditions (Benson and Meer 2009, Rowe et al. 2016). However, even a single cycle of bentonite desiccation can substantially degrade the low permeability of the GCL after it has undergone cation exchange (Benson and Meer 2009). Although some early investigators suggested that a 0.75 m thickness of soil cover may be adequate to protect the GCL, later investigations have shown that this thickness may be inadequate, and recommendations that 1.5 m (or potentially thicker in an arid or semiarid climate) is a better minimum thickness exist, unless long-term field studies, recommended for a minimum period of 5 years, can prove otherwise (Henken-Mellies 2010). The performance record of single GCL covers, determined either in the field via large-scale pan lysimeters or in the laboratory on field-exhumed samples, indicates a large scatter in performance over service times up to 10 years, with various degrees of degradation over time (Meer and Benson 2007, Rowe et al. 2016). Thus, the authors recommend that any proposed use of single-GCL barrier-layer final covers be given careful consideration for project-specific impacts that will be caused by the potential deleterious effects of long-term cation exchange, severe degradation with even one wet–dry cycle, the rate and adequacy of initial hydration, the effects of needle punching on patterns of hydration and cation exchange, and the long-term impacts of roots.

9.4.3 Single Geomembranes

Single-geomembrane barrier layers are often considered the most cost-effective single barrier-layer solution if they are properly selected and constructed. A common material that has been used is a 1.5 mm-thick high-density polyethylene (HDPE) because of long-term aging and durability studies indicating that this material can have a substantial lifetime in buried environments, its ability to be manufactured with an aggressively textured surface to improve veneer slope stability, and its relatively low cost. In some instances, a 1 mm-thick material has been accepted for projects. In other cases, the greater durability of a 2 mm-thick material has been judged to offset the higher additional cost; 2 mm-thick material is commonly used for covers in Germany as well as some other countries, but less frequently in the United States. Thicker geomembranes will be more durable and will last longer. The advantages of using a lower thickness material are ease of installation and lower cost. Concerns for the use of HDPE include low impact and puncture resistance; potential for stress-cracking over time if not properly formulated; and formation of wrinkles during construction that can create stress concentrations affecting long-term performance.

Linear low-density polyethylene (LLDPE), which is similar to HDPE in many respects, is often preferred in cover system applications because LLDPE has greater elongation and flexibility compared to HDPE and thus better accommodates differential settlements and stress concentrations.

One of the primary reasons that polyethylene (HDPE or LLDPE) is more widely used than other types of polymers for landfill cover geomembrane applications is that it is the only type of resin to which a durable textured surface can be readily created during manufacture. This textured surface provides a higher interface shear strength compared to smooth geomembranes to resist cover soils sliding off the geomembrane at the low normal stress levels encountered in cover systems.

Flexible unreinforced polyvinyl chloride, which was available as a liner material prior to polyethylene, has been used for landfill covers with typical thicknesses ranging from 0.75 to 1.5 mm. Other types of geomembranes that have been used for landfill covers include the following: polypropylene (PP), available in both a reinforced version for strength and an unreinforced version for high elongation and flexibility, with typical thicknesses ranging from 1 to 1.5 mm; ethylene propylene diene monomer (EPDM), a synthetic rubber, which has good elongation and conformability, good resistance to weathering and UV exposure, and makes for easy installation, with typical thicknesses ranging from 1 to 1.5 mm; and bituminous geomembrane, which is typically reinforced with a glass fleece, exhibiting good UV and oxidation resistance, has among the highest levels of construction durability, and has been used in rugged environments and cold climates, with typical thicknesses ranging from 3.5 to 5.6 mm. Other polymeric geomembranes are available, such as ethylene interpolymer alloy or chlorosulfonated polyethylene, but they are not common for landfill cover applications because their specialized formulations were developed more for exposed pond liners, and their extra cost and lack of texturing generally do not justify their use when covered with soil for final covers. In general, all types of geomembrane materials have relative advantages and disadvantages. For any project, each type of material should be evaluated with due consideration for the limitations and potential advantages related to short- and long-term performance, durability, constructability, aging characteristics, and cost. Detailed information on geomembrane material properties, engineering properties, design, and installation guidelines are provided in Scheirs (2009). General material properties and design approaches are also provided in Koerner (2012). Best practices for proper installation of geomembranes to reduce formation of defects and wrinkles and to prevent stress concentrations and high stresses during use are included, for example, in Giroud and Morel (1992), Scheirs (2009), Koerner and Koerner (2013), and Narejo (2016). Probabilistic analysis tools for polyethylene geomembrane seams in landfill covers are provided in Foye et al. (2016). A comparative analysis of the use of geomembranes for mine waste final cover systems is presented in Yesiller et al. (2018).

9.4.4 Composite Covers

Resistive composite covers typically consist of a geomembrane overlying a low saturated hydraulic conductivity barrier layer containing soil (CCL or GCL). The technical performance requirements for the geomembrane and soil layers are usually prescribed by regulations. For example, the USEPA RCRA Subtitle C performance-based regulation prescribes that the mineral component of the composite cover for a hazardous waste landfill be constructed with 0.6 m of compacted low-permeability soil having a maximum saturated hydraulic conductivity of 1×10^{-9} m/s. The RCRA Subtitle D prescriptive minimum criteria for an MSW landfill cover include a composite cover with 0.45 m of compacted soil having a maximum saturated hydraulic conductivity of 1×10^{-7} m/s.

The compaction requirements for constructing clay layers on top of landfills and waste piles to achieve low permeabilities can be demanding. Often, the low modulus-of-reaction foundation layer and waste below the final cover makes compaction of the soil liner difficult. The long-term integrity of these barriers can be compromised under low-normal loads because of desiccation and settlement-induced cracking. For these construction and performance reasons, as well as cost, GCLs are often preferred as an alternative for the soil-containing component of a composite final cover. Many GCL-based composite covers have been constructed. Scalia and Benson (2011) identified the measured in-service condition of four composite GM/GCL final covers that were exhumed approximately 5 to 7 years after construction. The results indicate that divalent cation exchange occurs, even though the GCL is covered by a geomembrane. They suggested that to ensure good long-term performance, the construction conditions should promote rapid hydration of the bentonite to at least 50% moisture content, after which the GCLs should not be allowed to desiccate, as discussed previously. This prevention of desiccation is more readily attainable when the GCL is overlain by a geomembrane than in a single GCL. Rowe et al. (2014) demonstrated that bentonite within GCLs overlain by a geomembrane on slopes can internally erode because of the flow of water that condenses beneath the geomembrane when left exposed (i.e., the GM is not covered by soil) even for relatively short periods (e.g., several months) while awaiting cover soil placement. Hanson and Yesiller (2019) observed significant cation exchange in GCLs after long periods where the overlying geomembrane was left uncovered. In addition, GCLs beneath uncovered geomembranes have been shown to shrink and lose their seam overlaps over a number of wet/dry cycles that are caused by diurnal heating and cooling (Rowe et al., 2018). Thus, timely covering of composite barriers with overlying soil shortly after geomembrane placement is critical to system success, unless the GCL is encapsulated between two geomembranes. If the recommended procedures are followed, existing evidence on composite cover barriers with GCLs indicates that these covers are likely to be effective in controlling percolation.

The USEPA guidelines for the selection of geomembrane materials for composite cover barrier layers are nonspecific in nature. The authors note that the most common types of geomembranes that have been used for composite final covers are the same materials discussed previously for single geomembrane barriers.

9.4.5 Exposed Geomembrane Covers

Exposed geomembrane covers for landfills started to be used in the United States in the mid-1990s in Delaware with a local preference to perform final closures of landfills with an EGC (Gleason et al. 2001). The replacement cycle for the exposed material was considered to provide opportunities to mine the landfill for resources prior to the next cycle of covering. A significant issue for EGCs is construction survivability under wind forces with specific designs available in the literature (Giroud et al. 1995, 2006). Early designs focused on using a reinforced PP geomembrane (with a nominal thickness of 1 mm). Additional materials, including HDPE, LLDPE, EPDM, and bituminous geomembranes, as well as designs with attached solar panels for renewable energy production (e.g., Boyd 2012), were also used. Early designs typically used large anchor trenches (e.g., 2 m wide by 1.3 m deep) spaced according to the design calculations, with spacing on the order of 30 to 50 m being common depending on the material, slope, and design wind speed. Although periodic large anchor trenches are still used, other methods have been developed, such as regularly spaced point-load soil anchors with tension hold-down cables and top plates. Exposed bituminous geomembranes have the advantage of a relatively high mass per unit area that can sustain large wind forces with no anchorage or ballasting. Although EGCs have withstood hurricanes in Louisiana and Florida, wind damage caused by stress concentrations at fixed penetrations (e.g., well heads) can occur. Damage owing to animals, destruction caused by fire, and aging failure of the geomembrane have occurred. The authors are not aware of an EGC project that was damaged because of hail, but this failure mechanism has occurred in ponds with exposed geomembranes.

A variation of an EGC that incorporates an artificial grass turf over the geomembrane (polyethylene and EPDM) has become available with different mechanisms used for keeping the geomembrane in place (e.g., broadcast sand or soil wedge anchors with tension hold-down cables and top plates). The artificial turf covering provides enhanced aesthetics and increases UV protection of the underlying geomembrane, thus increasing geomembrane life. The designs consider exposure effects and stormwater runoff. These covers also need periodic replacement similar to single-layer EGCs. A life-cycle analysis that considers financial assurance and engineering redundancy must include provisions that these covers will eventually degrade and need to be replaced.

The exposed geomembranes must be installed on a prepared earthen subgrade that covers the waste. To provide a reliable physical separation of the waste from the EGC and from the atmosphere in the event that the EGC becomes displaced or ripped because of aging, wind forces, animals, fire, and so on, a recommended guidance is to provide 0.6 m of soil between the EGC and the underlying waste. Additional considerations for management of side-slope seeps and landfill gas collection below EGCs are also often necessary.

9.5 ADDITIONAL TECHNICAL ISSUES FOR THE DESIGN OF FINAL COVERS

This section describes additional technical design, construction, and operational issues for final covers that have not been described previously.

9.5.1 Erosion Control and Surface Water Management

Control of stormwater runoff from precipitation events is a critical aspect of final cover design. Inadequate design and attention to construction details in this regard has been a cause of failures and postclosure maintenance costs for final covers.

The geometry of many waste containment facilities is characterized as a mound with significant slopes. For reasons of veneer stability and erosion control, final covers are typically designed to incorporate intermittent interruptions in the downslope drainage, usually in the form of side-slope ditches. The ditches can be minimally designed solely for the purpose of carrying stormwater flow, or they can be widened to incorporate benches to allow access for vehicles, landfill gas infrastructure, or other features.

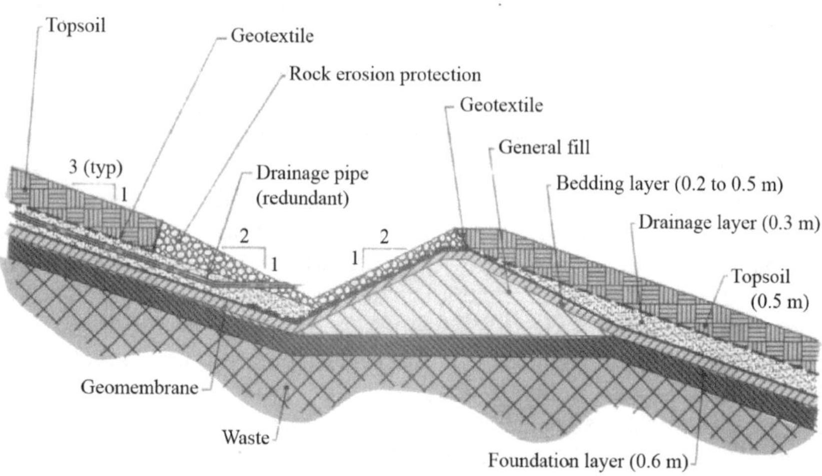

Figure 9-7. An example of a landfill cover section with a single geomembrane barrier at an intermediate side slope bench/ditch to control surface and subdrain water.
Source: Thiel and Stewart (1993).

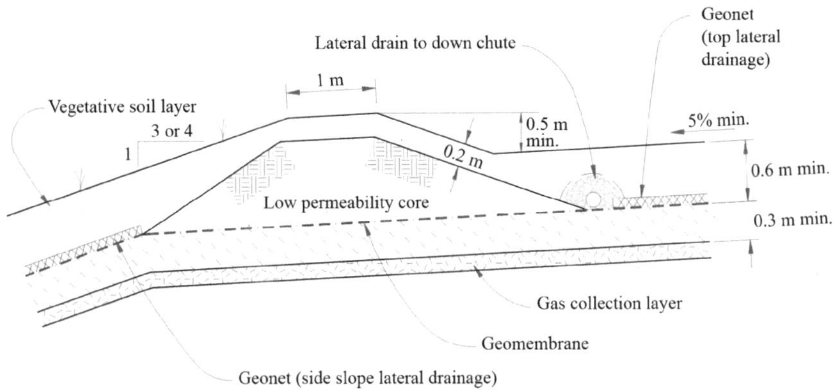

Figure 9-8. An example of surface and subdrain water control at a landfill crest to reduce water loading on slopes.
Source: Courtesy of Greg Richardson.

A typical bench/ditch design for a resistive cover with a geomembrane barrier layer, a drainage layer, and surface layer (topsoil) is presented in Figure 9-7. An example design of stormwater control at the top crest of a landfill mound, to reduce both surface and underdrainage water loading on the slopes, is presented in Figure 9-8.

Periodic down-drains are typically used to collect the water from the side-slope ditches and transfer this water to the perimeter toe. Various configurations have been used for this, including down-drain pipes and open-channel ditches. The need to accommodate waste settlement and maintenance of inlets are unique considerations for landfill covers. Minimum longitudinal ditch grades in the range of 3% to 5% are common to accommodate future total and differential settlement. Very stiff structures, such as concrete, may not perform well on landfills because of settlement, and thus, more flexible systems such as articulated mattresses, riprap, or in-filled geocells are common for open channels. Various surface water design considerations for final covers are presented in USEPA (2004).

9.5.2 Control of Fluid Pressures below and above Barrier Layers to Preserve Veneer Slope Stability

Control of fluids (both liquids and gases) below final covers, especially those incorporating a barrier layer, is important for geotechnical and environmental reasons. The environmental reason was discussed previously in the general discussion of the gas-control layer. The geotechnical issue relates to slope stability and is described subsequently. Similarly, control of liquids above a final cover-resistive layer also affects slope stability and is described herein.

The engineering approaches used to preserve the stability of soil layers placed on top of a landfill cover barrier layer are in accordance with basic principles that have been developed in geotechnical engineering. Because of the low effective stresses, the veneer stability of cover systems is sensitive to relatively small changes in pore pressures. Small changes in effective stress, caused by gas or liquid pressures (above or below the barrier layer), can destabilize a veneer system on a cover slope and cause a slide. Numerous reported slide failures highlight the importance of this issue.

9.5.2.1 Fluid Pressures below the Barrier Layer. Both landfill gas and liquid leachate seeps can potentially accumulate below a final cover barrier layer and create destabilizing pore pressures. MSW landfill gas pressures as high as 4 kPa have been reported immediately below a final cover (Thiel 1998). This gas pressure is sufficient to negate the effective normal stress of approximately 0.25 m of cover soil and is severe for a veneer cover system, potentially leading to failures. Although this level of gas pressure may not be common, such an occurrence points to the criticality of designing a gas-pressure relief layer below the final cover barrier for landfills that actively produce gas.

The gas collection layer can be constructed from either granular material or geosynthetic drainage layers. If the gas collection layer is constructed from granular soils, it is important to recognize that soils that are relatively saturated because of capillary moisture and condensate moisture will have a greatly reduced transmissivity to gas flow. Fine sands have been shown to be ineffective because of this, and thus, clean medium-to-coarse grained sands, or gravels, are required to transmit gas. Care must also be taken to avoid placement of angular gravels adjacent to a geomembrane because of the risks of puncture. If geosynthetics are used, the same concern for transmissivity reduction by capillary-held moisture exists. Standard heavy (550 g/m^2) nonwoven needle-punched geotextiles have been shown to have their gas transmissivity severely impeded by capillary water, and thus, geocomposite drainage layers (geonet with geotextiles on both sides) are preferred. For both cases, a network of lateral and header pipes, often with some trenched into the waste, is usually constructed in a manner to be in physical connection with the gas collection layer to provide the venting or suction outlet for the gas. For facilities with active gas control systems, design and operational provisions might be considered to allow the gas collection layer below the final cover to passively vent in the event that the active gas collection system is shut down for maintenance or other reasons.

The same layer that is used to collect landfill gas can also be used to collect and relieve pressures from side-slope leachate seeps. If a waste containment facility contains significant moisture and has the potential to produce considerable side-slope seepage, some form of subdrainage network below the final cover barrier layer should be considered to relieve pore pressures and maintain stability.

9.5.2.2 Liquid Pressures above the Barrier Layer. Infiltration of meteoric water through cover soils tends to pond above a low-permeability barrier layer, such as a geomembrane, causing the water in the overlying saturated soil to flow parallel to the slope. The effective normal stresses in the soil may be reduced by up to approximately 50% with full-cover saturation, reducing the slope stability factor of safety by approximately one-half compared to the dry condition.

The design solution for this problem is to construct a drainage layer on top of the barrier layer, below the protective soil layers, of adequate transmissivity, and limited slope length, to accept the infiltrated water and direct the water away from the cover without building excessive pore pressures. As with gas collection layers, both granular soils (gravels) and geosynthetics (typically geocomposite drainage layers) can be used for this lateral drainage layer above the barrier layer. A complete design methodology to estimate the rate of percolation into the drainage layer, the required drainage layer transmissivity, the maximum allowable slope length for the drainage layer, and the calculation of the impact of an unsaturated or saturated granular drainage layer on the veneer stability is presented in Thiel and Stewart (1993). The unit-gradient approach presented for estimating the liquid-supply rate to the drainage layer provides a conservative approach for these sensitive systems in response to storm events.

9.5.2.3 Other Slope Stability Considerations for Veneer Cover Systems. Once assessment and control of the potentially destabilizing pore pressures are completed, the remainder of the slope stability evaluation for veneer cover systems can generally follow standard geotechnical approaches. Some considerations that are uniquely related to final covers include the following:

- When the length of the slope exceeds approximately 15 times the thickness of the veneer cover, which is common for final cover designs, the slope can be considered an infinite slope from the point of view of geotechnical stability assessment. This simplifies the analysis without being overly conservative.
- Many regulations include maximum allowable landfill slopes of 3(*H*):1(*V*), which is supported by case history data showing higher frequencies of failure for final cover slopes steeper than 3:1. Projects with steep slopes may require specialized materials or specialized approaches such as reinforced slopes.

- Interface shear strength parameters between soils and geosynthetics need to be measured at the low normal loads in the final cover design, consider all internal and interface failure planes, and not be extrapolated from higher-normal load test data because of nonlinearity of failure envelopes at low stresses (leading to unconservative analysis).
- Drainage benches and access roads on final covers may have oversteepened portions where localized slumping may be of concern. If the final cover soils are comprised of overconsolidated high-plasticity silts and clays, then cyclic wetting and drying can reduce the shear strength of these types of compacted soils to the normally consolidated condition. Therefore, the long-term shear strength of these types of soils on slopes, and interfaces with these soils, should be modeled on the basis of normally consolidated densities representative of a fully softened condition, rather than compacted conditions. For cohesive soils, estimating shear strength for this condition at low normal stresses representative of covers is provided by Gamez and Stark (2014).
- Significant interface shear strength degradation can occur during soil spreading operations on a veneer slope. The critical conditions during construction are typically directly below the equipment tracks or wheels, rather than the entire slope length, unless the design is dependent on a tensile reinforcing layer. The maximum size of soil piles being pushed by dozers on a slope should be specified on the basis of site-specific materials, geometry, and construction equipment (Kerkes 1999, Paruvakat and Richardson 1999, Jones et al. 2000, Thiel and Giroud 2023). For localized equipment inertial loadings, a common assumption is 30% of the equipment weight as the dynamic force parallel to the slope.
- For seismic loading, calculated slope deformations on the order of 1 m have been referenced as being acceptable (Kavazanjian 1998), given that cover systems are generally accessible for repair, and that localized (veneer) failures are not expected to result in immediate endangerment of human health or the environment.
- As with most geotechnical slope stability evaluations, the commonly accepted long-term static target minimum factor of safety is 1.5. Evaluation of landfill stability in terms of probability of failure has also become more common. Probability analyses provide an indication of sensitivity and risk that complements the standard factor-of-safety approach. The acceptable level of risk is a project-specific decision between the designer, owner, and potentially other stakeholders and regulators.

A detailed discussion of veneer-stability slope stability analysis methods, including pseudo-static seismic stability and deformation evaluations, and the selection of interface shear strengths for design is presented in USEPA (2004). Additional considerations for potential interface shear strength degradation during construction should be considered as described by Paruvakat and Richardson (1999), Kerkes (1999), and Thiel and Giroud (2023).

9.5.3 Settlement

The impact of waste settlement on the performance of the cover system barrier layer and other layers as well as the infrastructure components is a common concern for all types of final covers. Two general technical issues related to settlement need to be considered: (1) the need to continuously provide positive stormwater drainage off the cover system and avoid ponding and (2) the need to avoid excessive stress/strain in the cover system that compromises performance of the barrier layers through tears, cracks, and similar mechanisms and other layers through tears, cracks, and damage to piping and similar components.

Two common types of settlement affect landfill final covers: (1) total settlement caused by the average long-term compression (and for organic wastes) decomposition of the waste. This is the most common type of settlement and typically has a relatively low impact on the cover systems. Drainage structures such as down-drains and associated inlet structures can be subjected to excess stresses. Vertical penetrations in the cover system, usually related to landfill gas wells, need to be designed or maintained to accommodate total settlement over time. (2) differential settlement represents localized vertical distortions of the cover system components (Scalia et al. 2017); these distortions may cause unacceptable tensile stresses or strains or the change of slopes, which may affect cover system performance. Differential settlement occurs in most MSW landfills because of the presence of zones that receive large volumes of highly compressible waste and differences in biochemical degradation and decomposition of waste. Extreme examples are sudden collapse/crushing of large constituents or underground smoldering, both of which can manifest as sinkholes. Areas around gas wells with excessive oxygen intrusion can create zones of differential settlement because of below-surface waste combustion. Differential settlement typically affects the cover system more than total settlement. Localized depressions, such as shown in Figure 9-9, can cause reverse ditch flow grades, ponding, erosion initiation sites, stresses and strains in the final cover system layers, and potential disruptions in down-drains or other structures on the landfill surface. When an unacceptable magnitude of differential settlement occurs, postconstruction/postclosure maintenance will likely be required to adjust the grades and repair the cover system in the affected area. For final covers with geomembrane components, a coupled analysis for geomembrane durability (through crack formation)

Figure 9-9. An example of differential settlement at a Florida landfill. Source: Courtesy of Bob Mackey.

and cover performance (through differential settlement and ponding over geomembrane defects) is presented in Foye and Soong (2017).

There are multiple design strategies for accommodating settlement for final covers:

- Delaying construction of the final cover system until some or most landfill settlements have occurred as established either by the initial design or by field settlement monitoring. However, delaying construction of the final cover system may result in significantly increased landfill gas emissions from an MSW landfill. In this case, an intermediate cover layer (e.g., soil with at least 12% high plasticity fines or exposed geomembrane) with a shallow gas collection system can be installed over the waste until it is time to place the final cover system.
- Providing adequate minimum final cover slopes (minimum values of 5% to 10% are common on MSW landfills) to accommodate some amount of differential settlement and still avoid the potential for ponding.
- Providing drainage benches with a minimum longitudinal slope (minimum values of 3% to 5% are common on MSW landfills) to accommodate some amount of differential settlement and still avoid the potential for reverse ditch grades.
- Designing down-drains, landfill gas wells, landfill gas headers, and other structures on the landfill to be able to accommodate settlement or accommodate corrective maintenance and still be able to function.

9.5.4 Landfill Cover Thermal Regime

The magnitude and variations of temperatures in a final cover system can affect the performance of the resistive layer (material properties and transport processes) as well as the other layers and components through high and/or low temperatures, wet–dry cycles, and freeze–thaw cycles. For example, compacted clay liners should be covered with adequate depth of soil to avoid freeze–thaw-driven degradation of permeability. Frozen water in the outlets of drainage layers have been known to result in pore pressure buildup in the drainage layer leading to veneer stability failures of the cover systems. Temperatures in the cover systems vary seasonally similar to air temperatures; demonstrate amplitude decrement and phase lag with depth; and for facilities such as MSW landfills and certain industrial waste landfills where heat generation occurs in the waste mass, typically are elevated compared to unheated conditions (i.e., no underlying heat-producing materials). Design charts for cover temperature ranges with depth and frost depth in landfill covers for different climatic regions are available in Yeşiller et al. (2008).

9.5.5 Landfill Gas Control in Final Cover Systems

The collection, control, and regulation of landfill gas is important to limit environmental, climate, and human health impacts and nuisance issues. For final covers designed and constructed on facilities that generate landfill gas, key considerations are summarized as follows:

- Placement of a final cover will change the flux of the gas from the landfill surface and may cause changes to the gas pressures and flows within the landfill. At unlined landfills, cases have occurred that had the unintended effect of causing downward and lateral gas migration, creating potential explosive hazards beyond the landfill perimeter, and groundwater impacts. For lined landfills, cases have occurred where this phenomenon has caused unexpected gas hotspots and condensate excursions around the perimeter where the bottom liner and cover systems meet. Gas collection, control, and monitoring systems should be considered as part of the final cover design to avoid these undesirable impacts.
- For resistive cover designs, the landfill gases that collect below the final cover system can generate excess pore pressures that create a significant destabilizing effect causing veneer stability failure of the final cover systems. A gas control layer is commonly incorporated in the final cover design to mitigate this, as described previously in this chapter.
- Water balance covers potentially allow for higher fluxes of landfill gas into the atmosphere than resistive covers. In a comparative field study of a conventional compacted clay cover and a monolithic water balance cover with similar underlying wastes at a landfill in the Western United States, Hanson et al. (2020) identified significantly

higher (orders of magnitude) methane emissions through the monolithic cover than the compacted clay cover, with less variation observed for nitrous oxide and relatively similar emissions for nonmethane trace organic compounds. Although methane oxidation through cover soils and reduced emissions have been demonstrated in several studies, the direct applicability of these results to water balance covers have not been fully established. Significantly higher landfill methane emissions from a monolithic water balance cover than a conventional compacted clay cover with similar underlying wastes were also obtained using numerical analyses (Hanson et al. 2022). Thus, depending on site conditions and special cover soil and botanic engineering, water balance covers lacking adequate fines content may not be suitable for controlling outward gas emissions. Water balance covers can be effective in preventing or treating low-gradient gas fluxes, especially when special design considerations are incorporated.

- Compacted clay covers can be effective at reducing landfill gas emissions if they are of adequate thickness, compaction, and fines content. Recommendations by Hanson et al. (2020) and Hanson et al. (2023), based on correlations between measured geotechnical index properties of covers at five landfill sites and measured landfill gas fluxes, are to provide a soil cover thickness of at least 1.5 m to control emissions. Their work suggests that higher-plasticity soils (with a plasticity index [PI] of more than 20%) are more effective in controlling emissions compared to soils with lower plasticity, and that a fines content for soil (% passing 0.075 mm) of more than 30% is more effective in controlling emissions compared to soils with lower fines content. They recommend using soil with at least 60% fines content and 12% clay content in covers with compacted clay barriers to control landfill gas emissions in the long term.
- Similar to compacted clay barriers, GCLs serve to minimize emissions of a wide variety of gas species from landfills. In tests performed by Bouazza et al. (2006) and Rouf et al. (2016), gas permeability has been shown to decrease with increasing GCL volumetric moisture content and increase significantly because of cation exchange coupled with wet–dry cycling.
- Cover systems that incorporate geomembranes are considered by USEPA (2011) to provide the greatest potential for landfill gas collection efficiency, thereby reducing emissions. It is recognized that gas hotspots may occur wherever there are penetrations (e.g., gas wells), defects in the geomembrane, and around the perimeter. HDPE and LLDPE geomembranes do not completely prevent the diffusive flux of VOCs and methane. For applications where strict control of diffusive gas fluxes of these constituents is required, coextruded

polyethylene geomembranes are available with a thin central layer of ethylene-vinyl alcohol copolymer that have very low gas diffusion characteristics for organic compounds (Eun et al. 2018).

- In all final cover systems incorporating soil layers, some amount of landfill gas is oxidized as it passes through the cover soil (Huber-Humer and Lechner 2005). The amount of landfill gas oxidation that occurs as the gas moves through cover soil has been the subject of extensive studies for the last 30 years and is difficult to predict. Methane oxidation is a microbially mediated process, by a group of bacteria called *methanotrophs*, whose efficiency in cover soils depends on a complex series of parameters, including climatic conditions, soil type (granulometry and organic content), moisture content, cover temperature, nutrients, density and type of vegetation, and biogas loadings (Huber-Humer and Lechner 2005, Spokas and Bogner 2011). Oxidation occurs within the soil layers near the surface with sufficient amounts of air/oxygen. The use of biocovers (compost-amended covers) and inclusion of specially blended soil/compost layers or windows in the final cover profile can enhance methane oxidation and reduce emissions from a cover system (Rower et al. 2011, Abichou et al. 2009).

9.6 POSTCLOSURE MAINTENANCE

Closed landfills create a legacy for future stakeholders, and good stewardship dictates that appropriate follow-through inspections and maintenance should occur for some years after final closure to verify that closure goals are being met. Depending on the project, the closure goals can include a simple physical isolation of the waste; leachate reduction; collection and treatment of landfill gas and leachate including control of unwanted migration of these fluids; vegetative and/or other erosion-control measures; stormwater management; resistance against accidental intrusion by humans or animals; and performance in a future land use such as an open space, recreational area, building site, or other. USEPA requires 30 years as a minimum postclosure care period for MSW landfills. Basic considerations for a complete postclosure maintenance program are described subsequently with an emphasis on technical details that relate to the final cover system.

9.6.1 Documentation

At the end of final closure, as-built documentation should record what was actually constructed and an operations and maintenance (O&M) manual must be provided. The documentation should generally include as-built record surveys of the constructed project showing all ditches, gas vents, surface water drain outlets, the sizes and materials of all culverts, manholes, vaults, structures, air relief valves, cleanouts, the location of buried utilities, and final contours on a 0.5 m contour interval. In addition, a manual should be provided that clearly describes the operations of any electrical/mechanical equipment, including the makes and models of all equipment, and a recommended inspection and maintenance schedule for all items must be included on the drawings or in the manual.

9.6.2 Inspections

Regular inspections of the final cover systems and all ancillary facilities and systems should be performed and documented. The frequency of the inspections may depend on regulatory requirements and the sophistication of the closure. Facilities with active landfill gas and leachate collection systems can be set up with automated monitoring, callouts, and reported in real time, often with user-friendly dashboards. Facilities with less-sophisticated systems may require regular (e.g., weekly) inspections for electrical/mechanical equipment adjustments and readings and seasonally adjusted inspections for general visible issues related to settlement and drainage, gas and leachate seeps, vegetation, stormwater, and erosion control. The frequency of inspections may decrease as conditions stabilize. A minimum of one complete annual inspection is recommended until the postclosure care program is terminated. Details regarding typical inspections include the following.

9.6.2.1 Vegetation. Vegetation on landfill covers can be actively or passively managed. Many cover systems are required to be mowed to facilitate inspection and for aesthetic purposes. Woody vegetation such as bushes and trees are often discouraged and should be removed as appropriate to avoid having the roots compromise the final cover system. Signs of weak vegetation can be indicative of either poor subsoils or zones of landfill gas displacing the soil oxygen. Signs of overly lush vegetation can be indicative of leachate seeps or surface water management problems. These issues should be investigated and measures taken to reestablish healthy vegetation.

9.6.2.2 Cover Soil Cracking, Perforations, and Slumping. Burrows in the landfill cover soils can be caused by animals, which can also create points of increased stormwater infiltration (California Water Boards 2014). On a final resistive cover that had numerous rodent burrows, the primary author observed an 80 mm deep erosion gully that appears to have been caused by a series of rodent burrows lined up in the slope direction. A study by DOE (2004) of natural analog sites for evaluation of long-term performance of water balance covers in the Mojave Desert indicated a likelihood that burrow activity by both mammals and insects would develop over time and suggested that the mammal burrows may be beneficial for the long-term performance of the water balance covers. Such burrows, conversely, may have negative impacts on resistive covers by allowing excessive water infiltration or gas emissions.

High-clay content cover soils may desiccate and crack, where crack widths of 10 mm and depths in excess of 150 mm have been observed by the authors, which can lead to increased infiltration during and immediately following precipitation events. Cracks in the cover soil can also be indicative of cover soil movement and slides, or differential settlement. Localized infiltration into the final cover can lead to erosion, slope instabilities, drainage layer clogging, and increased leachate production. These occurrences should not be ignored and should be investigated to determine the causes of the problems and addressed by appropriate means to eliminate the cracking, sliding, or settlement-related problems as well as potential increased gas emissions.

9.6.2.2.1 Settlement. As described previously, differential settlement is the most deleterious form of settlement and may cause ditch grade reversals, ponding, and interruption of surface drainage features and structures. Landfill gas wells that penetrate the final cover and surface piping may need to be adjusted periodically to accommodate settlement. If the settlement is severe, the situation may warrant corrective action to relieve strains on the barrier layer elements and ensure the performance of the cover.

9.6.2.3 Stormwater Control. Ditches and inlets are a prime location for problems to occur because of the energy that is focused by the concentrated stormwater flow. Signs of ditch instability evidenced by movement of soil and rocks, erosion, overtopping, undercutting, and gullying should be addressed immediately by upgrading the level of armoring or the capacity of the various elements under consideration. Clogging of drainage outlets, particularly those associated with gravel and pipe underdrains, should be checked, and maintained. A checklist of all stormwater structures (e.g., inlets, pipes, vaults) should be used as a guide for inspection.

9.6.2.4 Vandalism and Intrusion by Humans or Animals. Depending on the level of sophistication of the final cover and the level of site security, disturbances to final cover facilities by unauthorized people or by animals should be monitored. Corrective action measures may include upgrading the level of security.

9.6.2.5 Electrical/Mechanical Systems. The operation of landfill gas wells, blowers, condensate traps, pipelines, vents, leachate collection pumps, flow meters, air-release valves, and other types of valves should be performed and maintained in accordance with the final O&M manual.

9.6.2.6 Buildings, Roads, and Security. The conditions of the site access, fencing, and any buildings should be periodically verified.

9.6.2.7 Environmental Monitoring. Many landfills require a monitoring of potential landfill gas and leachate impacts on the environment. In addition to monitoring the wells and sumps within the landfill for the sampling and collection of these fluids, other types of environmental monitoring may include surface emissions tests for landfill gas, perimeter gas probes for detection of lateral gas migration in soils, and sampling groundwater monitoring wells to detect the status of contaminant transport coming from the landfill into groundwater.

9.6.2.8 Permits, Deed Restrictions, and End Use. Often, landfills will continue to be operated under a jurisdictional permit even during the postclosure period, and some form of restriction or notation may be required on the land deed alerting potential future owners of the landfill boundary and any associated land-use restrictions. Many successful, and some not-so-successful, end uses have been attempted and performed on closed landfills including open space, public trails and parks, sports facilities, parking areas, solar farms, and commercial and industrial building sites. End-use proposals for landfills should consider all geoenvironmental consequences of having the landfill as a foundation.

REFERENCES

Abdolahzadeh, A. M., B. L. Vachon, and A. Cabral. 2008. "Hydraulic barrier and its impact on the performance of cover with double capillary barrier effect." In *Proc., GeoEdmonton'08: 61st Canadian Geotechnical Conf.* B.C. Paper 154. Richmond: BiTech Publishers.

Abichou, T., K. Mahieu, L. Yuan, J. Chanton, et al. 2009. "Effects of compost biocovers on gas flow and methane oxidation in a landfill cover." *Waste Manage.* 29 (5): 1595–1601.

Albright, W. H., C. H. Benson, and W. J. Waugh. 2010. *Water balance covers for waste containment, principles and practice*. Reston, VA: ASCE.

Bareither, C. A., J. C. Foley, and C. H. Benson. 2016. "Using surrogate meteorological data to predict the hydrology of a water balance cover." *J. Geotech. Geoenviron. Eng.* 142 (4): 04015092.

Benson, C. H., and S. R. Meer. 2009. "Relative abundance of monovalent and divalent cations and the impact of desiccation on geosynthetic clay liners." *J. Geotech. Geoenviron. Eng.* 133 (7): 814–827.

Benson, C. H., and C. Bareither. 2012. "Designing water balance covers for sustainable waste containment: transitioning state of the art to state of the practice." *Geotechnical engineering state of the art and practice: Keynote lectures from ASCE Geocongress 2012*, 1–33.

Bouazza, A., T. Vangpaisal, and S. Jefferis. 2006. "Effect of wet–dry cycles and cation exchange on gas permeability of geosynthetic clay liners," *J. Geotech. Geoenviron. Eng.* 132 (8): 1011–1018.

Boyd, R. 2012. "Light on landfills: Solar energy covers turn maxed-out landfills into solar farms." Accessed February 29, 2024. https://blogs.scientificamerican.com/plugged-in/light-on-landfills-solar-energy-covers-turn-maxed-out-landfills-into-solar-farms/.

California Water Boards. 2014. "Rodent population impacts at Berkeley Landfill, Alameda County." Accessed February 29, 2024. https://www.waterboards.ca.gov/sanfranciscobay/press_room/documents/BerkeleyLandfill_Burrows_FAQ_031214.pdf.

Daniel, D. E. 1995. "Soil barrier layers versus geosynthetic barriers in landfill cover systems." In *Landfill Closures, Environmental Protection and Land Recovery*, Geotechnical Special Publication 53, edited by R. J. Dunn and U. P. Singh, 1–18. New York: ASCE.

DOE (US Department of Energy). 2004. *Coupled environmental processes and long-term performance of landfill covers in the northern Mojave Desert*. Publication No. 45203. Las Vegas, NV: DOE, Nevada Site Office National Nuclear Security Administration.

Dwyer, S. F. 1995. "Alternative landfill cover demonstration." In *Landfill Closures, Environmental Protection and Land Recovery*, Geotechnical Special Publication 53, 19–34. New York: ASCE.

Eun, J., J. Tinjum, C. Benson, and T. Edil. 2018. "Equivalent transport parameters for volatile organic compounds in coextruded geomembrane-containing ethylene-vinyl alcohol." *J. Geotech. Geoenviron. Eng.* 144 (7): 04018040.

Fayer, M. 2000. UNSAT-H version 3.0: Unsaturated soil water and heat flow model—Theory, user manual, and examples. *Rep. No. PNNL-13249*. Richland, WA: Pacific Northwest National Laboratory.

Foye, K. C., and T.-Y. Soong. 2017. "Geomembrane cracking and final cover performance." In *Geosynthetics, forging a path to bona fide engineering materials*, Geotechnical Special Publication 275, 1–13. Reston, VA: ASCE.

Foye, K. C., G. W. Toepfer, and T.-Y. Soong. 2016. "Evaluation of geomembrane destructive seam testing frequencies and risk management." In *Geosynthetics, Geotechnical Frontiers*, Geotechnical Special Publication 276, 97–106. Reston, VA: ASCE.

Gamez, J. A., and T. D. Stark. 2014. "Fully softened shear strength at low stresses for levee and embankment design." *J. Geotech. Geoenviron. Eng.* 140 (9): 06014010.

Giroud, J. P., and N. Morel. 1992. "Analysis of geomembrane wrinkles." *Geotext. Geomembr.* 11 (3): 255–276.

Giroud, J. P., T. Pelte, and R. J. Bathurst. 1995. "Uplift of geomembranes by wind." *Geosynth. Int.* 2 (6): 897–952. Errata: 1997, 4 (2): 187–207, and 1999, 6 (6): 521–522.

Giroud, J. P., R. B. Wallace, and C. J. Castro. 2006. "Improved methodology for geomembrane wind uplift design." In *Proc., 8th Int. Conf. on Geosynthetics*, 225–230. Yokohama, Japan: International Geosynthetics Society. www.geosyntheticssociety.org.

Gleason, M. H., P. E. Houlihan, and J. R. Palutis. 2001. "Exposed geomembrane cover systems: Technology summary." In *Proc., Geosynthetics 2001*, 905–918. Roseville, MN: IFAI.

Hanson, J. L., D. C. Manheim, and N. Yeşiller. 2023 "Geoenvironmental assessment of climate impacts from landfill gas emissions," *Soils and Foundations* 63 (2): 101279, 1–24.

Hanson, J. L., D. C. Manheim, G. E. Swarbrick, and N. Yesiller. 2022. "Comparison of gas emissions from conventional and alternative landfill final covers: Modeling estimates." In *Proc., Geo-Congress 2022: Geoenvironmental Engineering Unsaturated Soils and Contemporary Topics in Erosion, Sustainability, and Coal Combustion Residuals*, Geotechnical Special Publication 335, 209–218. Reston, VA: ASCE.

Hanson, J. L., and N. Yeşiller. 2019. "Assessment of condition of an uncovered geosynthetic landfill bottom liner system." In *Proc., Geosynthetics 2019*, 1–9. Roseville, MN: IFAI.

Hanson, J. L., N. Yesiller, and D. C. Manheim. 2020. *Estimation and comparison of methane, nitrous oxide, and trace volatile organic compound emissions and gas collection system efficiencies in California landfills*. Final Report Submitted to CalRecycle and California Air Resources Board. Sacramento, CA: CARB. ww2.arb.ca.gov.

Henken-Mellies, W. U. 2010. "GCL in a landfill cover: 10-year record of a lysimeter field test." In *Proc., 3rd Int. Symp. on GCLs*. Neckar, Germany: SKZ.

Huber-Humer, M. 2005. "Extended effects of methane oxidation and their relevance for biocover application in the field." In *Proc., Sardinia 2005, 10th Int. Waste Management and Landfill Symp*. Padlova, Italy: CISA.

Hutson, J. L., and R. J. Wagenet. 1992. *LEACHM: Leaching estimation and chemistry model: A process-based model of water and solute movement transformations, plant uptake, and chemical reactions in the unsaturated zone*. Version 3.0. Ithaca. NY: Cornell University, Dept. of Soil, Crop, and Atmospheric Sciences.

Jones, D. R. V., N. Dixon, N., and A. Connell. 2000. "Effect of landfill construction activities on mobilized interface shear strength." In *Proc., EuroGeo 2000*, 581–586. Bologna, Italy: International Geosynthetics Society Italian chapter. www.geosyntheticssociety.org.

Kampf, M., T. Holfelder, and H. Montenegro. 1999. "Designing capillary barriers." In *Proc., Sardinia 99, 7th Int. Waste Management and Landfill Symp.*, 381–388. Padova, Italy: CISA.

Kavazanjian, E., Jr. 1998. "Current issues in seismic design of geosynthetic cover systems." In *Proc., 6th Int. Conf. on Geosynthetics*, 219–226. Roseville, MN: IFAI.

Kerkes, D. J. 1999. "Analysis of equipment loads on geocomposite liner systems." In *Proc., Geosynthetics '99*, 1043–1054. Roseville, MN: IFAI.

Koerner, R. M. 2012. Vols. 1 and 2 of *Designing with geosynthetics*. 6th ed. Bloomington, IN: Xlibris.

Koerner, R. M., and G. R. Koerner. 2013. *The intimate contact issue of field placed geomembranes with respect to wave (or wrinkle) management*. GSI White Paper #27. Folsom, PA: Geosynthetic Institute.

Krahn, J. 2004. *Vadose zone modeling with VADOSE/W*. Calgary, AB: GEO-SLOPE International.

Meer, S. R., and C. H. Benson. 2007. "Hydraulic conductivity of geosynthetic clay liners exhumed from landfill final covers." *J. Geotech. Geoenviron. Eng.* 133(5): 550–563.

Narejo, D. 2016. "Geotextiles use for cushioning." Chap. 17 in *Geotextiles–From design to applications*, edited by R. Koerner, 395–411. Cambridge, UK: Woodhead Publishing.

NRC (National Research Council). 2011. "Engineered covers for waste containment: Changes in engineering properties and implications for long-term performance assessment." NUREG/CR-7028. Washington, DC: National Academies of Sciences, Engineering, and Medicine.

NRC. 2022. *Evaluation of in-service radon barriers over uranium mill tailings disposal facilities*. NUREG/CR-7288. Washington, DC: National Academies of Sciences, Engineering, and Medicine.

Ogorzalek, A., G. Bohnhoff, C. Shackelford, C. Benson, et al. 2008. "Comparison of field data and water balance predictions for a capillary barrier cover." *J. Geotech. Geoenviron. Eng.* 134 (4): 470–486.

Paruvakat, N., and G. Richardson. 1999. "Landfill cover failure prompts standards upgrade." *GFR Mag.* 17 (7): 1–7.

Risken, J. L., J. L. Hanson, and N. Yesiller. 2015. "Estimation of engineering behavior of geosynthetic clay liners using moisture–suction relationships." In Vol. 2 *Proc., 15th Asian Regional Conf. on Soil Mechanics and Geotechnical Engineering*, 2373–2378. Fukuoka, Kyushu Japan: Japanese Geotechnical Society. http://www.jgskyushu.net/uploads/15ARC/.

Ross, B. 1990. "The diversion capacity of capillary barriers." *Water Resour. Res.* 26 (10): 2625–2629.

Rouf, M. A., A. Bouazza, R. M. Singh, W. P. Gates, et al. 2016. "Gas flow unified measurement system for sequential measurement of gas diffusion and gas permeability of partially hydrated geosynthetic clay liners." *Can. Geotech. J* 53 (6): 1000–1012.

Rowe, R. K., L. E. Ashe, A. W. Take, and R. W. I. Brachman. 2014. "Factors affecting the down-slope erosion of bentonite in a GCL." *Geotext. Geomembr.* 42 (5): 445–456.

Rowe, R. K., R. W. I. Brachman, M. S. Hosney, W. A. Take, et al. 2016. "Insight into hydraulic conductivity testing of geosynthetic clay liners (GCLs) exhumed after 5 and 7 years in a cover." *Can. Geotech. J.* 54 (8): 1118–1138.

Rowe, R. K., R. W. I. Brachman, and W. A. Take. 2018. "Field measurements of overlap reductions for two reinforced fabric-encased geosynthetic clay liners (GCLs)." *Can. Geotech. J.* 55 (5): 631–639.

Rower, I. U., J. Gebert, J. Streese-Kleeberg, A. Grongroft, et al. 2011. "Design, implementation, and operation of soil-based methane oxidation windows for the remediation of gas emission hotspots in landfill cover soils." In *Proc., Sardinia 2011, 13th Int. Waste Management and Landfill Symp.* Padova, Italy: CISA.

Scalia, J., and C. H. Benson. 2011. "Hydraulic conductivity of geosynthetic clay liners exhumed from landfill final covers with composite barriers." *J. Geotech. Geoenviron. Eng.* 137 (1): 40–49.

Scalia, J., C. H. Benson, W. H., Albright, B. S. Smith, et al. 2017. "Properties of barrier components in a composite cover after 14 years of service and differential settlement." *J. Geotech. Geoenviron. Eng.* 143 (9): 0001744.

Scheirs, J. 2009. *A guide to polymeric geomembranes*. London: Wiley.

Schroeder, P. R., N. M. Aziz, C. M. Lloyd, and P. A. Zappi. 1994. *The hydrologic evaluation of landfill performance (HELP) model: User's guide for version 3*. EPA/600/R-94/168a. Washington, DC: US Environmental Protection Agency, Office of Research and Development.

Šimůnek, J., M. Sejna, and M. T. van Genuchten. 1999. *HYDRUS: Simulating water flow, heat, and solute transport in two-dimensional variably saturated media*. IGWMC–TPS53. Golden, CO: International Groundwater Modeling Center, Colorado School of Mines.

Spokas, K. A., and J. E. Bogner. 2011. "Limits and dynamics of methane oxidation in landfill cover soils." *Waste Manage.* 31 (5): 823–832.

Thiel, R. 1998. "Design methodology for a gas pressure relief layer below a geomembrane landfill cover to improve slope stability." *Geosynth. Int.* 5 (6): 589–617.

Thiel, R., and M. Stewart. 1993. "Geosynthetic landfill cover design methodology and construction experience in the pacific northwest." In *Proc., Geosynthetics '93 Conf.* Vancouver: IFIA.

Thiel, R., and J. P. Giroud. 2023. "Localized shear strength mobilization at geosynthetic interfaces caused by spreading soil upslope." In *Proc., GeoAfrica23, 4th African Regional Conf. on Geosynthetics,* Cairo, Egypt. 368: 02002. E3S Web of Conferences. https://doi.org/10.1051/e3sconf/202336802002.

USDA (US Department of Agriculture). 2017. *Soil survey manual*. Agricultural Manual No. 18. Washington, DC: USDA.

USEPA (US Environmental Protection Agency). 2004. *EPA (draft) technical guidance for RCRA/CERCLA final covers*. EPA 540-R-04-007, OSWER 9283.1-26. Washington, DC: USEPA, Office of Solid Waste and Emergency Response.

USEPA. 2011. *Available and emerging technologies for reducing greenhouse gas emissions from municipal solid waste landfills*. Washington, DC: USEPA, Office of Air Quality Planning and Standards.

Yesiller, N., J. L. Hanson, B. Bussiere, T. Pabst, et al. 2018. "Use of geomembranes in reclamation covers for reactive mining waste disposal sites." In *Proc., GeoEdmonton 2018, 71st Canadian Geotechnical Conf.*, 1–10. Vancouver: Canadian Geotechnical Society.

Yeşiller, N., J. L. Hanson, N. K. Oettle, and W.-L. Liu. 2008. "Thermal analysis of cover systems in municipal solid waste landfills," *J. Geotech. Geoenviron. Eng*. 134 (11): 1655–1664.

Yesiller, N., J. L. Hanson, J. L., Risken, T. Abichou, et al. 2019. "Hydration fluid and field exposure effects on moisture-suction response of geosynthetic clay liners." *J. Geotech. Geoenviron. Eng*. 145 (4): 04019010.

CHAPTER 10

STATIC SLOPE STABILITY OF CONTAINMENT SYSTEMS

Kevin Foye, Ph.D., P.E., M.ASCE

10.1 INTRODUCTION

Static slope stability refers to a category of geotechnical engineering analyses whose purpose is to assess the relative likelihood of unacceptable soil and rock movements under the action of gravity loads (especially self-weight) and seepage. The defining characteristic of these analyses is the geometry of the soil and rock, which typically include a ground surface inclined to the horizontal. The term *static* is typically used to distinguish these calculations from "seismic" slope stability analyses, which consider effects from earthquakes and is described in Chapter 11. Static slope stability of containment systems is a special category of these analyses attributed to the additional material properties, potential slip surfaces, and pore pressure considerations introduced by waste containment liner systems, waste materials, cover systems, and liquid management.

The goal of this chapter is to present guidelines and principles to guide geotechnical engineers in the planning, execution, and critique of slope stability calculations with an emphasis on slopes and modes of failure encountered in containment systems. For an exhaustive treatment of analytical methods and training in the use of available computer software programs used to conduct slope stability analyses, the reader should seek out textbooks, manuals, and other technical literature related to the specific methods selected for analysis, for example, Duncan et al. (2014). Suggested additional reading is included in the references section. References such as GeoRG (2004) and Qian et al. (2002) offer broad guidance for slope stability analyses of landfills, which is applicable to all containment systems. The other references offer more detailed guidance on specific topics introduced in this chapter. This chapter assumes that the geotechnical engineer conducting the analysis has already completed coursework that includes some slope stability analyses and is conversant in the engineering properties of soils.

10.2 TYPICAL APPLICATIONS

Slope stability analyses are conducted to support the design of slopes and embankments, assessment of landslide hazards, and remediation of existing landslides. In general, any time an engineer offers an opinion about the acceptability of an inclined ground surface, some quantitative consideration of slope stability is warranted. For example, slope stability analyses are often conducted in the following circumstances:

- Evaluation of existing slopes to assess their hazard in connection with nearby infrastructure,
- Evaluation and design of alterations to existing slopes to determine the acceptability of the designed changes,
- Evaluation of new embankment designs, and
- Forensic evaluation and stabilization design of past landslides or landslides in progress.

In the context of containment systems, additional reasons to conduct a slope stability analysis include

- Evaluation of liner system or cover system components for inclusion in a containment system design;
- Proposed changes to the operational filling plans (slopes, maximum heights, rate of placement) for materials placed in a containment system;
- Consideration of waste materials with varying properties for placement within a containment system;
- Observed changes to the pore pressure conditions within a containment system—for example, piezometer measurements of subgrade, leachate collection, or waste mass pore pressures; and
- Engineering demonstrations to regulatory authorities.

Several different slope stability analysis methods are available to quantify the relative likelihood of unacceptable movement. These methods vary in their underlying assumptions and the complexity of the calculations involved. Methods can vary in the difficulty involved in obtaining the required engineering parameters to conduct the analyses. Depending on the site-specific conditions, different degrees of precision and complexity may be required to conduct the analyses with satisfactory confidence in the results. For example, in many design cases, a simple analysis may determine that the chances of unacceptable movement are sufficiently remote to preclude additional effort. In contrast, for some forensic evaluations where unacceptable movement has already occurred, significant effort may be needed to conduct analyses with sufficient confidence that the causes of movement have been fully understood.

10.3 LIMITATIONS

The analyses described in this chapter are focused on the evaluation of slopes encountered in containment system engineering practice, particularly embankments founded on soil and incorporating geosynthetics. Because of the complexity of these systems and the wide range of waste material properties encountered in practice, this chapter cannot address all possible scenarios that could lead to a landslide in these facilities. Accordingly, this chapter presents a set of principles for the engineer to consider when conducting slope stability analyses. Depending on the complexity of the system, the safety margin against failure, and/or the consequences of failure, additional expertise and information should be sought. A number of references are included at the end of this chapter to aid the engineer.

Like other geotechnical engineering problems, slope stability analyses necessarily involve significant simplifications of the real-world system when modeling the ground and its properties. Although complex surface geometries are relatively easy to obtain and program, subsurface geometries, soil/rock properties, and complex groundwater flow regimes are not. The engineer must pay attention to the sensitivity of the analysis to various aspects of the problem, particularly those that contain significant uncertainty.

10.4 ROLES

Slope stability analyses are typically conducted by geotechnical engineers who are knowledgeable in soil mechanics. A knowledge of the geology of landslides and groundwater flow is also highly beneficial, depending on the application. For waste containment systems, a knowledge of the physical and chemical processes affecting the generation, disposition, and degradation of wastes is important. Other important roles may include

- Geologists, engineers, and technicians who provide information about existing ground features, subsurface conditions, and soil properties;
- Various technical professionals skilled at installing and monitoring instruments that measure ground movements and groundwater conditions;

- Surveyors, design engineers, and other design professionals who provide geometric data about the existing ground surface and proposed ground surface for design projects;
- Supervising or independent geotechnical engineers who provide technical reviews of the inputs, analyses, and conclusions;
- Property owners, neighbors, and other observers with historical and eye-witness information about existing slopes and landslides;
- Landfill operators who will be responsible for implementing the recommendations of the engineer and who also have direct observation of site conditions;
- Government regulators, who are responsible for administering the rules associated with waste containment facilities and also have experience with engineering analyses and observations and multiple facilities;
- Other parties with a knowledge of the materials placed in the landfill, such as industrial waste generators, especially when engineering evaluations of materials placed in the waste containment facility were a condition for disposal;
- Geosynthetic manufacturers, who typically test their products under a range of potential operating conditions for engineering properties such as strength; and
- Testing agencies that are responsible for directly evaluating the engineering properties of waste, construction materials, and in situ soils through in situ and laboratory testing.

10.5 WORKFLOW

Slope stability analyses require a proper selection of inputs and critical evaluation of the results obtained. The formulation and solution of slope stability problems is well established and conducted using computer software. However, for simplified geometries, hand calculations can also be used. The following steps are recommended for most slope stability analyses:

1. Determine project analysis needs/objectives,
2. Determine available data,
3. Collect missing data,
4. Construct an engineering model,
5. Perform calculations,
6. Postprocess and critique calculations and results, and
7. Make engineering recommendations.

Note that the steps may be iterative should the initial findings be unacceptable. In the engineering of waste containment systems, slope stability analyses typically proceed from simple analyses to more sophisticated analyses to interrogate which failure modes are most critical to the design of the system. Unlike many other civil engineering applications of slope stability, waste containment systems are constantly changing during their operational life and therefore often require a reevaluation of slope stability as conditions develop that were not previously analyzed. Examples of conditions that can change in the life of a waste containment system include

- Materials accepted for placement in the facility;
- Construction of new or revised portions of the containment system;
- Materials used for the construction of the facility liner system or cover system;
- Measured or suspected changes in pore pressures resulting from liquid or gas accumulation; and
- Changes to the plans for fill placement such as sequence, slope, or height.

In each of these cases, the engineer should be consulted to reevaluate slope stability and advise personnel constructing, operating, and monitoring the waste containment system to ensure that target margins of safety are still met.

10.6 METHODS OF ANALYSIS

Most slope stability analyses used in practice model a cross section of the ground oriented along the steepest slope of the ground surface under consideration. However, more than one cross section may be needed to model the problem (e.g., slope where external loads such as stockpiles, structures, and so on are present). These calculations are considered two-dimensional because the cross-section model changes in the vertical and horizontal directions (i.e., plane strain). Most analyses assume a unit width out of plane for computational convenience.

10.6.1 Limit Equilibrium

Most slope stability analyses used in practice are limit equilibrium analyses that statically analyze the driving forces (or moments) and the available resisting forces (or moments) acting on a slope at the initiation of movement to compute a factor of safety (FS):

$$FS = \frac{\sum \text{available resisting forces}}{\sum \text{driving forces}} \quad \text{or} \quad FS = \frac{\sum \text{available resisting moments}}{\sum \text{driving moments}} \tag{10-1}$$

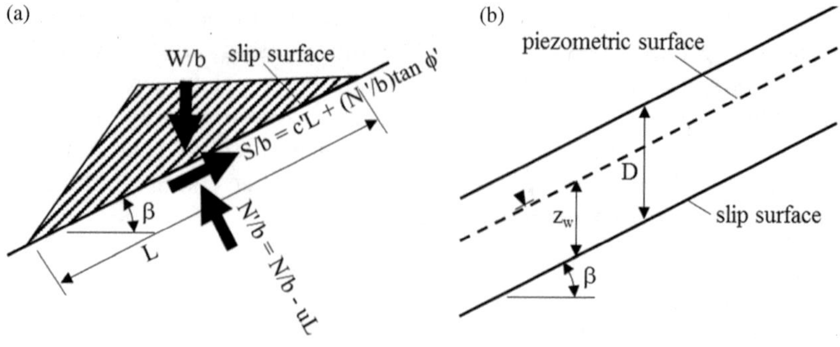

Figure 10-1. Simple planar slope stability analysis: (a) finite slope length, (b) infinite slope length.

Types of limit equilibrium analyses differ in the geometry of the assumed failure plane (slip surface), the degree to which the problem geometry is discretized in the computations and the assumptions required to obtain a numerical result. The resisting forces include the soil's shear strength, normally quantified using Mohr–Coulomb. The simplest type of limit equilibrium analysis models the potentially sliding soil mass as a single volume of soil with a planar slip surface. The slip surface is the interface of the potentially moving soil mass with the relatively static ground. In this type of analysis, the modeled soil mass can be of either finite or infinite extent. Figure 10-1 illustrates the basic geometry, forces, and symbols for this analysis.

For limit equilibrium analysis of the sliding masses shown in Figure 10-1, Equation (10-1) is rewritten to compute the FS with respect to force equilibrium as

$$\mathrm{FS} = \frac{c'L + \left(\left(\frac{W}{b}\right)\cos\beta - uL\right)\tan\phi'}{\left(\frac{W}{b}\right)\sin\beta} \tag{10-2}$$

where

W/b = Weight W of the sliding mass per cross-section unit width b,
S/b = Available shear resistance to sliding per unit width b at the base of the sliding mass,
N/b = Normal reaction force per unit width b at the base of the sliding mass,
β = Slip surface inclination angle to horizontal plane (also ground surface inclination in the case of infinite slope stability),

c' = Effective stress Mohr–Coulomb strength intercept (also called *cohesion*) representing the shear strength of the material undergoing shear at the base of the sliding mass,
ϕ' = Effective stress Mohr–Coulomb strength slope angle (also called *friction angle*) representing the shear strength of the material undergoing shear at the base of the sliding mass,
u = Pore pressure acting at the base of the sliding mass $= \gamma_w z_w$,
γ_w = Unit weight of water,
z_w = Height of the piezometric surface above the base of the sliding mass, and
L = Length of the base of the sliding mass in contact with the static ground.

For infinite slopes, Equation (10-2) can be more conveniently written

$$\mathrm{FS} = \frac{(\gamma D - \gamma_w z_w)\tan\phi'}{\gamma D \tan\beta} \tag{10-3}$$

where γ is the unit weight of soil in the sliding mass, and D is the thickness of the sliding mass above the slip surface.

Infinite slope stability analyses are an important tool to evaluate the stability of potential slides along inclined layers of discontinuous materials, especially when the upper layers can accumulate pore pressure. Waste containment covers and exposed bottom liner systems are ideal applications for infinite slope stability analyses. Infinite slope stability analyses are also a useful tool for engineers to check the results of more complex analyses.

Engineers commonly use the method of slices to analyze slip surface geometries with both regular and irregular shapes. The method of slices involves the division of the modeled ground into a number of discrete slices that are then collectively analyzed for moment and/or force equilibrium. Because these slices are not statically determinant, simplifying assumptions about the orientation and magnitude of interslice forces are made to enable a solution. Accordingly, the method of slices includes a number of distinct computational methods depending on the simplifying assumptions made. In this chapter, only equations for ordinary method of slices and modified Bishop's method are provided because these methods can be readily implemented by hand to check computer solutions. Other methods, such Morgenstern–Price (MP), Spencer, and Janbu, are widely available in common slope stability software packages. See Duncan et al. (2014) or Duncan (1996) for a discussion of slope stability analysis methods. As with any engineering software, the engineer should review the slope stability software's documentation and try the demo, prior to purchase.

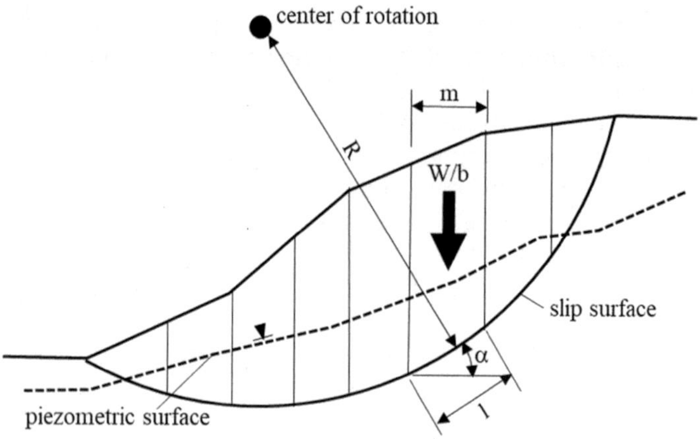

Figure 10-2. Method of slices slope stability analysis.

The engineer should also validate simple analyses with hand calculations to verify that the software is working correctly.

Figure 10-2 illustrates the basic geometry and symbols for analysis using the method of slices with an assumed circular slip surface. Note that the slip surface need not be circular. At the beginning of a slope stability analysis using method of slices, the geometry of the slip surface is assumed. The resulting FS computed for this trial slip surface may not be the minimum value possible for a given cross section. Accordingly, different trial slip surfaces with different geometries must be evaluated to estimate a minimum FS value. The slip surface corresponding to the minimum value of FS is called the *critical slip surface* for the cross section evaluated. Because of the interaction of slope geometry, soil profile, and pore pressure, the critical slip surface may have an irregular shape. This possibility should be considered by the engineer when selecting trial slip surfaces for analysis.

Ordinary method of slices (e.g., Coduto 1999) ignores interslice forces and solves the system for moment equilibrium only, computing the FS using the following summations of resisting and driving moments for all slices:

$$\mathrm{FS} = \frac{\sum R\left[c'l + \left(\left(\frac{W}{b}\right)\cos\alpha - ul\right)\tan\phi'\right]}{\sum R\left[\left(\frac{W}{b}\right)\sin\alpha\right]} \tag{10-4}$$

where

W/b = Weight w of a slice per cross-section unit width b,

R = Radius from the assumed center of rotation to the bottom of a slice (for circular arc analyses, R is constant for all slices),
l = Length of the base of a slice, and
α = Inclination angle of the base of a slice to the horizontal.

A number of references will not include the R term in Equation (10-4) because this variable cancels out of the FS equation for analyses of circular slip surfaces with constant radius. However, it is necessary to retain this formulation when considering noncircular slip surfaces, which are treated by more generalized computational methods, and when considering the effects of slope reinforcement, which is discussed subsequently. The ability to model noncircular slip surfaces is especially important for slope stability analyses that include material interfaces such as liner systems featured in many waste containment systems.

The modified Bishop's method, a commonly used method, assumes equal shear forces and colinear normal forces at the sides of each slice. Using the modified Bishop's method, the FS is computed iteratively using the following summations of resisting and driving moments for each slice:

$$FS = \frac{\sum R\left[\dfrac{mc' + \left(\left(\dfrac{W}{b}\right) - um\right)\tan\phi'}{\psi}\right]}{\sum R\left[\left(\dfrac{W}{b}\right)\sin\alpha\right]} \tag{10-5}$$

where m is the width of a slice, and

$$\psi = \cos\alpha + \frac{\sin|\alpha|\tan\phi}{FS} \tag{10-6}$$

Simple limit equilibrium calculations, such as infinite slope and two-slice analyses, are often carried out by hand or by spreadsheet. These simple calculations are valuable tools for engineers to estimate the FS prior to attempting more complex calculations. For more complex calculations, there are many computer software implementations of limit equilibrium methods (e.g., SLIDE, SLOPE/W, SVSlope). Computer software implementations allow the rapid iteration of multiple slip surface geometries to attempt to identify the critical slip surface for a given cross section and search parameters. These software packages also allow a computation of the FS for complex geometries, including noncircular slip surfaces, which are of interest for many problems, especially slopes with weak or soft zones, or other preferential slip surfaces such as liner systems in modern landfills.

10.6.2 Limit Analysis

Limit analysis differs from limit equilibrium analysis in that this form of analysis explicitly models stress or deformations in the soil mass to bound the modeled conditions at failure according to the lower and upper bound theorems of plasticity theory. For a discussion of limit analysis and how it compares to limit equilibrium methods, see Yu et al. (1998). Although these bounds can be calculated by hand for simple geometries, limit analysis is not typically applied in practice because of the relative novelty of plasticity theory to geotechnical engineers. Accordingly, most practitioners will encounter limit analysis solutions in the form of either published charts or computer software.

10.6.3 Finite-Element Analysis

Finite-element analysis (FEA) numerically estimates a solution to a system of differential equations describing a discretized field of deformations over the modeled slope cross section. Finite-element analysis can estimate ground movement and predict concentrated shear zones (representing slip surfaces). Because of the relative complexity of FEA, it is performed by computer. An FEA of slopes is possible using different commercially available FEA programs. However, a detailed discussion of the method depends on the specific software that is used.

In addition to quantifying possible ground movement directly, methods have been devised to calculate slope stability factors of safety using FEA. For a discussion of methods to apply FEA to slope stability analysis, refer to Griffiths and Lane (1999) and Fredlund and Scoular (1999). These methods include the following:

- Using FEA to estimate the most likely slip surface (zone of concentrated shear deformation) and then calculating driving and resisting forces along the slip surface.
- Applying a strength reduction factor to the soil strength parameters (c' and ϕ') until a threshold value of deformation is obtained.
- Applying a portion of the gravity loads in the model until a threshold value of deformation is obtained.

For waste containment systems, FEA can be useful to evaluate the effects of evolving conditions within the system on slope stability. Example scenarios where FEA offers unique benefits include consideration of coupled development of pore pressures, volumetric changes, and shear strains within a waste mass and displacement of materials with significantly different stiffnesses—for example, municipal solid waste and soft clay require significantly greater displacement to reach peak shear strength than many geosynthetic interfaces and stiffer soils.

10.6.4 Other Computational Methods

Discrete element methods (DEMs) simulate movement and forces within soil and rock as a massive collection of individual particles that apply forces to each other at their boundaries according to the models of contact forces. These methods require advanced computer software to perform. DEM can be useful to understand the mechanics of rock falls, debris flows, and other landslide phenomena that do not fit well within the assumptions of limit equilibrium methods (e.g., Zhang and Rock 2011). The potential for rock falls is also evaluated in practice using various correlations developed from geological survey methods (e.g., Michoud et al. 2012). Practitioners considering the stability of rock masses are encouraged to consult these methods.

10.6.5 Three-Dimensional Analyses

Limit equilibrium analysis software that models the soil mass as a collection of three-dimensional (3D) soil columns instead of two-dimensional (2D) soil slices is available (e.g., CLARA, SVSlope, Slide3). As for other software, the engineer should carefully evaluate the documentation and performance of the software before and during use.

This software allows a more exhaustive consideration of combinations of slope geometry, soil stratigraphy, and pore pressure distributions that exist in the considered problem than is possible with 2D analyses. However, because engineers have experience using 2D limit equilibrium methods for many years, 3D methods are less used in practice. Applications in which 3D methods may offer greater confidence include

- Back-analysis of landslides with irregular slip surfaces that vary outside the critical cross section;
- Analysis of slopes with zones of weak soil or elevated pore pressures that do not coincide with the anticipated critical slip surface cross section;
- Analysis of narrow excavations, canyons, or other slope geometries where the slope geometry differs significantly outside the critical cross section; and
- Analysis of containment systems where waste materials with varying strength, density, and pore pressure exist in various volumes within the overall system geometry (e.g., pockets of soft sludge embedded within a larger waste mass).

In these cases, a 2D analysis of any particular cross section would not be representative of the material properties present throughout. Moreover, the relative effect of differences in materials or geometry outside of a particular cross section is difficult to evaluate using 2D analyses alone.

The engineer may estimate different factors of safety using 3D methods than using 2D methods. In these cases, the engineer should critique the results to understand what out-of-section properties are affecting the results to determine their applicability to the problem analyzed.

10.6.6 Probabilistic Analyses

The aforementioned methods of analysis are called *deterministic* because the soil properties used to compute the FS values or ground movement are assumed by the analyses to have known values. Although engineers recognize that there is significant uncertainty in the "exact" values of various soil properties in the ground, it is commonplace to allow for such uncertainty by setting a target static FS for acceptable slopes that is greater than 1, typically in the range of 1.3 to 1.5. The reader is cautioned that the selection of the FS depends on many considerations, including the uncertainty of inputs and the consequences of failure. Duncan (1992) and Duncan (2000) discuss these considerations.

Alternatively, methods are available to model the effects of soil parameter uncertainty directly (Baecher and Christian 2003). These methods are called *probabilistic* because they often result in the calculation of a probability of failure. Low et al. (1998) proposed a simple spreadsheet method for probabilistic limit equilibrium analysis using the method of slices. This method uses the built-in Solver tool in Excel to optimize the critical slip surface according to the probability of failure rather than the FS. Commercial 2D limit equilibrium method software also has this capability.

The probabilistic approach discussed addresses the uncertainty in a set of soil properties as it is applied uniformly over the entire model cross section. An alternate approach is to model the apparently random fluctuations in soil properties that vary spatially over the cross section. This type of probabilistic method is known as "random field modeling." Software to perform this type of analysis in conjunction with either FEA or limit equilibrium methods is also available, for example, RFEM, presented by Fenton and Griffiths (2008), but should be used by engineers who are well-versed in its limitations. This type of analysis is occasionally useful to engineers when there is a need to directly account for material property heterogeneity in slope stability analyses. Many waste materials, especially municipal solid waste, exhibit significant heterogeneity because of variability in material composition, placement, and history. For many slope stability analyses, the scale of the potential movement is large enough to minimize the practical effect of this heterogeneity on analysis results—especially when considering bulk waste properties obtained through back-analysis. However, when analyses show that localized differences in material properties could result in unacceptable performance, more sophisticated analyses should be considered (e.g., Kumar and Reddy 2022).

As an advanced approach, probabilistic methods are not typically applied in practice, except for high-consequence projects such as earth dams, where estimates of probabilities of failure are needed as part of larger system reliability studies. For a simplified probabilistic evaluation method, refer to Duncan (2000). For additional background on probabilistic methods, refer to USACE (1997, 2006).

10.7 MODES OF FAILURE IN CONTAINMENT SYSTEMS

All natural slopes, cut slopes, and constructed embankments have the potential to develop movement and slip surfaces through all or part of the slope and in various geometries. However, containment systems offer even more potential slip surface configurations because of the construction of multiple layers of materials with differing properties. Figure 10-3 shows an example waste containment bottom liner system under construction where multiple layers are visible.

In general, the occurrence of interfaces within the subsurface geometry that can slide over one another deserves special consideration. Examples of where this phenomenon occurs regularly in practice include joints in rock, the interface between soil and rock, in discrete layers of weak soil, between geosynthetic layers or geosynthetic and soil layers (especially in

Figure 10-3. Waste containment bottom liner system under construction, including sand drainage media, geotextile, geomembrane, and subgrade.

landfills), and along historic landslide slip surfaces. It is important in these cases that the interfaces in question be investigated and the relevant interface shear strengths be represented in the model. Often multiple models considering different candidate interfaces within the ground are needed to conduct an exhaustive set of analyses.

Methods available to directly evaluate stability along interfaces vary by the software package used. For some programs, the analyst must define discrete material layers and then define candidate critical slip surfaces that pass through these layers. In other programs, it is possible to limit the definition of the cross section to explicitly place the interface under consideration at the bottom of possible slip surface searches, ensuring their inclusion in the model. Regardless of the technique used, it is important that the analyst review the software output to verify that the shear strengths used to compute the FS along slices in contact with key interfaces are using the correct interface strength parameters. Engineers can also conduct simplified slope stability analyses directly using infinite slope, sliding wedge, and two-slice methods to investigate the effects of an interface on stability.

Landfills for the containment of solid wastes are a special class of embankment because of the number of interfaces created by their construction. Figure 10-4 presents a conceptual cross section that illustrates the different candidate slip surfaces that typically need to be evaluated in the engineering design of landfills. Final cover and liner systems during construction (Figure 10-5) are best analyzed using infinite slope stability calculations because they are long slopes with relatively thin soil layers. Possible slip along the liner system after waste has been placed is typically analyzed using the method of slices where the bottom slip surface is constrained to the liner system. Critical slip surfaces that pass through the foundation soils are also possible and must be

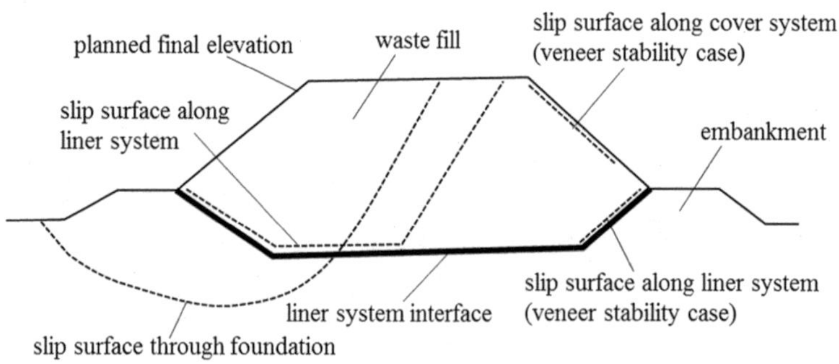

Figure 10-4. Conceptual cross section showing potential slip surfaces within a landfill.

Figure 10-5. Side slope of a waste containment bottom liner system under construction. Shallow "veneer" slides can develop along the various interfaces created between the soils and the geosynthetics.

considered. Qian et al. (2002) offer detailed guidance for the evaluation of landfill slope stability.

Many causes of landslides in landfills have been deduced from case histories. Bonaparte et al. (2020) and Koerner and Soong (2000) offer insights into key causes. These publications should also be consulted for a wide range of lessons learned from landslide case histories affecting slope stability analyses, as well as operation of these facilities. Insights include the following:

- Development and underappreciation of liquid and gas pressures within the waste mass. In addition, measures to control these pressures are not always effective and vary with waste properties and fill geometry. Practices that introduce additional liquids must be carefully considered.
- Introduction of wastes with properties that vary from typical and/or those assumed during design.
- Weak interfaces with geosynthetics and the weakening of geosynthetic interfaces caused by movements during and after construction.
- Mobilized strength incompatibility between waste, geosynthetic interfaces, and subgrade soils. The strength of these different materials develops at a different value of displacement and therefore analyses assuming that peak values occur simultaneously will be unconservative.

- Soft subgrade soils that dissipate load-included pore pressures and gain strength slowly over time.
- Differences between assumed and actual facility construction or fill placement sequence and geometry.

10.8 INPUTS

Determining the input parameters to a slope stability analysis is one of the most important steps in the overall process. The quantity and quality of data will determine the reliability of any results obtained. In the case of slopes with marginally acceptable stability, it may be necessary to conduct preliminary analyses, including sensitivity studies, to determine the most important data for follow-up investigation. The primary inputs to slope stability analyses are

- Geometry, including existing surface topography, slope height, inclinations, benches, and design cut and fill zones;
- Soil, rock, and/or waste unit weights;
- Soil, rock, waste, and/or interface strength parameters, including discrete zones, such as geosynthetics, soil layers, and historical slip surfaces with differing properties;
- Soil stress–strain response parameters (in the case of FEA models);
- Pore pressures, determined from either a direct observation of piezometers, modeled flow conditions, or projected extreme events;
- Exterior forces, including structures, transient surface loads, hydraulic pressures, and so on;
- Inertial forces (in the case of seismic analyses);
- Reinforcing elements (in the case of reinforced earth embankments or stabilization design); and
- Discrete interfaces, such as liner systems or cover systems, zones of fill with different properties, and/or contacts with in situ soil or rock layers.

As discussed in Section 10.4, obtaining reliable data for each of these inputs typically involves the collaboration of many different professionals drawing upon different sources of information. The different sources of information are used to construct a simplified engineering model of the physical slope system. Figure 10-6 illustrates an example plan and cross-section view of a slope for analysis. In many cases, preliminary analyses may be required to determine the most important data for follow-up investigation.

For preliminary analyses, many soil properties may need to be assumed on the basis of typical values from the engineer's experience or technical literature. Because the interaction of these inputs determines whether the slope will be stable to an acceptable degree, it is often useful in a slope

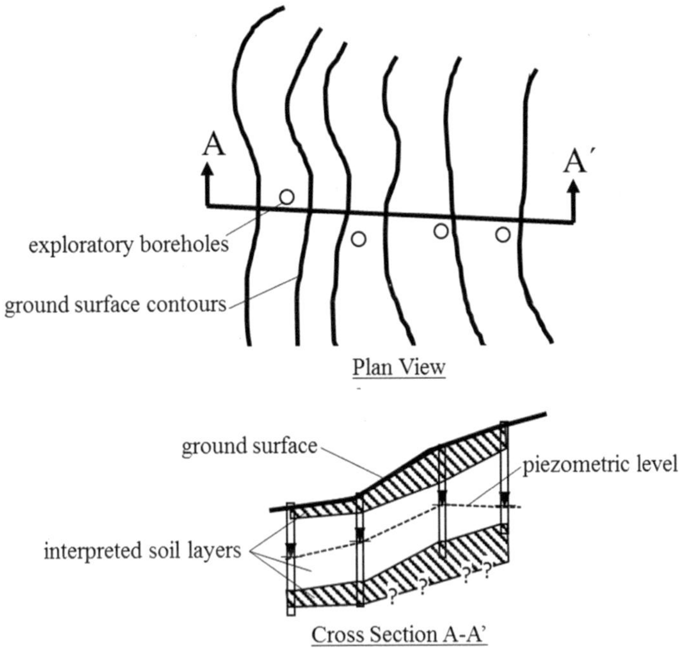

Figure 10-6. Example slope plan view and cross section.

stability analysis to vary the input parameters over the range of possible values. This approach is known as a sensitivity study. Parameters that have a significant impact on FS values should be investigated further, possibly including in situ investigations, sampling, and testing.

For waste containment systems, the engineer may specify acceptable strength parameters for the acceptance of liner system components or waste materials for disposal. In these cases, the design geometry is known, and the acceptable strength parameters are varied to produce the target FS value. If the required strength parameters are less than anticipated or tested values, a simple analysis may be sufficient for design purposes. Otherwise, more sophisticated analyses incorporating actual tested strength values may be needed and/or the design geometry may need to be revised.

10.8.1 Slope Geometry

Depending on the application, slope geometry data will be obtained from a combination of existing topographic surveys and design plans for altering the existing topography.

Typically, the most important cross section for analysis will be the cross section aligned with the steepest, longest slope of the investigated topography. Because FS values computed using limit equilibrium methods offer a relative assessment of stability, proposed changes to the slope geometry can be assessed with respect to their effect of either increasing or decreasing the estimated FS. This relative assessment is especially important for landslide remediation projects where the goal is to make improvements to increase the FS value. In some cases, the goal of the slope stability analysis will be to recommend limiting values of slope inclination for construction. In those cases, the slope geometry is a design variable to be changed, similarly to a sensitivity study.

10.8.2 Soil Shear Strength/Interface Shear Strength

For the typical limit equilibrium methods presented previously, soil strength is modeled using the Mohr–Coulomb shear strength envelope. The choice of shear strength parameters depends on the conditions being modeled:

- For long-term conditions, where sufficient time has elapsed for pore pressures to equilibrate with boundary conditions, effective stress analyses may be the most appropriate. Equations (10-2) to (10-6) are written with respect to effective stress strength parameters.
- For short-term conditions (often during construction), where there is insufficient time for changes in pore pressure to equilibrate with boundary conditions, total stress analyses using undrained strength parameters are the most appropriate.
- For staged loading of soft subgrade soils, where included pore pressures may slowly dissipate over time, total stress analyses using undrained strength parameters are also appropriate, with modification to account for changes in strength with time.
- For the case of rapid drawdown, often applied to dams and levees, pore pressures accumulated within the soil have not yet equilibrated with boundary conditions, similar to short-term conditions. Depending on the available soil stress and strength data, the engineer may model this case using either total or effective stress parameters, provided the engineer selects them accurately to model the stresses acting on the soil during drawdown.

For effective stress analyses (applicable to long-term conditions), the shear strength τ of a soil is expressed as follows:

$$\tau = c' + \sigma' \tan\phi' \tag{10-7}$$

where

c' = Effective stress Mohr–Coulomb strength intercept,

ϕ' = Effective stress Mohr–Coulomb friction angle,
σ' = Effective normal stress = $\sigma - u$,
σ = Total normal stress, and
u = Pore pressure.

Effective stress strength parameters c' and ϕ' can be obtained from consolidated-undrained triaxial testing (ASTM D4767) (ASTM 2011) of undisturbed or remolded samples of fine-grained soils, reconstituted samples of coarse-grained soils, and, if relatively undisturbed samples of the in situ soils are not available, from correlations with various in situ tests (e.g., standard penetration, cone penetration, and pressuremeter) and index tests (e.g., Atterberg limits). These sources of parameters are not equally reliable and the analyst must carefully consider the source of data for key parameters affecting the stability of the slope under consideration.

Engineers typically perform effective stress slope stability analyses for slopes of coarse-grained soils. For fine-grained or other poorly draining soils, effective stress analyses can be difficult to perform without careful modeling of the changes in pore pressure that occur over time as these soils are loaded and subsequently consolidate. However, for some applications, especially the gradual loading of embankments, it may be necessary to conduct this modeling to obtain relevant slope stability results.

For total stress analyses (applicable to short-term conditions), the shear strength τ of a soil is expressed as follows:

$$\tau = c + \sigma \tan\phi \quad \text{or} \quad \tau = c_u \tag{10-8}$$

where

σ = Total normal stress,
c = Total stress Mohr–Coulomb strength intercept (also called *cohesion*),
c_u = Undrained shear strength, and
ϕ = Total stress Mohr–Coulomb friction angle.

Total stress strength parameters c, c_u, and ϕ can also be obtained from consolidated-undrained triaxial testing of undisturbed or remolded samples of fine-grained soils. Undrained shear strength c_u can also be estimated from in situ tests such as cone penetrometer testing (ASTM D5778) (ASTM 2020b) and vane shear testing (ASTM D2573) (ASTM 2008), as well as undrained laboratory shear strength testing such as unconfined compression (ASTM D2166) (ASTM 2006) and unconsolidated-undrained triaxial compression (ASTM D2850) (ASTM 2023) or constant volume simple shear (ASTM D6528) (ASTM 2017).

Total stress analyses using undrained shear strength can yield valuable results, especially when considering the sudden change in loading or inclination of slopes in normally or lightly overconsolidated clays. A total stress analysis is especially useful for the analysis of end-of-construction stability of embankments over soft clay or cut slopes into soft clay. Engineers should obtain and select undrained shear strength parameters that are relevant to the problem under consideration. In the case of end-of-construction stability, the engineer must be careful to select total stress strength parameters that represent the state of stress and consolidation for the soils at the point in time represented by the analysis geometry.

Many containment systems are founded on fine-grained soils, either because of local geology or because of technical siting criteria that favor low-permeability subgrades. As a consequence, placement of large embankments typical of waste containment systems may result in excess pore pressures in the subgrade that reduce its undrained strength. Moreover, the development and filling of containment systems may take place over several years, allowing a partial consolidation of the subgrade soils and increase in undrained strength. This sequence of events is best addressed using an appropriate staged loading analysis methodology such as that presented by Ladd (1991). Following this methodology, subgrade pore pressures are modeled at each stage of fill development, the corresponding subgrade shear strengths are estimated, and the resulting factors of safety are computed. Depending on the results, the engineer may need to adjust the waste containment system geometry or construction schedule to maintain an acceptable FS against slides through the foundation soils.

The undrained shear strengths along the slip surfaces of current or historical landslides in clays, especially sensitive clays, can be significantly less than the undrained shear strengths of undisturbed clays. This phenomenon is attributed to the reorientation of clay particles at the base of the slip surface induced by previous movement, called *slickensides*. Therefore, engineers must identify zones of previous movement in applicable terrain (Highland and Bobrowsky 2008) and apply residual strengths to analyses of these zones.

Equations (10-2) through (10-6) can be adapted for use with total stress strength parameters by substituting the total stress parameters for the effective stress parameters and omitting the pore pressure (u) terms.

Containment systems constructed for the placement of waste materials (i.e., landfills) are a common geoenvironmental engineering application. As shown in Figure 10-4, potential critical slip surfaces can pass through the waste mass. Accordingly, appropriate strength parameters are needed to analyze the stability of these materials. The Mohr–Coulomb strength envelope expressed using Equation (10-7) is often applied to waste materials. The following bullet points summarize typical waste material types and applicable strategies to obtain strength parameters for use in slope stability analyses:

- **Municipal Solid Waste (MSW) (Chapter 5):** Frequently encountered in landfill engineering practice because of the common need for MSW disposal, this material is highly heterogenous, varies in composition regionally, and is difficult to test. Strategies that have been used to quantify strength parameters for MSW include large-scale laboratory testing, in situ tests, test embankment failures, and back-calculation from full-scale failures (Matasovic et al. 2011, Bray et al. 2011, Gourc and Zekkos 2018). For preliminary analyses, $\phi' = 36$ and $c' = 5$ kPa may be used with caution. Depending on the FS values obtained or site-specific variations in the MSW profile (such as layers of weak material), the strength parameters should be refined. CPT soundings and soil borings can be effective to locate low-strength layers within an existing waste mass. Specification of waste placement techniques can prevent the development of low-strength layers for newly constructed facilities.
- **Construction and Demolition Debris (C&DD):** This material includes concrete fragments, masonry, wood, metal, plastic, and other products used in building construction. If the composition is more concrete, bricks, and stone, this material may behave like a rock fill. If wood, metal, and plastic and more prevalent, MSW strength properties may be more appropriate.
- **Coal Combustion Residual (CCR) (Chapter 6):** These materials are the by-products of coal-burning industrial processes, especially power plants, and include a wide range of potential grain sizes, chemical compositions, and mechanical properties. Depending on origin, these materials may behave more like fine-grained soils or more like coarse-grained soils. CCR is more amenable to laboratory and in situ soil testing than MSW and engineers should therefore perform site-specific strength testing (e.g., Jin et al. 2022). Care is needed to recognize spatial variations in placement technique and in-place composition.
- **Other Industrial Waste (Chapter 7), Hazardous Waste (Chapter 4), and Radioactive Waste:** A number of industrial processes generate waste that is disposed in landfills. Unlike MSW landfills, industrial landfills often receive waste from fewer generators and therefore may be more amendable to material testing. The engineer should consider the origin and mechanical behavior of the materials landfilled when selecting strength parameters. Some hazardous and radioactive waste treatment practices include chemical stabilization techniques that can affect the strength parameters of the waste. Some hazardous waste streams include contaminated soils, which generally

behave like other soils and may be amenable to laboratory soils testing with adequate precautions to protect worker health. When analyzing the stability of these facilities, the engineer should investigate the origin, treatment practices, and placement practices for the waste in place. For proposed waste containment facilities, the engineer may consider the range of potential properties for planned waste streams and set acceptable limits to be used in specifications for waste acceptance.

Interface strength is typically also quantified using Equation (10-7), with interface friction angle δ and C_a taking the place of ϕ and c, respectively. Interface shear strength parameters for rock joints (obtained from tests such as ASTM D5607, ASTM 2016), weak soils (Figure 10-7), and soil/rock interfaces can be estimated from the shear strength parameters for the weakest soil at the interface. Interface shear strength parameters for historic landslide slip surfaces will typically need to be back-calculated because of the difficulty in reliably obtaining samples from these interfaces. Shear strength parameters for geosynthetic interfaces can be determined through direct shear strength testing of the materials at the interface. ASTM D5321 (ASTM 2012) provides the standard procedures for conducting direct shear tests for this purpose. For interfaces involving geosynthetic clay liners (GCL), ASTM D6243 (ASTM 2020a) is used.

For waste containment systems, special consideration must be given to interface shear strength at contacts between dissimilar materials, particularly between different layers of geosynthetics and between

Figure 10-7. A soft clay layer responsible for a landslide.

geosynthetics and soil. The selection and specification of appropriate values of interface friction for analysis and design requires careful consideration of how sliding may preferentially develop and along which interfaces. Stark and Choi (2004) discuss considerations affecting the selection of interface shear strength values for design, particularly the displacements that geosynthetic interfaces experience during their service life. Because peak- and large-displacement shear strength develop at different rates depending on the specific interface, the engineer must consider which interface will reach peak stress first and progress to large displacement values when sheared. This consideration is complicated by the variation in the interface shear behavior of geosynthetics at different normal stresses.

An analysis of interface friction in containment systems may be performed in stages, with an increasing degree of sophistication and accuracy at each stage. The purpose of the preliminary stages of analysis is to determine which interfaces and slip surfaces have the lowest FS and to determine whether additional information or design modifications are needed before expending additional engineering effort. Accordingly, database values of interface strength properties are often consulted at the preliminary stage. Table 10-1 presents some typical interface strength parameters for commonly used geosynthetic interfaces. Following this preliminary analysis, the engineer may perform direct shear testing of the actual geosynthetic products specified or, alternatively, specify the required interface strength parameters for construction. If specifying required parameters, the engineer should be careful to determine whether the specified values are attainable with available products and also to specify test conditions (method of installation, normal stresses, and moisture conditions) that are representative of the expected conditions in service.

10.8.3 Pore Pressure

For many types of slope stability problems, pore pressure can be the most significant factor affecting acceptable or unacceptable outcomes. For analyses of existing slopes, piezometers and other types of groundwater instrumentation are needed to accurately estimate existing piezometric levels. Engineers should be cautious about seasonal variations and other transient effects on these data and consider the resulting effects on FS values. For evaluations of existing slopes where anticipated variations in pore pressure are outside of the range of the available monitoring data, groundwater flow modeling is necessary. For design slopes, flow modeling will be necessary, particularly for embankments near water such as levees and dams. In these cases, it is essential that the boundary conditions selected for modeling flow capture the full range of boundary conditions expected for the slopes. One possible outcome of these analyses is a set of

Table 10-1. Database Values of Interface Shear Strength Parameters for Selected Geosynthetic Interfaces.[a]

Material 1	Material 2	Interface friction angle (δ, degrees) – peak/LD[b]	Interface shear strength intercept (C_a, kPa) – peak/LD[b]
Textured[c] HDPE geomembrane[d]	Nonwoven needle-punched geotextile[e]	25/17	8/0
Textured LLDPE geomembrane[d]	Nonwoven needle-punched geotextile[e]	26/17	8/9
Faille PVC[f]	Nonwoven needle-punched geotextile[e]	27/23	0/0
Reinforced GCL (internal shear strength)		16/6[g]	38/12[g]

Source: Koerner and Narejo (2005).

[a] These values are useful for preliminary analyses. Actual laboratory test values should be obtained from the specific materials under consideration.

[b] Large displacement (LD) refers to a condition near the end of a direct shear test where the interface has been sheared sufficiently to measure a reduction in shear strength. In some cases, this value will reach an asymptote. The engineer is advised to consider displacement values anticipated in service (Stark and Choi 2004).

[c] The use of smooth geomembranes was more common historically than in current waste containment practice and resulted in landslides attributed to this weak interface as noted by Bonaparte et al. (2020). Actual test values should be sought when possible to analyze interfaces between smooth geomembranes and other geosynthetics, as well as between smooth geomembranes and soil.

[d] Smooth geomembranes have been installed in a significant number of waste containment facilities historically. However, the use of textured geomembranes has become more commonplace because of their improved interface friction properties. When considering smooth geomembrane installations, material samples should be obtained for site-specific testing.

[e] Nonwoven needle-punched geotextiles are incorporated into a number of geocomposite products, including geocomposite drainage layers and geocomposite clay liners (GCLs).

[f] *Faille* refers to a technique to apply an embossed surface to PVC geomembranes.

[g] Studies of the internal shear strength of GCLs (e.g., Zornberg et al. 2005) indicate that the internal shear strength of GCLs tends toward LD values of friction angle for bentonite, suggesting a complete destruction of the internal reinforcement.

recommendations for groundwater control measures, such as drains, relief wells, and other improvements to control the possible range of pore pressures to within acceptable limits.

For waste containment systems, pore pressures can be affected by environmental conditions, groundwater conditions, and chemical/ biological processes within the contained materials. Figure 10-8 illustrates the hydrology of typical multilayer cover systems. Pore pressures may accumulate from surface water infiltration to the cover soils, water trapped above the barrier layers, or water that infiltrates the materials below. Koerner and Soong (1998) present a methodology to analyze the veneer stability of cover systems. Stability of a landfill interior considering infiltration should incorporate unsaturated flow modeling, such as HELP analysis (Schroeder et al. 1994), as described in Chapter 9.

Ongoing chemical and biological degradation of waste, particularly MSW, can result in the accumulation of landfill gas pressure, pore-water (leachate) pressure, and excess heat. These processes and products interact with one another dynamically within a landfill (e.g., Reddy et al. 2017, Khire et al. 2022). Figure 10-9 shows an example of landfill leachate being expelled by gas pressure at an MSW landfill. Leachate collection and removal as well as landfill gas extraction systems are important engineered systems to limit the potential impact of these effects on stability. The presence of both liquid and gas within MSW under pressure can hinder the effectiveness of these systems. For planned facilities, the engineer

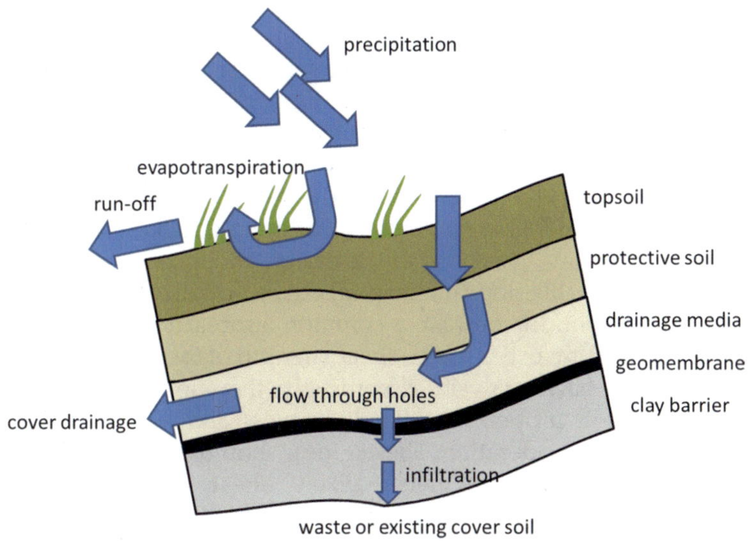

Figure 10-8. Cover system hydraulics. Pore pressures can accumulate above and below the barrier layers.

Figure 10-9. Leachate being expelled from the surface of a landfill by gas pressure, evidenced by escaping bubbles.

should consider acceptable limits on accumulated pore pressures within the landfill and design liquid and gas control systems to achieve these limits. For existing facilities, investigation and measurement on in situ pressures should be used to develop appropriate pore pressure distributions for slope stability analyses.

10.9 FORENSIC EVALUATIONS

Analysis of slopes with known movement is a special type of challenge and demands engineering approaches different from those for design scenarios. In particular, the rationale for the selection of geometry, shear strength, pore pressure, and loading parameters is largely inverted. For example, in the case of a design scenario, it is highly desirable to model a conservative set of adverse parameters that are more likely to produce a landslide. In this way, the aspect of the design under consideration (such as maximum allowable slope inclination) can be optimized conservatively to avoid movement. In contrast, a common approach when evaluating landslides is to estimate the in situ shear strength of the soil along the slip surface through back-analysis. Accordingly, the engineer must test or assume some soil properties ("fixed" values) and vary the strength parameter under consideration. The engineer should assign as close to the actual service values of geometry, "fixed" shear strengths, and pore pressure so as not to overestimate the shear strength value in question. In the case of "fixed" input parameters with uncertainty, the engineer may need to overestimate these parameters so that the soil strength obtained through back-analysis is not overestimated.

Another aspect of forensic evaluations that differs significantly from analyses supporting design is the role of field investigations to define the slip surface. Forensic investigations typically include significant instrumentation, in situ investigations, and surveys to identify the horizontal and vertical limits of movement as well as the slip surface. This information is used to define a fixed slip surface within the analyses rather than the typical critical slip surface search. Cornforth (2005) offers detailed guidance regarding the analysis and remediation of landslides. Bonaparte et al. (2020) present a number of landslide observations specific to landfill case histories. In waste containment systems, especially landfills, a visual inspection of the landslide surface as well as subsurface features accessible with excavation equipment can reveal significant information about the location of the slip surface and the materials involved. Figures 10-10 and 10-11 present some example visual observations of landslides at landfills showing various features, including evidence of sliding along a bottom

Figure 10-10. Visual evidence of sliding along geosynthetic interfaces in a bottom liner system exposed after movement occured. The pronounced wrinkles in the geomembrane visible in the photograph indicate movement of overlying materials and a slip surface between the geomembrane and the underlying materials.

Figure 10-11. Diagonal "echelon" cracks at one side of a landslide in clay.

liner system (Figure 10-10) and surface cracks at the sides of a clay embankment slide (Figure 10-11).

Similarly to all slope stability analyses, forensic studies must define the landslide geometry, material properties, and pore pressures. The more accurately each of these aspects can be quantified, the more accurate the resulting analyses will be, and consequently, the more confident the engineer can be in subsequent recommendations to mitigate the landslide. In waste containment practice, it is typically easier to obtain high-quality information about landslide geometry and pore pressure than about material properties. For this reason, the shear strengths of soils and especially wastes involved in landslides are often back-calculated during forensic evaluations.

10.10 SOIL REINFORCEMENT

This section is not guidance for mechanically stabilized earth (MSE) retaining walls but rather for the stabilization of slopes using tensile members.

Soil reinforcement refers to constructing inclusions within the soil mass to improve the resistance of the slope to movement. Typical inclusions may be metal strips, metal rods, cables, geogrids, geotextiles, or other materials that primarily resist deformation through tension. These materials transfer tensile forces to the ground through interface friction with the surrounding soil. Other ground improvements may include steel and concrete piles that primarily resist deformation through shear strength and bending stiffness. Accordingly, design of soil reinforcement not only considers the contribution of tension or shear resistance to resisting slope movement but must also consider whether the reinforcement is sufficiently embedded or anchored in surrounding materials to transfer load to portions of the slope that are not moving or otherwise beyond the critical slip surface. Figure 10-12 illustrates a modification to the method of slices often used to analyze soil reinforcement.

The contributions of soil reinforcement to the available resistance to slope movement can be quantified by adding a moment term to the limit equilibrium numerator. For example, for ordinary method of slices, Equation (10-4) is modified as follows:

$$FS = \frac{\sum R\left[c'l + \left(\left(\frac{W}{b}\right)\cos\alpha - ul\right)\tan\phi'\right] + \sum\left(\frac{T}{b}\right)y}{\sum R\left[\left(\frac{W}{b}\right)\sin\alpha\right]} \tag{10-9}$$

where

T/b = Allowable resisting force (tension for geosynthetics),
T = Reinforcing element, per slope unit width b, and
y = Moment arm to the center of rotation (often vertical distance) for a reinforcing element.

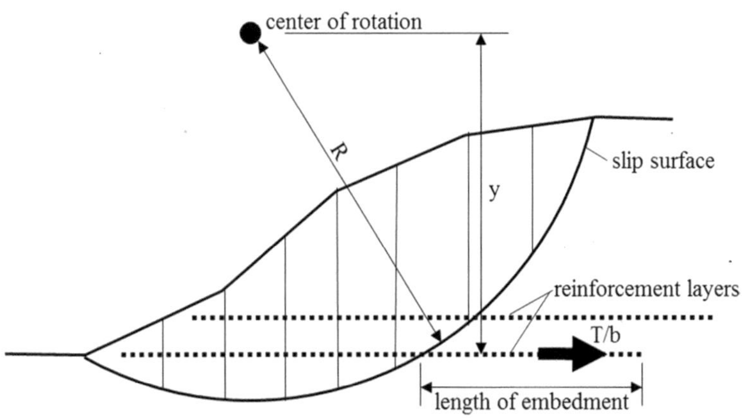

Figure 10-12. Method of slices incorporating tensile reinforcement.

Note that (T/b) is dependent on the out-of-section spacing of the reinforcing elements. The more widely spaced the reinforcement, the smaller the relative value of (T/b). For reinforcement that is a continuous out of section, such as layers of geotextile, allowable tension is typically expressed in units of force per unit width, allowing convenient use in these calculations.

For slopes that have unsatisfactory FS values without reinforcement, multiple design iterations may be required to obtain a combination of reinforcement strength and geometry that both satisfies the target FS value and provides the required embedment beyond the critical slip surface. For constructed embankments, reinforcement may be introduced by embedding tensile members such as steel strips and geosynthetics as part of fill placement. For stabilization of excavated slopes, tensile members must be introduced from the outside, such as soil nails and ground anchors. A detailed discussion of methods to analyze slopes reinforced with geosynthetics is provided by Koerner (2012).

10.11 REVIEW GUIDELINES

A key component of any slope stability analysis is a critique of the inputs used to develop the analysis and of the results obtained. The guidelines that follow are primarily directed to critiques of limit equilibrium analyses conducted using the method of slices.

In general, slope stability analysis software programs are good at conducting a large number of candidate slip surface computations. However, this capability does not ensure the validity or exhaustiveness of the cases considered. Accordingly, engineers should thoroughly critique the inputs used, the slip surface geometries considered, and the critical combinations of factors obtained. The skillful application of slope stability analysis software requires an appreciation of the factors most likely to result in adverse conditions in the modeled problem and the adjustment of the corresponding parameters in the analyses to interrogate the effects of these parameters on FS.

With regard to the choice of strength parameters, total stress analyses will tend to produce deeper critical slip surfaces when constant undrained shear strength profiles are used for soil layers. This result is obtained because, as deeper slip surfaces are analyzed, the slice weights, hence driving forces tend to increase, whereas the shear strengths do not increase to compensate. This consequence of total stress analysis is useful to evaluate the short-term (end-of-construction) stability of embankment placement and the effects of weaker layers within the profile. Therefore, engineers must ensure that they have sufficiently discretized the soil layers to account for differences in strength at different depths. Effective stress analyses tend to produce shallower critical slip surfaces by comparison because of the increase in slice shear strength with depth. Therefore, these results are applicable only in long-term analyses where strength gain can occur. Engineers should be aware of these effects to ensure that they define soil layers and select strength properties that are representative of the soil conditions under consideration. As discussed previously, when effective stress analyses are attempted, an understanding of the pore pressure conditions is especially important.

In the case of design scenarios, elements of the slope geometry, strength, pore pressure, or reinforcement that are found to have significant influence on the FS should be reported and used as the basis for the engineering recommendations for construction. In the case of forensic studies and hazard analyses, sensitive parameters may require follow-up data collection or research to reduce uncertainty in the findings.

The following bulleted lists are a recommended checklist to assist the reviewer and analyst in their critique of a slope stability analysis and to help determine whether the analyses performed are adequate for the engineering problem considered.

Problem Setup

- Is the purpose of the analysis adequately defined? What kind of problem? What are the expected types of conclusions and recommendations?
- What codes, regulations, or client directives govern the analysis?
- What is the required FS? Do the conditions for this FS apply to the type of analysis?
- What are the inputs to analysis (geometry, materials, pore pressure) that are fixed by the problem under consideration?
- What inputs are subject to change as part of an engineering recommendation? (i.e., what parts of the geometry or materials are under design control or are under investigation?)

Geometry

- Are all relevant material contacts modeled?
- Is the existing topography represented and credible?
- Is the design excavation/fill defined?
- Does the design excavation/fill conform to the correct stage of the project (phased construction, etc.)?
- Is the most critical section/slope direction represented by the design geometry (longest slope, steepest slope, weakest material)?
- For effective stress analysis, are potential piezometric surfaces defined?
- If optimizing for design geometry, is the impact of the design variable promulgated to all affected inputs?

- Is the analysis domain extended sufficiently in the horizontal and vertical directions to prevent unrealistic constraints on the slip surface search?
- If applicable, are tension cracks modeled?

Material Properties

- Are all relevant materials modeled?
- For geosynthetic interfaces and waste materials, are the correct peak- or large-displacement values being used? Do these values take into account the differences in strength development with displacement for each material?
- For design, are conservative (low) strength values represented?
- For back-analysis, are the "fixed" strength values conservative (high or precise values)?
- Is the stratigraphy and range of strength values representative and credible?
- If optimizing strength values to achieve a target FS, is the goal to determine the minimum acceptable strength for proposed fill or for verification of existing materials?
- Do minor changes in shear strength result in significantly different critical slip surface geometries?
- In the case of effective stress analyses, are shallow slip surfaces that are not of interest being evaluated instead of more pertinent deeper mechanisms? If so, it may be necessary to revisit the strength of these materials under low confining stress to remove irrelevant critical slip surface searches near the ground surface.
- For landfills, has the rate of material placement been considered? Is a staged loading analysis required? How well defined is the fill placement schedule? What are the consequences of changes to this schedule? How is this information communicated to operators and other engineers?

Pore Pressure

- Are the pore pressures modeled representative of appropriately adverse conditions and have their effect on shear strength been accordingly modeled in the analysis?
- Are the factors of safety obtained sensitive to pore pressure over the range of plausible values?
- Are changes in pore pressures anticipated from construction activities? Are they modeled?
- Are the long-term pore pressures defined after construction pore pressure has dissipated? What time is required to achieve long-term conditions? Are there intermediate conditions where additional changes in pore pressure may occur?

- For dikes, levees, dams, and treatment ponds: have steady-state seepage conditions been considered? Has rapid drawdown been considered?
- For landfills, have future liquid and gas pore pressure developments resulting from waste decomposition and liquid movement been considered? Are current pore pressure conditions from direct measurements included?

Other loads

- Is the internal reinforcement placed and defined correctly?
- Are surface surcharges properly defined (location, magnitude, direction)?
- For design, are limiting surface loads obtained and reported?

Slip Surface Definition

- Is the analysis domain sufficiently large to encompass the critical slip surfaces?
- Are the possible combinations of search parameters in the search window (centers of rotation, tangents, defined slip surfaces) sufficiently broad and sufficiently subdivided to capture the critical slip surfaces?
- Was the range of candidate slip surfaces considered inclusive of combinations of geometry, strength, and pore pressure that are likely to be less stable?
- Are there weak materials or interfaces that can form preferential slip surfaces? Are they explicitly defined in the analysis?
- Have each of the typical waste containment slip surface geometries been considered?
- Was the range of candidate slip surfaces considered inclusive of surface loads that are central to the design problem?
- After running the analysis, is there a part of the analysis domain that requires a more detailed investigation? Have all credible critical slip surfaces been exhausted? (are excluded weak zones, steep slopes, or combinations of slope and materials more likely to slide present?)

Analysis

- Is the correct analysis type being used?
- Is the solution converging, or are there numerical problems?
- Is there sufficient resolution in the discretization of the problem to avoid artificially low/high factors of safety?
- Are the critical slip surfaces obtained credible in terms of the trajectories of movement and changes in those trajectories over the critical slip surface?

- Is the critical slip surface unnecessarily shallow because of problems with material parameters?
- Is the critical slip surface unusually deep because of an improper profiling of undrained strength properties?
- Are pore pressures included? Are they applied to relevant layers that would experience these pressures?

Postprocessing and Documentation

- Is the critical slip surface clearly defined and visible?
- Can the analysis output for each slice be listed and is it credible?
- Are errors or unexpected results reported for the shear strength at the base of each slice?
- Are the FS and relevant design parameters reported?
- Can the problem's geometry and material properties be traced back to their applicable sources and/or proposed design features?

10.12 DOCUMENTATION

The following items are recommended for the documentation of slope stability analyses:

- Name of the analyst and date of analysis;
- Objectives;
- Description of the problem geometry, source of geometric data;
- Description of the stage of construction represented by the analysis;
- Complete list of material strength values, source of strength data;
- Pore pressure data, source of data;
- Surface loads, source of load data;
- For assumptions, a description of the input parameters that are acceptable and/or subject to verification;
- Cross-section plan view with traces and graphical depiction of analyzed cross sections;
- Analysis methods used;
- Record of quality control process, comment resolution, independent reviewer, and dates; and
- Engineering recommendations and items for follow-up.

REFERENCES

ASTM International. 2006. *Standard test method for unconfined compressive strength of cohesive soil*. ASTM D2166. West Conshohocken, PA: ASTM.

ASTM. 2008. *Standard test method for field vane shear test in cohesive soil*. ASTM D2573. West Conshohocken, PA: ASTM.

ASTM. 2011. *Standard test method for consolidated undrained triaxial compression test for cohesive soils*. ASTM D4767. West Conshohocken, PA: ASTM.

ASTM. 2012. *Standard test method for determining the shear strength of soil-geosynthetic and geosynthetic-geosynthetic interfaces by direct shear*. ASTM D5321. West Conshohocken, PA: ASTM.

ASTM. 2016. *Standard test method for performing laboratory direct shear strength tests of rock specimens under constant normal force*. ASTM D5607. West Conshohocken, PA: ASTM.

ASTM. 2017. *Standard test method for consolidated undrained direct simple shear testing of fine grain soils*. ASTM D6528. West Conshohocken, PA: ASTM.

ASTM. 2020a. *Standard test method for determining the internal and interface shear strength of geosynthetic clay liner by the direct shear method*. ASTM D6243. West Conshohocken, PA: ASTM.

ASTM. 2020b. *Standard test method for electronic friction cone and piezocone penetration testing of soils*. ASTM D5778. West Conshohocken, PA: ASTM.

ASTM. 2023. *Standard test method for unconsolidated-undrained triaxial compression test on cohesive soils*. ASTM D2850. West Conshohocken, PA: ASTM.

Baecher, G. B., and J. T. Christian. 2003. *Reliability and statistics in geotechnical engineering*. Hoboken, NJ: Wiley.

Bonaparte, R., R. C. Bachus, and B. A. Gross. 2020. "Geotechnical stability of waste fills: Lessons learned and continuing challenges." *J. Geotech. Geoenviron. Eng*. 146 (11): 05020010.

Bray, J. D., D. Zekkos, and S. M. Merry. 2011. "Shear strength of municipal solid waste." In *Geotechnical characterization, field measurement, and laboratory testing of municipal solid waste*, Geotechnical Special Publication 209, edited by D. Zekkos. Reston, VA: ASCE.

Coduto, D. P. 1999. *Geotechnical engineering: Principles and practices*. Upper Saddle River, NJ: Prentice Hall.

Cornforth, D. H. 2005. *Landslides in engineering practice*. Hoboken, NJ: Wiley.

Duncan, J. M. 1992. "State-of-the-art: Static stability and deformation analysis." In *Proc., Stability and Performance of Slopes and Embankments II, A 25 Year Perspective*. Geotechnical Special Publication 31. New York: ASCE.

Duncan, J. M. 1996. "State of the art: Limit equilibrium and finite-element analysis of slopes." *J. Geotech. Geoenviron. Eng*. 122 (7): 577–596.

Duncan, J. M. 2000. "Factors of safety and reliability in geotechnical engineering." *J. Geotech. Geoenviron. Eng*. 126 (4): 307–316.

Duncan, J. M., S. G. Wright, and T. L. Brandon. 2014. *Soil strength and slope stability*. 2nd ed. Hoboken, NJ: Wiley.

Fenton, G. A., and D. V. Griffiths. 2008. *Risk assessment in geotechnical engineering*. New York: Wiley.

Fredlund, D. G., and R. E. G. Scoular. 1999. "Using limit equilibrium concepts in finite element slope stability analysis." In *Proc., Int. Symp. on Slope Stability Engineering, Sichikoku '99*, 31–47. Rotterdam, Netherlands: Balkema-Rotterdam.

GeoRG (Geotechnical Resource Group). 2004. *Geotechnical and stability analyses for Ohio waste containment facilities*. Guidance Document 660. Columbus, OH: Ohio Protection Agency.

Gourc, J.-P., and D. Zekkos. 2018. "Waste mechanical properties." In *Solid waste landfilling: Concepts, processes, technology*, edited by R. Cossu and R. Stegman, 193–210. Amsterdam, Netherlands: Elsevier.

Griffiths, D. V., and P. A. Lane. 1999. "Slope stability analysis by finite elements." *Géotechnique* 49 (3): 387–403.

Highland, L. M., and P. Bobrowsky. 2008. *The landslide handbook: A guide to understanding landslides*. Circular No. 1325. Reston, VA: US Geological Survey.

Jin, L., A. Fuggle, and C. P. Armstrong. 2022. "Determining the in situ state of coal combustion residuals using cone penetration testing." In *Geo-Congress 2022*, 463–473. Reston, VA: ASCE.

Khire, M. V., T. Johnson, and R. Holt. 2022. "Heat loss in landfills as a function of landfill size." In *Geo-Congress 2022*, 83–91. Reston, VA: ASCE.

Koerner, G. R., and D. Narejo. 2005. *Direct shear database of geosynthetic-to-geosynthetic and geosynthetic-to-soil interfaces*. GRI Report 30. Folsom, PA: Geosynthetic Research Institute.

Koerner, R. M. 2012. *Designing with geosynthetics*. Bloomington, IN: Xlibris.

Koerner, R. M., and T.-Y. Soong. 1998. "Analysis and design of veneer cover soils." In *Proc., 6th Int. Conf. on Geosynthetics*, 28–49. Leeds, UK: Emerald Publishing.

Koerner, R. M., and T.-Y. Soong. 2000. "Stability assessment of ten large landfill failures." In *GeoDenver 2000*, Geotechnical Special Publication 103, 1–38. Reston, VA: ASCE.

Kumar, G., and K. R. Reddy. 2022. "Reliability assessment of bioreactor landfill performance using coupled thermo-hydro-bio-mechanical model." *J. Geotech. Geoenviron. Eng.* 148 (5): 04022023.

Ladd, C. C. 1991. "Stability evaluation during staged construction." *J. Geotech. Geoenviron. Eng.* 117 (4): 540–615.

Low, B. K., R. B. Gilbert, and S. G. Wright. 1998. "Slope reliability analysis using generalized method of slices." *J. Geotech. Geoenviron. Eng.* 124 (4): 350–362.

Matasovic, N., R. El-Sherbiny, and E. Kavazanjian Jr. 2011. "In-situ measurements of MSW properties." In *Geotechnical characterization, field measurement, and laboratory testing of municipal solid waste*, Geotechnical Special Publication 209, edited by D. Zekkos. Reston, VA: ASCE.

Michoud, C., M.-H. Derron, P. Horton, M. Jaboyedoff, et al. 2012. "Rockfall hazard and risk assessments along roads at a regional scale: Example in Swiss alps." *Nat. Hazards Earth Syst. Sci.* 12 (3): 615–629.

Qian, X., R. M. Koerner, and D. H. Gray. 2002. *Geotechnical aspects of landfill design and construction*. Upper Saddle River, NJ: Prentice Hall.

Reddy, K. R., R. K. Giri, and H. S. Kulkarni. 2017. "Modeling coupled hydromechanical behavior of landfilled waste in bioreactor landfills: Numerical formulation and validation." *J. Hazard. Toxic Radioact. Waste* 21 (1): D4015004.

Schroeder, P. R., T. S. Dozier, P. A. Zappi, B. M. McEnroe, et al. 1994. *The hydrologic evaluation of landfill performance (HELP) model: Engineering documentation for version 3*. EPA/600/R-94/168b. Washington, DC: US Environmental Protection Agency, Office of Research and Development.

Stark, T. D., and H. Choi. 2004. "Peak versus residual interface strengths for landfill liner and cover design." *Geosynth. Int.* 11 (6): 491–498.

USACE (US Army Corps of Engineers). 1997. *Introduction to probability and reliability methods for use in geotechnical engineering*. Engineer Technical Letter 1110-2-547. Washington, DC: USACE.

USACE. 2006. *Reliability analysis and risk assessment for seepage and slope stability failure modes for embankment dams*. Engineer Technical Letter 1110-2-561. Washington, DC: USACE.

Yu, H. S., R. Salgado, S. W. Sloan, and J. M. Kim. 1998. "Limit analysis versus limit equilibrium for slope stability." *J. Geotech. Geoenviron. Eng.* 124 (1): 914–918.

Zhang, R., and A. Rock. 2011. "Rockfall and mitigation evaluation with 3-D discrete element modeling." In *GeoRisk 2011: Geotechnical Risk Assessment and Management*, edited by C. H. Juang, K. K. Phoon, A. J. Puppala, R. A. Green, and G. A. Fenton. Reston, VA: ASCE.

Zornberg, J. G., J. S. McCartney, and R. H. Swan. 2005. "Analysis of a large database of GCL internal shear strength results." *J. Geotech. Geoenviron. Eng.* 131 (3): 367–380.

CHAPTER 11

SEISMIC STABILITY OF WASTE CONTAINMENT SYSTEMS

Edward Kavazanjian Jr., Ph.D., P. E., D.GE;
Dimitrios Zekkos, Ph.D., P. E., M.ASCE; and George Zalachoris, Ph.D.

11.1 INTRODUCTION

Seismic stability of waste containment systems may be assessed by a hierarchical set of analyses of increasing complexity. At each hierarchical level, the seismic stability assessment consists of five elements: (1) determination of the design ground motion, (2) evaluation of the seismic response of the system, (3) evaluation of system element performance under seismic loading, (4) assessment of the performance of system elements relative to a seismic performance standard, and (5) modification of the design and reanalysis in case the seismic performance standard is not met. Element 4 may be implicit based on the calculation of a seismic performance index. Elements 2 and 3 may be integral parts of the same analysis. Seismic stability analyses are generally conducted once the static stability of the system under consideration has been established. In this regard, seismic analysis is sometimes referred to as a "check" on system performance. Although seismic stability is often evaluated only for final configuration, it may also be evaluated at the end of the discrete phases of containment system development. However, the seismic loading and performance criteria for intermediate configurations of the containment system may differ from that of the final configuration.

11.2 HIERARCHY OF SEISMIC STABILITY ANALYSES

The hierarchical level that is required for the seismic stability assessment of a particular facility depends on the regional seismotectonic environment, the seismic performance standard, and the desired level of precision.

In general, the level of detail of the characterization of the design earthquake, the complexity of the analysis, the required level of detail regarding the seismic performance, and the effort associated with the analysis all increase as the hierarchical level increases. At the lowest level, the factor of safety (FS) may be calculated as an index of seismic performance on the basis of a simple pseudo-static limit equilibrium analysis using a seismic coefficient, k_s, defined as some fraction of the peak ground acceleration (PGA) from a seismic hazard map or specified by code, regulation, or the owner (Bray 2007). In this type of analysis, the calculated FS is compared to a target value (often 1.0 or 1.1). However, failure to meet or exceed the target value does not mean that the design is unacceptable: more sophisticated analyses may still demonstrate that seismic performance is acceptable. At the highest level, a numerical stress–deformation analysis (e.g., finite-element or finite difference analysis) is performed in which a suite of acceleration time histories is used to explicitly calculate the forces and strains exerted on the elements of the waste containment system as well as the displacements of the waste mass. Calculated forces, strains, and displacements are then compared to the seismic resistance of individual system elements to determine whether they constitute unacceptable threats to the integrity of the containment system.

The sophistication and fragility of the waste containment system may also influence the level of the analysis. For example, the seismic stability of a municipal solid waste (MSW) landfill without an engineered liner system and subject to a performance standard corresponding to no-unconstrained release of contaminants and repairable damage may be satisfactorily addressed using the simple pseudo-static limit equilibrium analysis approach described previously. Conversely, the seismic stability of a geosynthetic-lined hazardous waste landfill constructed on soft clay in a region of high seismicity subject to a performance standard of no damage to the containment system elements may require detailed seismic hazard and site-specific response analyses and sophisticated numerical modeling.

Table 11-1 summarizes the hierarchical levels of seismic stability analyses for waste containment systems. The details of these analyses are discussed subsequently. Note that the accuracy and precision of these analyses increases with their complexity, and, thus, with the level of analysis. Simpler analyses are intentionally developed to be conservative (and thus not necessarily conclusive), but in many cases, they may be sufficient to demonstrate seismic stability. However, as noted previously, failure to demonstrate that a waste containment system is stable using one of the simpler levels of analysis does not indicate that the containment system is seismically unstable, but it is merely an indication that a higher level of analysis may be needed to demonstrate acceptable seismic performance.

Table 11-1. Hierarchical Levels of Seismic Stability Assessment.

Type of analysis	Level[a]	Sublevel	Performance criteria
Pseudostatic limit equilibrium	1.1	PGA from chart (e.g., Bray and Rathje 1998)	FS > 1.0 or 1.1 for k_s = MHA or PGA/2 (Kavazanjian et al. 2011)
	1.2	MHA from chart (e.g., Anderson et al. 2008)	
	1.3	PGA from site response analysis	
Newmark-type seismic displacement[b]	2.1	Correlation with spectral accelerations (e.g., Anderson et al., 2008)	Allowable calculated seismic displacement
	2.2	Time history analysis	
Stress–deformation analysis (e.g., finite-element or finite difference)	3.1	Time history analysis without geosynthetic elements	Calculated displacements
	3.2	Time history analysis with geosynthetic elements (e.g., geomembranes)	Calculated displacements, forces, and strains

[a]Levels reflect a progressively decreasing conservatism.

[b]Pseudostatic limit equilibrium analysis (for yield acceleration) is also required.

11.3 SEISMIC PERFORMANCE STANDARD

The seismic performance standard is a key factor in determining the adequacy of a waste containment system subject to seismic loading. Experience in moderate to high seismicity regions has shown that it is often difficult to show that a geosynthetic-lined waste containment system is unconditionally stable subject to the design earthquake. That is, from a practical point of view, a demonstration that the containment system can resist the design earthquake shaking without any damage may be technically unattainable and/or economically burdensome. Fortunately, a demonstration of unconditional stability is often unnecessary for effective environmental protection, as observations of the performance of solid waste landfills during earthquakes has shown that some waste containment system elements can sustain damage because of seismically induced forces, strains, and displacements without a harmful discharge of contaminants (Augello et al. 1995, Matasovic et al. 1998a).

Prevention of a discharge of contaminants that are harmful to human health or the environment (i.e., "harmful discharge") provides a rational basis for economically and environmentally sound seismic design of waste containment systems. This performance standard often allows for limited damage to containment system elements under the design earthquake scenario if the damage is repairable, including complete replacement of some system elements in some cases. For critical waste containment facilities (e.g., high-level radioactive waste containment facilities), and for containment system elements where damage may not be detectable or cannot be repaired even if detected (e.g., liner systems buried under tens of meters of waste and overlying useable aquifers), a "no-damage" performance standard may be required. These performance standards are consistent with the current state of practice regarding the seismic design of bridges and other critical facilities that allows for repairable damage, while maintaining serviceability for critical or important facilities and for complete replacement of less important facilities, provided that life safety is maintained (AASHTO, 2015). Therefore, seismic performance requirements should be established considering: (1) the potential for detecting damage that could lead to a harmful discharge of contaminants, (2) the ease and cost of interim measures to maintain containment system effectiveness, and (3) the ease and cost of complete restoration of system integrity.

Seismic design of a waste containment system with a goal to prevent harmful discharge requires an evaluation of the seismic performance of individual components and subsystems. Performance requirements should be established on a system- and component-specific basis. Figure 11-1 shows elements of the containment system for both a modern, lined, solid waste landfill and an older, unlined, facility. Appurtenant facilities (e.g., access roads, surface water drainage channels, and gas and leachate treatment systems) (not shown in Figure 11-1) are also essential elements of the overall system and must be considered in seismic design. Many of these systems can be designed to sustain damage during the design earthquake, provided that a postearthquake inspection and repair program is in place. One example of recommendations regarding the acceptable level of damage, expressed in terms of permanent seismic displacement and time required for interim repair and full restoration of noncritical solid waste landfill systems is provided in Table 11-2 (Kavazanjian et al. 1998).

Design practices such as the use of ductile materials and inclusion of redundant critical systems and components, as well as design details that

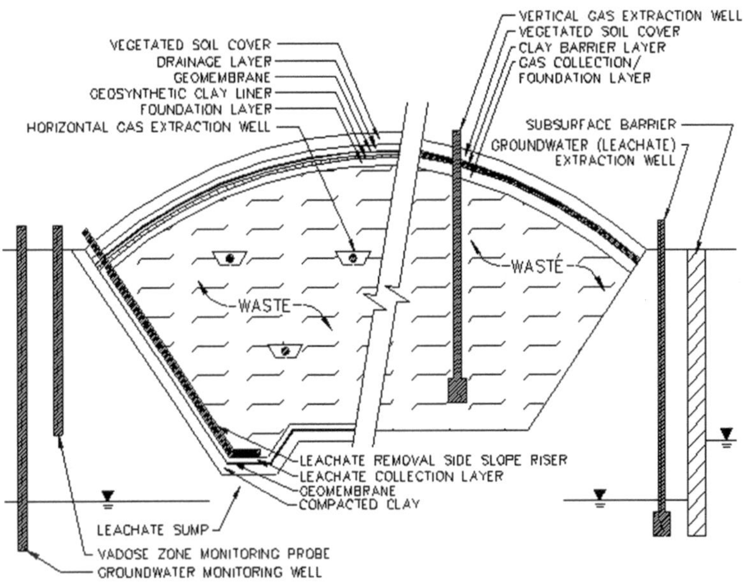

Figure 11-1. Containment system components for modern (left side) and older (right side) landfills.
Source: Kavazanjian et al. (1998).

can accommodate anticipated seismic forces and deformations, can facilitate compliance with a performance standard that simply requires prevention of harmful discharge. For instance, infiltration control can be provided by redundant subsystems for surface water management, including drainage layers and multiple infiltration barriers. Contaminated liquid conveyance and storage systems may include redundant features such as double containment for pipes and secondary containment for tanks. However, if the basis of the seismic design is to allow for limited damage without harmful discharge, a postearthquake inspection, repair, and recovery plan becomes an essential component of facility design. Kavazanjian et al. (1998) describe the necessary elements of an appropriate postearthquake response and recovery plan.

11.4 DESIGN CRITERIA

Consistent with current geotechnical practice for earth dams, pipelines, and other important facilities, the seismic design of landfills to prevent harmful discharge to the environment is most often implemented through the concept of "allowable calculated seismic displacement." In many common seismic analyses (e.g., in a decoupled Newmark-type

Table 11-2. Seismic Performance Standards for Ancillary Facilities at Solid Waste Landfills.

Cover system component	Design criteria and performance standard	Interim remediation to restore compliance	Repair to pre-earthquake condition
Final cover			
Soil monocover on side slopes	150 mm of soil deformation. Partial failure contained on site	3 months to strip vegetation, regrade and recompact areas of cracking	12 months to restore vegetation
Landfill gas control			
Collection wells	Up to 25% of wellheads broken	1 month to route headers around broken wellheads	12 months to repair/replace broken wells heads
Headers	Up to 25% of header pipes cracked or broken	1 month to bypass broken header pipes	3 months to repair/replace broken headers
Vacuum pumps	Power loss. No structural damage	None required	1 month to restore off- site power
Leachate transmission pipes	Acceptable breakage of pipes with double containment	1 month to bypass broken pipes	3 months to repair broken pipes
Surface water management			
Conveyance systems (bench channels, down drains, culverts)	Cracking and up to 300 mm of displacement	2 months to completely restore surface pathways	9 months to replace/rebuild surface pathways
Sedimentation basin	Minor cracking of concrete	2 weeks to 1 month to patch the cracks	9 months to rebuild the basin (if needed)
Access roads	300 mm displacement (cracking)	2 months to patch the cracks	12 months for full repair.

Source: Kavazanjian et al. (1998).

displacement analysis) that use a yield acceleration on the basis of shear strength parameters defined at large displacements, the calculated deformation is a conservative value that is really only an index of seismic performance and does not correspond to the actual anticipated displacement of the waste mass or other system elements under the design earthquake scenario. However, in more sophisticated analyses (e.g., finite-element and finite difference analyses) that explicitly simulate the behavior of the elements of the containment system under the design earthquake shaking, the calculated displacements and the seismically induced forces represent an estimate of the actual anticipated displacements and forces. The allowable calculated seismic displacement depends on a number of factors, including the relationship between the calculated displacement and the anticipated displacement, the ability of specific elements of the containment system to sustain deformation without loss of function, the impact of a possible release caused by a loss of function, the ability to detect any such loss of function, and the feasibility (and associated cost) of repair (or replacement) of the potentially damaged element.

It is difficult to establish generic criteria for allowable calculated seismic displacements of waste containment systems and containment system components as the allowable calculated displacement depends on many site-, system-, material-, and analysis-specific factors. However, several generalizations can be made. Many waste containment system components and subsystems can sustain significant deformation without loss of function. Furthermore, except for fluids (liquids and gases), contaminants discharged from a waste containment system may not be highly mobile, thus allowing for timely detection and repair of the damage without harmful consequences to human health or the environment. Damage to landfill covers and hazardous waste site caps, above ground pipes and tanks, surface water control systems, and ancillary facilities can, in general, be detected and repaired. However, a postevent inspection and repair program must be in place if the system elements are to be designed following this rationale. Damage to the elements of an engineered liner system that are buried under the waste mass, or to other subsurface components of the containment system, may be hard to detect and, even if detected, difficult to repair, except at great expense.

Many engineered waste containment systems have substantial built-in redundancy. Most final cover systems and liner systems provide multiple, redundant levels of protection against leachate migration to groundwater. For instance, the leachate collection and removal system and the composite liner in a modern solid waste landfill provide a redundant system for groundwater protection from landfill leachate and gas condensate. Any of these components may lose their functionality without compromising the ability of the liner system to contain liquids, particularly if the final cover system provides adequate infiltration control. However, because of difficulties in detecting contaminant releases through a breached liner system in a timely fashion, geomembranes in a single or composite liner system are generally designed according to a "no-damage" standard. Note that this does not mean that the calculated seismic displacement must be zero. When applying any seismic performance standard, the ability of the employed analytical method to accurately predict seismic deformations must also be considered. So, for instance, based on observations from past earthquakes, it is generally assumed that a seismic displacement of < 15 cm (6 in.) calculated in a decoupled Newmark analysis using large displacement properties for the soil and geosynthetic interfaces is considered to represent acceptable displacements for a liner system designed to a "no-damage" performance standard (Augello et al. 1995). Consideration of the seismic performance standard on a project-specific basis can lead to the development of rational and economical seismic design.

11.5 DESIGN GROUND MOTION

The extent of characterization of the ground motions considered during the seismic design of a waste containment system, as well as the necessity of adjusting the design motions to account for local site conditions (including the presence of the landfill), depends on the level of complexity of the seismic analysis. Design ground motions are typically specified for free-field, level-ground conditions and may require adjustment for local site conditions. Moreover, the effect of the waste mass and waste containment system on the characteristics of the design ground motions may also need to be considered.

The simplest seismic stability analysis (i.e., a pseudo-static limit equilibrium analysis) may only require the free-field, PGA corresponding to the design earthquake as input to the analysis. However, the free-field PGA design value, as defined by codes and standards, is typically specified for a reference ground condition. Therefore, as part of the analysis, the free-field PGA defined by a code or standard may need to be corrected to establish a site-specific PGA value (i.e., if site conditions differ from the reference ground condition). Figure 11-2 presents a map, produced by the US Geological Survey (USGS) as part of the National Earthquake Hazard Reduction Program (NEHRP), depicting the spatial variability of the PGAs corresponding to a 90% probability of not being exceeded in a 250-year period for a reference ground condition. This threshold is specified in the US Environmental Protection Agency (USEPA) regulations for the design of MSW landfill facilities and in the solid waste landfill regulations for many states.

The PGA values shown in Figure 11-2 are computed for a reference site condition characterized by an average shear wave velocity over the top

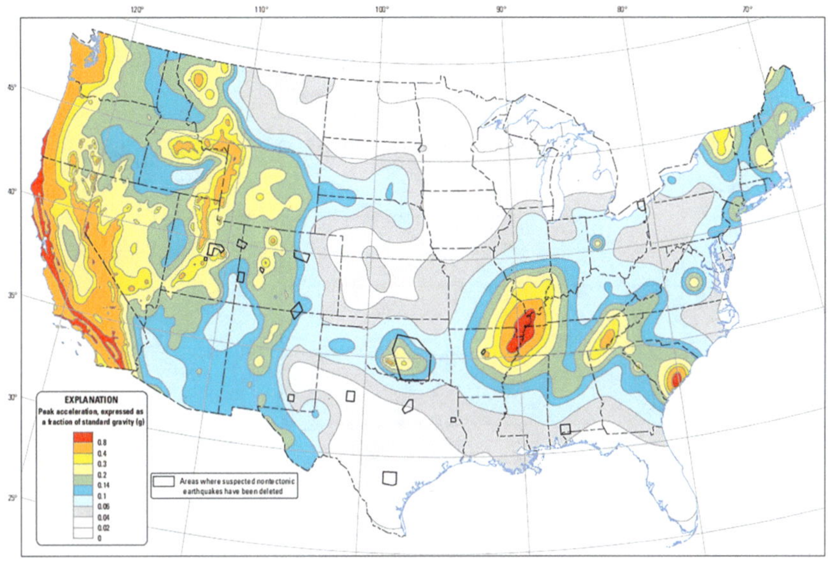

Figure 11-2. US national seismic hazard map for PGA with a 90% probability of not being exceeded in 250 years for the B/C boundary.
Source: USGS (2014).

30 m of the site, V_{S30}, of 760 m/s, a condition referred to as "the B/C boundary" (the boundary between Site Class B and Site Class C as defined by NEHRP). Peak ground acceleration values for probabilities of not being exceeded other than 90% and exposure periods other than 250 years for the contiguous United States, Alaska, and Hawaii are available from the USGS National Seismic Hazard Mapping Program (NSHMP) website (USGS 2024) and may also be available from other sources (e.g., regional seismic hazard analyses or site-specific studies).

NEHRP (2009) defines a series of five standard site classes (Site Classes A through E) on the basis of V_{S30}. One additional class of sites (Site Class F) is used to specify special study sites (i.e., site conditions requiring site-specific ground motion studies). The NEHRP provides "site class coefficients" that can be used to adjust a PGA value for the reference site condition to a PGA value for sites in Site Classes A through E. Figure 11-3 provides a graphical representation of the corrected PGA for these five standard site classes. For Site Class F, the NEHRP mandates that a site-specific response analysis be conducted to correct the reference PGA value for local ground conditions. Site Class F includes sites with organic and high-plasticity clay soils, which are particularly susceptible to ground motion amplification, and landfill sites belong to this class. Note also that a number of studies have shown that there are additional ground

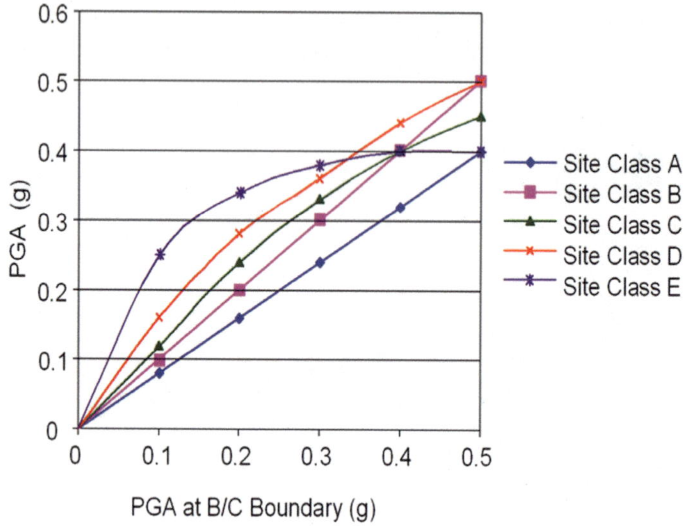

Figure 11-3. PGA amplification based on NEHRP site class.
Source: Kavazanjian et al. (2011).

conditions besides those specified in NEHRP as special study sites for which the NEHRP site class coefficients may not be appropriate (Hashash and Park 2001, Choi and Stewart 2005, Huang et al. 2010, Ghanat and Kavazanjian 2011, Pehlivan et al. 2017, Parker et al. 2019). These supplemental special study sites include sites within deep soil basins, basin edge sites, and sites with shallow impedance (shear wave velocity) contrasts.

As an alternative to using a value from a seismic hazard map or a code, the design PGA may be established on the basis of a site-specific seismic hazard analysis. US Environmental Protection Agency regulations explicitly allow for the use of a site-specific analysis in lieu of map values for a PGA with a 90% probability of not being exceeded in a 250-year period. Furthermore, the design PGA mandated by other regulatory agencies for other types of waste containment systems (e.g., hazardous waste on mixed waste landfills) or established by a facility owner may not be available from a recognized source. In such cases, a site-specific seismic hazard analysis may be required. If a site-specific seismic hazard analysis is conducted for reference site conditions, the resulting PGA should be adjusted for local site conditions. However, it is also possible to use ground motion prediction equations (GMPEs) appropriate for the region of interest that incorporate parameters describing local site conditions such as V_{S30} (e.g., NGA-West2 GMPEs for active tectonic regions such as California, NGA-East GMPEs for the central and eastern

North American regions) and calculate a PGA that directly accounts for local site conditions.

Although the PGA alone may be sufficient for simplified analyses, spectral acceleration values associated with the design earthquake may also be required for some types of analyses. Spectral acceleration values are sometimes used to adjust the site-specific free-field PGA to account for the presence of the waste containment system and the waste mass, for Newmark-type seismic displacement analyses, and for producing representative acceleration time histories for the design earthquake. In the United States, spectral acceleration values for spectral periods of 0.2 and 1.0 s are available from NEHRP maps and the USGS NSHMP website cited previously for the same probabilities of exceedance for which PGA values are available. NEHRP also provides site class coefficients for spectral accelerations at spectral periods of 0.2 and 1.0 s as well as guidance for evaluating spectral accelerations at other spectral periods. Spectral accelerations are also usually available from the results of a site-specific seismic hazard assessment and, depending on the GMPEs employed in the analysis, they may automatically correct for local site conditions (e.g., V_{S30}). However, corrections of map-, website-, or site-specific seismic hazard–derived reference spectral accelerations are subject to the same qualifications as the corrections of PGA values based on site class: special study sites require site-specific seismic response analyses.

For site-specific response analyses, for Newmark-type displacement analyses, and for advanced numerical simulations of the effect of the containment system and the waste mass on the design ground motion, a suite of representative acceleration time histories is typically required. It is generally recognized that the use of just a single time history in such analyses is not appropriate, even in cases where a spectrum-compatible motion (that uses a synthetic or modified time history that is matched to a smooth, statistically average response spectrum) is used as the input ground motion. Variability among ground motion time histories has been shown to be a major contributor to the variability in observed response, even for ground motions of the same magnitude and PGA (Zekkos and Carlson 2011, Harris et al. 2013). The suite of acceleration time histories should be representative of the design event in terms of magnitude, distance, and/or duration, and it should, on aggregate, encompass the spectral acceleration envelope from the seismic hazard analysis (Kottke and Rathje 2008). ASCE 7-05 and 7-10 (ASCE 2005, 2010) for seismic design of buildings call for the use of a set of at least seven two-component time histories if the average response is used in design, with that number increasing to 11 in ASCE 7-16 (ASCE 2017). However, if the maximum response is used (rather than the average), a suite of at least three two-component time histories is mandated for design. Figure 11-4 illustrates the large variability among the response spectra of 10 selected time

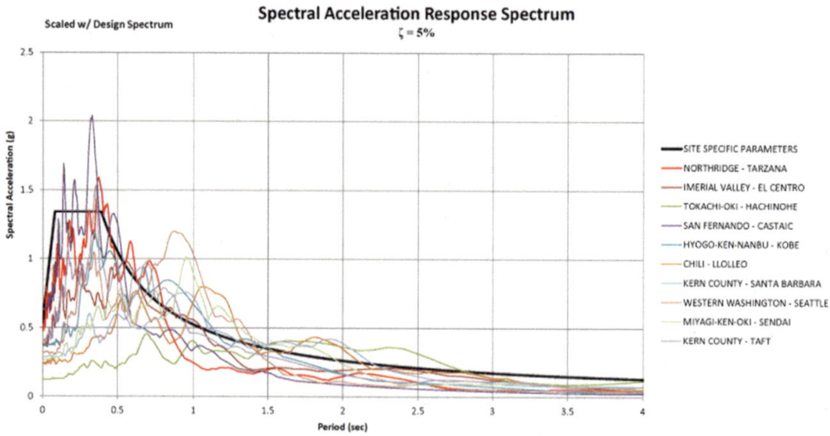

Figure 11-4. Variability in the elastic response spectrum from 10 ground motions of similar magnitude and distance.
Source: Harris et al. (2013).

histories, all of which are considered representative of a design event. Selection of representative time histories should always be subject to peer review from a professional with experience in this field (ASCE 7-16).

11.6 PSEUDO-STATIC LIMIT EQUILIBRIUM

In regions of low seismicity, and for ductile waste containment systems, seismic stability may be demonstrated by a simple pseudo-static limit equilibrium analysis. In such an analysis, either the FS is calculated using an appropriate seismic coefficient, k_s, and compared to a target FS value, or the yield acceleration, k_y, is compared to an appropriate peak acceleration. These two approaches are identical when the target FS is 1.0, which is generally considered an appropriate target in cases where the design ground motion is obtained on the basis of an extreme event (e.g., a PGA with a 90% probability of not being exceeded in 250 years). By convention, large displacement–compatible strength parameters for the waste masses, earthen materials, and geosynthetic materials and interfaces are used to provide a conservative basis for such analyses. Note that the site-specific PGA is not, in general, the appropriate PGA to use as a seismic coefficient, or for comparison to the yield acceleration, k_y.

As a first step in the pseudo-static limit equilibrium analysis, the site-specific free-field PGA may need to be adjusted to estimate the PGA for the waste containment system, e.g., the PGA at the top of the waste mass of the landfill. Observations and analyses of solid waste landfill response

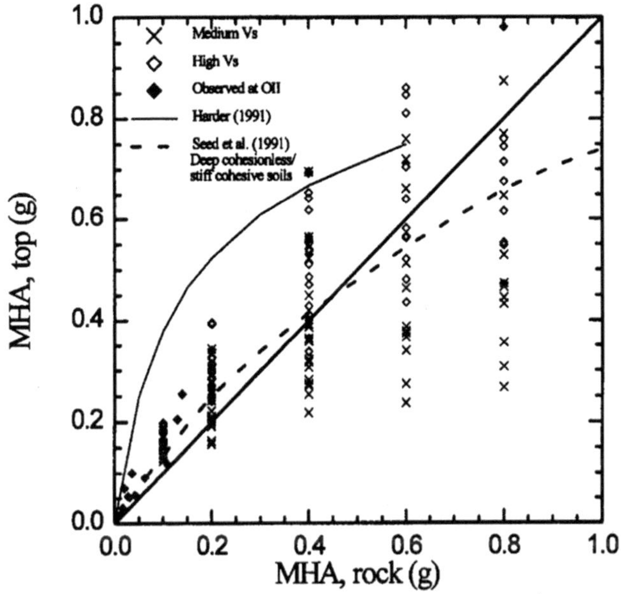

Figure 11-5. PGA amplification at solid waste landfills.
Source: Bray and Rathje (1998).

have shown that, like earthen embankments, the free-field acceleration may be significantly amplified by the waste mass (Bray and Rathje 1998). Figure 11-5, from Bray and Rathje (1998), illustrates the free-field site PGA (termed the *MHA, rock*) versus the PGA at the top of the waste mass (i.e., *MHA, top*), on the basis of multiple analyses of MSW landfill seismic response.

The "High V_S" and "Medium V_S" data points in Figure 11-5 are based on results from one-dimensional landfill response analyses. The bounding surface in Figure 11-5 [i.e., the "Harder 1991" curve] may be used in lieu of seismic response analyses when the MHA in rock is < 0.4 g to estimate the PGA (i.e., *MHA, top*) of the waste mass. However, using the PGA at the top of the waste mass, divided by the acceleration of gravity (g) as the seismic coefficient (k_s) or for comparison to the yield acceleration (k_y) is a very conservative means of assessing seismic stability, particularly when shear strength parameters at large displacements are used for the analysis. If this approach yields acceptable values (i.e., if the pseudo-static FS is > 1.0 or the yield acceleration is greater than the PGA of the waste mass), the waste containment system is unconditionally stable. Because of the exceedingly conservative nature of this approach, it is unlikely to indicate satisfactory performance unless the waste containment system has a high static FS and is in a region of low seismicity. In other cases, experience shows that a more sophisticated analysis may show the performance of the containment system when subject to the design seismic loading to be acceptable (depending on the performance standard), even if the pseudo-static FS is < 1.0 or the yield acceleration is less than the PGA of the waste mass.

A somewhat less conservative and more realistic approach than using the PGA of the potential failure mass is the use of the peak average acceleration of the potential failure mass, termed the MHA (maximum horizontal acceleration) by Bray and Rathje (1998), as the seismic coefficient (k_s) or as the limiting value for the yield acceleration (k_y). The MHA of the potential failure mass is proportional to the peak inertial force of the potential failure mass and is always less than the PGA of the waste mass because of spatial and temporal incoherence. By definition, as illustrated in Figure 11-6, the peak accelerations at other points within the potential failure mass are smaller than the PGA of the waste mass. The peak accelerations at other points within the potential failure mass will also typically occur at a different time from the PGA because of the propagation of the seismic wave. Thus, when accelerations are averaged over the entire potential failure mass, as illustrated in Figure 11-6, the failure mass MHA will be less than the PGA. The reduction from the PGA to the MHA becomes smaller with increasing size of the considered failure mass.

Although using the MHA of the potential failure mass as k_s, or as the limiting value for k_y, is less conservative than using the PGA of the waste mass, it is still a very conservative approach to pseudo-static analysis, considering that (1) the MHA occurs only one time, instantly, during an earthquake, (2) the waste containment system elements are commonly ductile, and (3) large displacement–compatible shear strength parameters are used in the limit equilibrium analysis. Ductility of system elements allows for the average acceleration of the waste mass to exceed the yield acceleration multiple times before significant displacements occur, whereas the use of the large displacement shear strength provides an additional margin of conservatism.

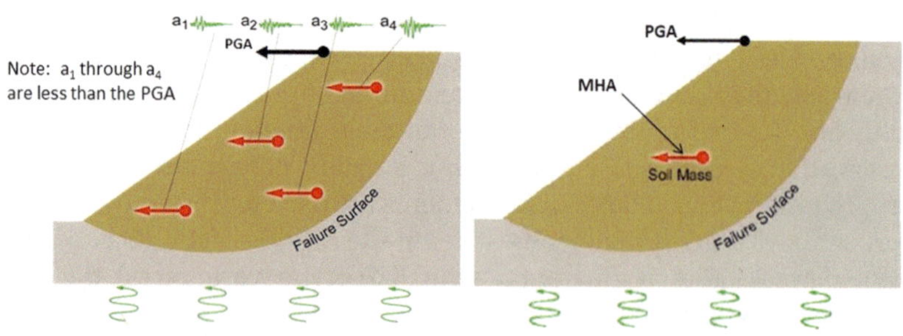

Figure 11-6. Peak average acceleration (MHA) of the waste mass. Source: Anderson et al. (2008).

Several approaches can be used to calculate the MHA of a potential failure mass from the PGA of the waste mass. Historically, the Makdisi and Seed (1978) approach has been used to calculate the MHA (termed k_{max} by the authors) in earth dams as a function of the depth of the sliding surface beneath the crest of the dam. The Makdisi and Seed (1978) approach indicates that, for sliding surfaces that penetrate the full height of the embankment, the MHA may be no more than 50% and as little as 20% of the PGA at the crest of the embankment. However, this approach was based on a limited number of ground motions, a limited number of analyses, and material properties that are different from those of landfills. The Makdisi and Seed (1978) approach also does not consider a number of important factors influencing seismic response, including the absolute height of the embankment and the characteristics of the design ground motion.

Anderson et al. (2008) developed an alternate approach for the calculation of the MHA using a scaling factor, α:

$$\text{MHA} = \alpha \times \text{PGA} \tag{11-1}$$

The value of α takes into account both the height of the potential failure mass and the characteristics of the design ground motion. The characteristics of the design ground motion, and thus the value of α, are captured by a parameter β that is defined as

$$\beta = \left(S_1/\text{PGA}\right)_{ff} \tag{11-2}$$

where S_1 is the spectral acceleration at a spectral period of 1 s, and the subscript *ff* denotes the free-field, site condition–corrected ground motion.

Anderson et al. (2008) suggest that Equation (11-3) can be used to calculate the value of α for slope heights of < 30 m:

$$\alpha = 1 + 0.033H(0.5\beta - 1) \tag{11-3}$$

where H is the slope height in meters. For slope heights > 30 m, Anderson et al. (2008) suggest that the value of α be limited to the value obtained using β for a slope height of 30 m.

Anderson et al. (2008) report that a β value of 1.5 is representative of large-magnitude earthquakes (e.g., $M > 7.5$), typical of Western US design earthquakes, whereas β equal to 0.5 is representative of moderate magnitude earthquakes (e.g., $M = 5.5$ to 6.0), typical of Eastern US design earthquakes. As shown in Figure 11-7, values of α range from 0.75 to 0.25 for a failure mass with a height of 30 m (100 ft). These values are consistent with the values reported by Makdisi and Seed (1978), discussed previously.

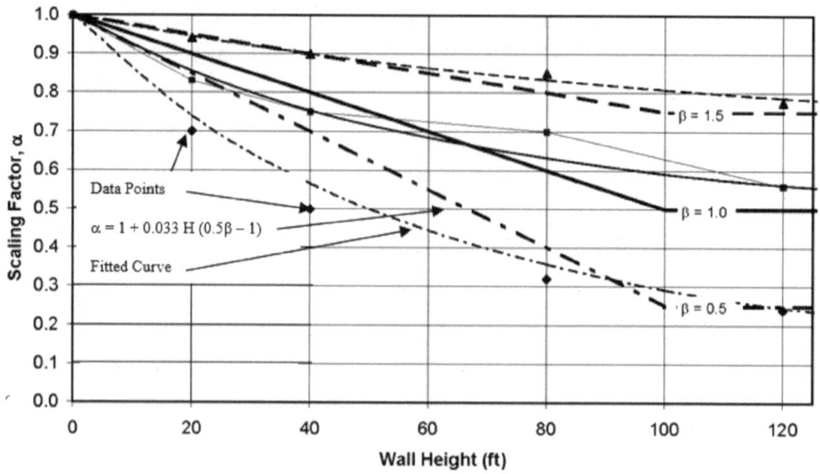

Figure 11-7. MHA scaling factor, α, for slopes and embankments. Source: Anderson et al. (2008).

The Anderson et al. (2008) approach, using the free-field spectral accelerations and the peak acceleration at the top of the potential failure mass, allows for a more refined analysis based on the characteristics of the slope or embankment, therefore providing a rational basis for the calculation of the MHA for a potential failure mass in a waste containment system.

A more sophisticated approach than the one embodied in Figure 11-7 is to use the results of a site response analysis to evaluate the MHA of a potential failure mass. As illustrated in Figure 11-8, the MHA (in terms of the acceleration of gravity, g) can be estimated by dividing the peak earthquake–induced shear stress at the base of a column representative of the failure mass, $(\tau_{base})_{max}$, by the total vertical stress at the bottom of this column, σ_v.

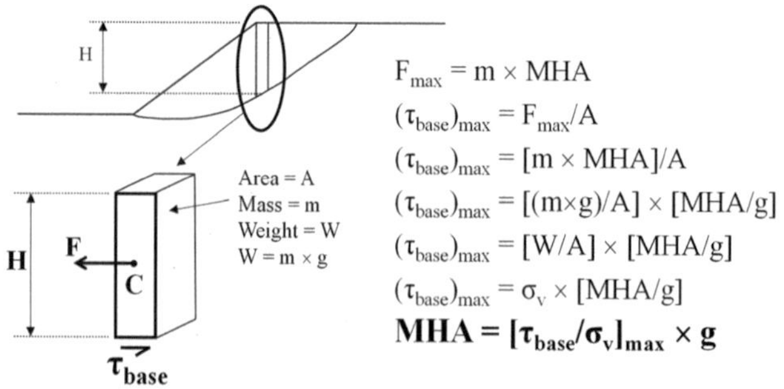

Figure 11-8. Relationship of the MHA to peak shear stress.

Newmark-type displacement analyses, conducted using a large number of acceleration time histories, suggest that an FS of 1.0 obtained using k_s equal to 50% of the MHA (or a limiting k_y equal to 50% of the MHA) corresponds to small-to-negligible calculated permanent seismic displacements (Anderson et al. 2008). This level of calculated seismic displacement indicates seismic performance of a waste containment system associated with a "minimal" or "no-damage" standard, provided that large displacement shear strengths are used in the analysis and that there are no brittle elements in the system (Kavazanjian 1999). Analyses indicate that even smaller fractions of the MHA may be used if a seismic performance standard corresponding to "measurable but limited" permanent seismic displacements is employed (Kavazanjian 1998).

11.7 NEWMARK-TYPE DISPLACEMENT ANALYSIS

If the simplistic and very conservative pseudo-static analysis approach using k_s equal to 50% of the PGA of the waste mass, or even a more sophisticated but still conservative analysis approach using the 50% of the MHA as k_s, does not yield satisfactory seismic performance (an FS > 1.0), a quantitative Newmark-type seismic displacement analysis may still yield acceptable performance. This type of analysis may be necessary in regions of high seismicity, except for cases with a very high static FS. Even in regions of low seismicity, if the seismic safety margin is low because of a low k_y, or if brittle system elements or hazardous wastes are present, or if a more stringent seismic performance standard is in place (e.g., a "no-damage" criteria), a quantitative Newmark-type analysis may be necessary.

A key parameter in a Newmark-type seismic displacement analysis is k_y, the yield acceleration of the potential failure mass. As in the pseudo-static method, the yield acceleration in a Newmark-type analysis is typically evaluated using the large displacement shear strength. Although analyses are available that allow the shear strength to be specified as a function of the cumulative seismic displacement (Matasovic et al. 1997, 1998b), recommendations for the allowable calculated seismic displacement are not calibrated on this type of analysis. Newmark-type permanent seismic displacement analyses may be conducted using charts, correlation equations, or acceleration time histories resulting from a containment system seismic response analysis.

11.7.1 Simplified Newmark-Type Analysis Chart Procedures

The Makdisi and Seed (1978) method is the classical chart-based approach to a Newmark-type seismic displacement analysis in which the permanent seismic displacement is obtained as a function of the ratio of the yield acceleration, k_y, to the MHA (termed k_{max} by the authors) for a given earthquake magnitude. However, as mentioned previously, this approach has limitations because of the number of ground motions, analyses, and material properties used to develop the Makdisi and Seed (1978) design charts.

Significant advances have been made since Makdisi and Seed (1978) that have led to improved simplified methods for calculating seismic displacements using Newmark-type analyses. These studies have demonstrated the significant influence of ground motion variability on the results of the predicted seismic displacements. Although the Bray and Rathje (1998) methodology was developed specifically for landfills and accounts for the nonlinear stress–strain response of the waste mass, it still is based on a decoupled one-dimensional seismic response analysis. Bray and Travasarou (2007) used a nonlinear coupled stick-slip sliding model in their seismic displacement analyses. The Bray and Travasarou (2007) nonlinear coupled stick-slip deformable sliding block model offers a more realistic representation of the dynamic response of a waste mass by (1) accounting for the deformability of the sliding mass and (2) considering the simultaneous occurrence of its nonlinear dynamic response and periodic sliding episodes along a potential sliding plane. The authors found that seismic displacements were significantly affected by the yield coefficient k_y, the considered failure mass initial fundamental period T_s, and the ground motion's spectral acceleration at a degraded period equal to $1.5T_s$. T_s is evaluated on the basis of a representative waste mass thickness, H_{wm}, and the average shear wave velocity of the waste mass, V_{wm}, as

$$T_s = \alpha \times H_{wm} / V_{wm} \tag{11-4}$$

where α is a coefficient based on the geometry of the potential sliding mass: $\alpha = 4$ is recommended for the case of a relatively wide potential sliding mass that is shaped like either a trapezoid or a segment of a circle where its response is largely 1D; $\alpha = 2.6$ for the case of a triangle-shaped sliding mass that largely has a two-dimensional (2D) response.

Anderson et al. (2008) presented correlation equations on the basis of thousands of Newmark-type displacement analyses relating the calculated permanent seismic displacement to k_y, the MHA, and the peak ground velocity (PGV) of the ground motion. The PGV of the ground motion was then correlated to S_1, the spectral acceleration at a spectral period of 1 s. The reader is referred to National Cooperative Research Program Report 611 (Anderson et al. 2008) for these equations. Rathje and Antonakos (2011) developed a predictive model for rigid sliding blocks (representative of shallow, stiff slopes) and flexible sliding masses (for deeper and softer slopes that have natural periods that are not close to zero) in which the seismic displacement is a function of the PGA, PGV, the natural period of the sliding mass (T_s), and the mean period of the earthquake motion (T_m) (Rathje et al. 2004). More recently, Bray et al. (2018) and Bray and Macedo (2019) used a fully coupled, nonlinear, deformable stick-slip sliding model and an expanded ground motion database to account for the influence of ground motion variability on the predicted seismic displacements. Bray et al. (2018) used 810 two-component subduction zone ground motion recordings, and Bray and Macedo (2019) used 6,711 two-component horizontal ground motion recordings to develop equations for the permanent seismic displacement. The Bray and Macedo (2019) analysis considers the system's yield coefficient k_y, its initial fundamental period T_s, the ground motion's spectral acceleration at a degraded period of the slope taken as $1.3T_s$, the earthquake magnitude as a proxy for duration, and PGV and fault-normal effects for near-fault pulse motions.

11.7.2 Site-Specific Newmark-Type Analysis

As an alternative to the design chart approach, a one-dimensional seismic response analysis of a column representative of the potential failure mass can be conducted and the results of the analysis (i.e., the average acceleration time history of the potential failure mass) can be used directly in a Newmark-type analysis to calculate the permanent seismic displacement. Consistent with the derivation of the MHA presented in Figure 11-8, the shear stress time history at the base of the column divided by the total vertical stress at the base of the column provides the average acceleration time history of the potential failure mass (in g's). Some commercial site response analysis programs have provisions for a Newmark-type displacement analysis built into them. Alternatively, a freeware program such as SLAMMER (Jibson et al. 2013) can be used.

Keeping in mind the required assumptions and the conservatism inherent to a decoupled Newmark analysis that employs large-deformation strengths, displacements calculated in this type of analysis should still be regarded merely as an index of seismic performance. The calculated permanent seismic displacement from a Newmark-type analysis must be compared to acceptable values for the various components of a waste containment system. Based on a comparison of results from Newmark-type analyses that used residual shear strength parameters to observations of the performance of earth structures under seismic loading (Augello et al. 1995), a calculated permanent seismic displacement of 5 cm or less is generally considered representative of "negligible" seismic displacements, and a calculated value of 15 cm or

Table 11-3. Relationship of Seismic Displacements Calculated from a Decoupled Newmark-Type Analysis to Field Displacements.

Calculated displacement	< 5 cm	5–15 cm	15–30 cm	30–1 m	> 1 m
Anticipated displacement	Negligible	Minimal (no damage)	Minor cracking	Constrained (small)	Unconstrained (large)

Source: Kavazanjian (1999).

less is generally considered representative of "minimal" displacements, not likely to damage containment system elements. Calculated seismic displacements of 30 cm to 1 m are considered representative of "constrained" (i.e., limited) displacements, whereas calculated displacements > 1 m are generally considered to be indicative of large displacements and the potential for unconstrained movement.

The relationship between calculated displacements from a Newmark-type analysis and the anticipated field displacements based on Augello et al. (1995) (as summarized by Kavazanjian 1999) is presented in Table 11-3. It should be emphasized that these criteria are based on conventional decoupled Newmark-type analyses, using large displacement strength values, in which the response of the potential failure mass is calculated assuming that there is no relative displacement (e.g., no slip at a liner interface) between the waste mass and the waste containment system elements. Criteria on the allowable displacements based on analyses that attempt to accurately calculate the seismic response of the landfill, including the forces, stains, and displacements induced in the waste mass and containment system elements, are likely to be substantially more rigorous than these values.

An important consideration in a limit equilibrium analysis is that the MHA, the yield acceleration, and the factor of safety may vary with the geometry of the potential failure surface. In general, the critical surface (i.e., the surface yielding the largest calculated seismic displacement) may be neither the surface with the largest MHA nor the surface with the lowest yield acceleration, but it is likely to be the surface with the smallest ratio of k_y/MHA. Therefore, it may be necessary to evaluate an *en echelon* family of failure surfaces within the framework of a seismic stability (and displacement) analysis. Moreover, different elements of the containment system have different levels of seismic fragility. Thus, as illustrated in Figure 11-9, a comprehensive displacement analysis must consider a family of potential failure surfaces.

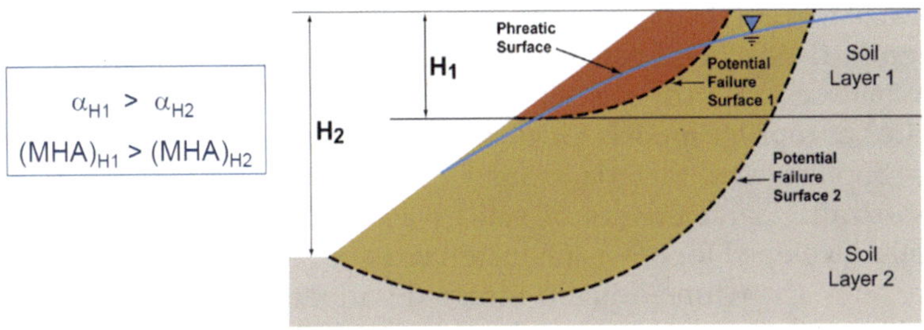

Figure 11-9. En echelon failure surfaces for seismic displacement analysis. Source: Anderson et al. (2008).

11.8 STRESS–DEFORMATION DYNAMIC ANALYSES

At the most sophisticated technical level, if a more precise assessment of seismic performance is desired, or if a Newmark-type displacement analysis does not yield acceptable seismic performance, a stress–deformation analysis of the seismic response of the waste containment system can be conducted. Both finite-element and finite difference analyses can be employed for this purpose, and there are still several levels of complexity and precision for these types of analyses. As noted previously, a seismic analysis of this type should never rely on just a single time history. The standard practice is to use the maximum computed response if a limited number of time histories (e.g., 3 to 5) is used in the analysis, or to use the average computed response if a relatively large suite of time histories (e.g., 10 or more) is used.

The simplest type of 2-D stress deformation analysis is a frequency domain analysis in which the geosynthetic elements of the waste containment system are not explicitly modeled. This type of analysis can be used to evaluate seismically induced stresses and deformations in the foundation and waste mass and, by exporting the average acceleration time history of a potential failure mass, to estimate permanent seismic displacements through a Newmark-type analysis. Note that a limit equilibrium analysis to estimate the yield acceleration of the potential failure mass as well as consider a family of potential failure masses is still required for this type of analysis. Matasović and Kavazanjian (1998) provide details regarding this type of analysis for evaluation of the stability of an unlined solid waste landfill. Matasovic and Kavazanjian (2006) describe using this type of analysis to assess the performance of the geosynthetic elements of a waste containment system.

The most sophisticated type of seismic stability analysis for a waste containment system is a time-domain, stress–deformation analysis conducted to estimate the seismically induced displacements, strains, forces, and stresses on the waste mass and/or the waste containment system. An example of a time-domain stress–deformation analysis for an old, unlined landfill is shown in Figure 11-10 (Rix et al. 2018). Examples of such analyses are rare. Part of the reason for this is that such analyses require appropriate models for all materials involved, including the waste mass. Sophisticated material behavior models that capture the dynamic stress–strain characteristics of solid waste do not presently exist, but models developed for other soil materials have been used (Rix et al. 2018). Even rarer than time-domain stress–deformation models of unlined landfills are the same type of analyses for landfills with geosynthetic liner systems, as these models also require advanced constitutive models for liner system elements and interfaces. Kavazanjian et al. (2014, 2018) and Arab et al. (2011) describe models that capture the dynamic characteristics of geosynthetic liner materials and liner interfaces, including postpeak strength degradation, and the use of the models in the performance-based seismic design (PBSD) of waste containment systems with geosynthetic elements. Wu (2017) applied this model to the prediction of liner strains measured in a dynamic centrifuge test of a geosynthetic-lined landfill (Kavazanjian and Gutierrez 2017).

The tensile strains in geosynthetic elements of the containment system calculated in a PBSD analysis can be compared to allowable values to assess the potential for barrier layer breaching and the subsequent release of contaminants. However, the allowable strain of geosynthetic liner elements is significantly less than the yield strain measured on an intact geomembrane because of a variety of factors, including strain

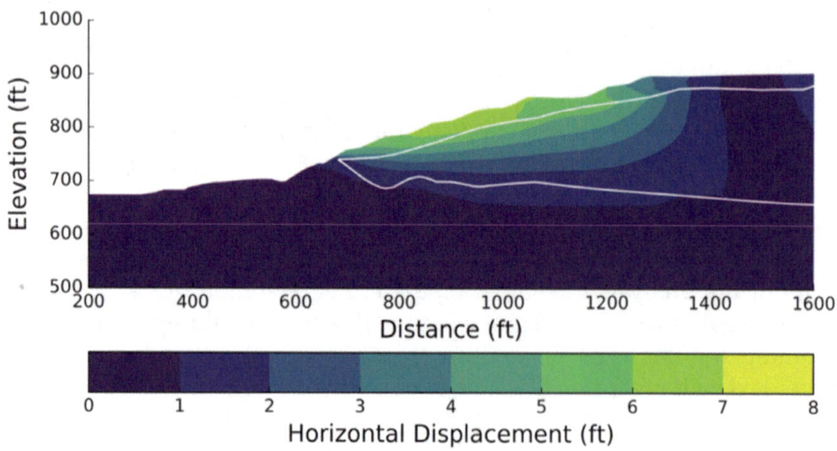

Figure 11-10. Example of calculated horizontal seismic displacement for an unlined hazardous waste landfill in Southern California.
Source: Rix et al. (2018).

concentrations caused by seams and scratches, wrinkles in the geomembrane, and penetrations of the membrane. Taking these factors into account, 4% tensile strain is often considered the maximum allowable tensile strain for a geomembrane with a yield strain of 11% to 15% in a uniaxial tension test (Peggs et al. 2005). Recent work on strain concentrations caused by seams and scratches (Kavazanjian et al. 2017, Andresen and Kavazanjian 2017) and wrinkles (Mitchell et al. 2016) suggests that even 4% may be too large of a value when allowable tensile strains in a geomembrane are considered. Parametric studies conducted by Kavazanjian et al. (2018) using a PBSD model suggest that critical areas of tension in geomembrane liners for landfills are on the benches and the top third to one-half of the side slope, as illustrated in Figure 11-11. These analyses also show that when the peak interface shear strength on the bottom of the geomembrane is greater than the peak interface strength on top of the geomembrane, settlement- and earthquake-induced tensile strains in the geomembrane may be minimized. More detailed numerical analyses reported by Gao et al. (2021) led to similar conclusions.

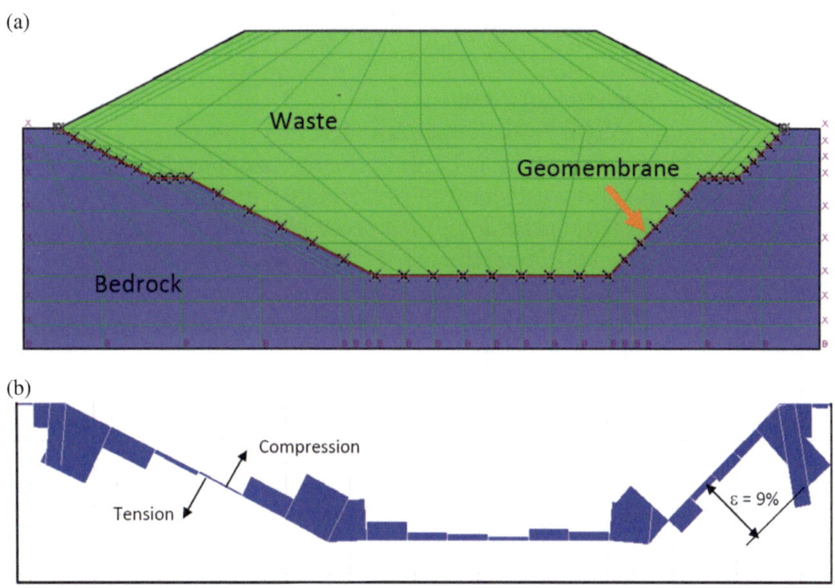

Figure 11-11. Numerical analysis of seismic performance of a geosynthetic-lined landfill: (a) finite difference mesh of a geomembrane-lined landfill, (b) maximum geomembrane strains between the base and the first bench during seismic loading. Source: Kavazanjian et al. (2018).

REFERENCES

AASHTO (American Association of State Highway and Transportation Officials). 2015. *Guide specifications for the LRFD bridge design, 2nd ed., with 2012, 2014, and 2015 interim revisions*. Washington, DC: AASHTO.

Anderson, D. G., G. R. Martin, I. Lam, and J. N. Wang. 2008. Vol. 1 of *Seismic analysis and design of retaining walls, buried structures, slopes, and embankments*. NCHRP Rep. 611. Washington, DC: Transportation Research Board.

Andresen, J., and E. Kavazanjian Jr. 2017. "Experimental evaluation of strain concentrations in HDPE geomembranes due to scratches." In *Proc., Sardinia 2017, 16th Int. Waste Management and Landfill Symp.* https://eu-refresh.org/sardinia-2017-16th-waste-management-and-landfill-symposium.html.

Arab, M. G., E. Kavazanjian Jr., P. J. Fox, and J. D. Ross. 2011. "Displacement-softening constitutive model for geosynthetic interfaces." In *Proc., 14th Pan American Conf. on Soil Mechanics and Geotechnical Engineering*. Vancouver: Canadian Geotechnical Society.

ASCE. 2005. *Minimum design loads for buildings and other structures*. ASCE 7-05. Reston, VA: ASCE.

ASCE. 2010. *Minimum design loads for buildings and other structures*. ASCE 7-10. Reston, VA: ASCE.

ASCE. 2017. *Minimum design loads and associated criteria for buildings and other structures*. ASCE 7-16. Reston, VA: ASCE.

Augello, A. J., N. Matasovic, J. D. Bray, E. Kavazanjian Jr., et al. 1995. "Evaluation of solid waste landfill performance during the Northridge earthquake." In *Earthquake design and performance of solid waste landfills*, Geotechnical Special Publication 54, 17–50. Reston, VA: ASCE.

Bray, J. D. 2007. "Simplified seismic slope displacement procedures." In *Earthquake geotechnical engineering*, edited by K. D. Pitilakis, 327–353. Dordrecht, Netherlands: Springer.

Bray, J. D., and J. Macedo. 2019. "Procedure for estimating shear-induced seismic slope displacement for shallow crustal earthquakes." *J. Geotech. Geoenviron. Eng.* 145 (12): 04019106.

Bray, J. D., J. Macedo, and T. Travasarou. 2018. "Simplified procedure for estimating seismic slope displacements for subduction zone earthquakes." *J. Geotech. Geoenviron. Eng.* 144 (3): 04017124.

Bray, J. D., and E. M. Rathje. 1998. "Earthquake-induced displacements of solid-waste landfills." *J. Geotech. Geoenviron. Eng.* 124 (3): 242–253.

Bray, J. D., and T. Travasarou. 2007. "Simplified procedure for estimating earthquake-induced deviatoric slope displacements." *J. Geotech. Geoenviron. Eng.* 133 (4): 381–392.

Choi, Y., and J. P. Stewart. 2005. "Nonlinear site amplification as function of 30 m shear wave velocity." *Earthquake Spectra* 21 (1): 1–30.

Gao, W., E. Kavazanjian Jr., and X. Wu. 2021. "Numerical study of strain development in high-density polyethylene geomembrane liner system in landfills using a new constitutive model for municipal solid waste." *Geotext. Geomembr.* 50 (2): 216–230.

Ghanat, S., and E. Kavazanjian Jr. 2011. "Site-specific response analysis at non-standard sites." In *Proc., 5th Int. Conf. on Earthquake Geotechnical Engineering*. Paper No. SSRGH. London: International Society for Soil Mechanics and Geotechnical Engineering.

Harris, K. D., N. D. Robinson, and E. L. Sammarco. 2013. "Seismic time histories." *Structure*. Accessed March 5, 2024. https://www.structuremag.org/?p=886.

Hashash, Y. M. A., and D. Park. 2001. "Nonlinear one-dimensional seismic ground motion propagation in the Mississippi embayment." *Eng. Geol.* 62 (1–3): 185–206.

Huang, Y. N., A. S. Whittaker, and N. Luco. 2010. "NEHRP site amplification factors and the NGA relationships." *Earthquake Spectra* 26 (2): 583–593.

Jibson, R. W., E. M. Rathje, M. W. Jibson, and Y. W. Lee. 2013. "SLAMMER—Seismic landslide movement modeled using earthquake records." Chap. B1 of *US Geological Survey techniques and methods*, Book 12. Reston, VA: US Geological Survey.

Kavazanjian, E. Jr. 1998. "Current issues in seismic design of geosynthetic cover systems." In Vol. 1 of *Proc., 6th Int. Conf. on Geosynthetics*, 219–226. West Roseville, MN: Industrial Fabrics Association International.

Kavazanjian, E. Jr. 1999. "Seismic design of solid waste containment facilities." In *Proc., 8th Canadian Conf. on Earthquake Engineering*, 51–68. Vancouver: Canadian Association for Earthquake Engineering and Seismology.

Kavazanjian, E. Jr., J. Andresen, and A. Gutierrez. 2017. "Experimental evaluation of HDPE geomembrane seam strain concentrations." *Geosynth. Int.* 24 (4): 1–10.

Kavazanjian, E. Jr., M. G. Arab, P. J. Fox, and N. Matasovic. 2014. "Performance based seismic design of geosynthetic barriers for waste containment." In *Earthquake geotechnical engineering design*, edited by M. Maugeri and C. Soccodato, 363–385. Berlin: Springer.

Kavazanjian, E. Jr., and A. Gutierrez. 2017. "Large scale centrifuge test of a geomembrane-lined landfill subject to waste settlement and seismic loading." *Waste Manage.* 68 (October): 252–262.

Kavazanjian, E. Jr., N. Matasovic, and J. Caldwell. 1998. "Damage criteria for solid waste landfills." In *Proc., 6th US National Conf. on Earthquake Engineering*. Oakland, CA: Earthquake Engineering Research Institute.

Kavazanjian, E. Jr., J.-N. Wang, G. R. Martin, A. Shamsabadi, et al. 2011. "LRFD seismic analysis and design of transportation geotechnical features and structural foundations." *Geotech. Eng. Circular No. 3, Report No. FHWA-NHI-11-032*. Washington, DC: Federal Highway Administration.

Kavazanjian, E. Jr., X. Wu, M. Arab, and N. Matasovic. 2018. "Development of a numerical model for performance-based design of geosynthetic liner systems." *Geotext. Geomembr.* 46 (2): 166–182.

Kottke, A., and E. M. Rathje. 2008. "A semi-automated procedure for selecting and scaling recorded earthquake motions for dynamic analysis." *Earthquake Spectra* 24 (4): 911–932.

Makdisi, F. I., and H. B. Seed. 1978. "Simplified procedure for estimating dam and embankment earthquake-induced deformations." *J. Geotech. Geoenviron. Eng.* 104 (GT7): 849–867.

Matasović, N., and E. Kavazanjian Jr. 1998. "Cyclic characterization of OII landfill solid waste." *J. Geotech. Geoenviron. Eng.* 124 (3): 197–210.

Matasovic, N., and E. Kavazanjian Jr. 2006. "Seismic response of a composite landfill cover." *J. Geotech. Geoenviron. Eng.* 132 (4): 448–455.

Matasovic, N., E. Kavazanjian Jr., and R. L. Anderson. 1998a. "Performance of solid waste landfills in earthquakes." *Earthquake Spectra* 14 (2): 319–334.

Matasovic, N., E. Kavazanjian Jr., and J. P. Giroud. 1998b. "Newmark seismic deformation analysis for geosynthetic covers." *Geosynth. Int.* 5 (1-2): 237–264.

Matasovic, N., E. Kavazanjian Jr., and L. Yan. 1997. "Newmark deformation analysis with degrading yield acceleration." In Vol. 2 of *Proc., Geosynthetic '97*, 989–1000. Albany, NY: North American Geosynthetics Society. https://geosyntheticsconference.com/proceedings-archive/.

Mitchell, T., G. J. Fowmes, R. Wackrow, N. Dixon, et al. 2016. "Assessing the planarity of containment liner systems using a terrestrial laser scanner." In *Proc., 6th European Geosynthetics Congress.* Austin, TX: International Geosynthetics Society.

NEHRP (National Earthquake Hazards Reduction Program). 2009. *NEHRP recommended seismic provisions for new buildings and other structures (FEMA P-750).* Washington, DC: Building Seismic Safety Council.

Parker, G. A., J. P. Stewart, Y. M. Hashash, E. M. Rathje, et al. 2019. "Empirical linear seismic site amplification in central and eastern North America." *Earthquake Spectra* 35 (2): 849–881.

Peggs, I. D., B. Schmucker, and P. Carey. 2005. "Assessment of maximum allowable strains in polyethylene and polypropylene geomembranes." In *GeoFrontiers 2015: Waste Containment and Remediation*, Geotechnical Special Publication 142. Reston, VA: ASCE.

Pehlivan, M., D. Park, S. Sadiq, and Y. M. A. Hashash. 2017. "Seismic site response of shallow sites in moderate seismicity regions." In *Proc., 19th Int. Conf. on Soil Mechanics and Geotechnical Engineering*, 1579–1572. London: International Society for Soil Mechanics and Geotechnical Engineering.

Rathje, E. M., and G. Antonakos. 2011. "A unified model for predicting earthquake-induced sliding displacements of rigid and flexible slopes." *Eng. Geol* 122 (1–2): 51–60.

Rathje, E. M., F. Faraj, S. Russell, and J. D. Bray. 2004. "Empirical relationships for frequency content parameters of earthquake ground motions." *Earthquake Spectra* 20 (1): 119–144.

Rix, G. J., R. C. Bachus, C. Conkle, and M. Schultheis. 2018. "Evaluation of the seismic performance of a class I landfill." In *Geotechnical earthquake engineering and soil dynamics V: Slope stability and landslides, laboratory testing, and in situ testing*, 56–66. Reston, VA: ASCE.

USGS (US Geological Survey). 2014. "National seismic hazard map, PGA 2% in 50 yrs." Accessed December 28, 2017. https://earthquake.usgs.gov/hazards/hazmaps/conterminous/.

USGS. 2024. "USGS national seismic hazard mapping program (NSHMP)." Accessed March 5, 2024. https://www.usgs.gov/natural-hazards/earthquake-hazards/design-ground-motions.

Wu, X. 2017. "Validation of a Numerical Model for Design of Geomembranes Subject to Extreme Loads." Ph.D. diss. Arizona State University, Dept. of Civil and Environmental Engineering.

Zekkos, D., and C. Carlson. 2011. "Scaling vs. modification of ground motions and impact on seismic analyses of a MSW landfill." In *Proc., 5th Int. Conf. on Earthquake Geotechnical Engineering*, 1–12. London: International Society for Soil Mechanics and Geotechnical Engineering.

PART 5

MINE WASTE GEOTECHNICS

CHAPTER 12

MINE WASTE GEOTECHNICS: CHARACTERISTICS AND MANAGEMENT

Christopher A. Bareither, Ph.D., P.E., M. ASCE

12.1 INTRODUCTION

Mine operations produce considerable quantities of waste materials as a result of ore extraction and refining processes. Two types of mine waste that present relevant challenges to practicing engineers and require on-site management are mine tailings and waste rock. Mine tailings are ubiquitous in mining operations and are typically fine-grained materials with high water contents (low solids contents) that are commonly disposed as slurry in tailing storage facilities (TSFs). Waste rock is usually managed in piles and predominantly contains boulder- and gravel-sized particles with some sand and fines (Bussière 2007, Blight 2010). Management of tailings and waste rock often requires large areas of land, may present challenges with acid rock drainage (ARD) if sulfide minerals (e.g., pyrite) are present, and may present water contamination concerns depending on leachate and seepage chemistry. Management of tailings also includes challenges related to large potential settlement and low shear strength, which are typical engineering characteristics of slurry-deposited mine tailings (Qiu and Sego 2001, Bussière 2007).

The objective of this section is to provide an overview of the characteristics, management techniques, and common challenges encountered by geotechnical engineers working on waste rock and tailing projects. This content has been prepared from a case study–based course developed at Colorado State University titled, *Mining Geotechnics*, which was created via a partnership with the Tailings and Mine Waste (TMW) Conference Committee and a research grant from the National Science Foundation. This section should be viewed only as a summary of the subject matter, and interested readers are referred to the proceedings of the

TMW Conference, relevant journal articles, forensic reports (e.g., Morgenstern et al. 2015, 2016; Robertson et al. 2019), textbooks (e.g., Vick 1990, Blight 2010), as well as other resources.

12.2 MINE WASTE ROCK AND TAILINGS

Mine waste rock and mine tailings are generated from ore extraction processes. Ore is the earthen material (typically rock) that contains a valuable commodity that can be extracted profitably. Common minerals and mined commodities include copper, gold, silver, zinc, iron, precious metals, phosphate, and hydrocarbons (e.g., coal, oil, and gas), among many others. Production of raw metals and minerals has increased over the past decades and will increase in the foreseeable future in response to resource demands from the green energy economy, vehicle electrification, and novel energy storage, among others (e.g., Herrington 2021, Song et al. 2022). Lithium in particular has proven reserves that may last only until 2080 because of the high demand associated with lithium-ion batteries in electrical vehicles (Luong et al. 2022). Considerations for mine development include ore grade, extent of the deposit, burial depth, site accessibility, mining technique, demand for the commodity of interest, environmental regulations, and closure/reclamation costs and requirements (Blight 2010). Mining companies are increasingly relying on large-scale exploitation of ever-lower grade ores, resulting in vast amounts of waste rock and tailings that must be managed in perpetuity.

Photographs of waste rock samples from two different gold mines are shown in Figure 12-1a, b. Waste rock is generated from rock containing uneconomical concentrations of the target commodity that exists around the ore body. Removal of this rock is essential to access the ore, and because there is no economic value in the rock, this material is a waste stream. The physical and chemical characteristics of waste rock depend on the composition of the parent rock, geologic history (e.g., extent of fracturing), blasting process, and transportation and disposal to the final waste containment location.

The photographs of waste rock in Figure 12-1 depict considerably different physical characteristics. The waste rock in Figure 12-1a contained predominantly crystalline rock with limited weathering potential and was nearly free of fine-grained particles (i.e., fines). In contrast, the waste rock in Figure 12-1b contained more weathered rock particles and had a fines content of approximately 14%. These photographs present a range of physical characteristics of waste rock and highlight the importance of not assuming characteristics prior to site characterization, sample collection, and/or laboratory analysis.

Figure 12-1. Mine waste: (a) waste rock that has strong particles with high weathering resistance; (b) waste rock that has weak particles, low weathering resistance, and fines content; (c) mine tailings discharged into a storage facility.

A photograph of slurry-discharged tailings into a TSF is shown in Figure 12-1c. Tailings are generated from ore processing. Typical ore processes include crushing and milling to grind the ore, hydrometallurgy to liberate target metals from the ore and form a concentrate, and separation and refining to subsequently treat the concentrate to meet end-point specifications. Tailings are the residuum of crushing, milling, and metallurgical processing and contain ground particles of the parent rock, unrecoverable minerals/metals, and process water. The physical and chemical characteristics of tailings from any given mine are unique, as these characteristics are dependent on the parent rock, crushing and milling, and metallurgical process. A typical tailing stream leaving the processing plant is a high-water-content material in slurry form. The slurry facilitates pipeline transport and direct disposal in a TSF (Figure 12-1c); however, dewatering tailings prior to or after disposal is also common to reclaim water for subsequent reuse, decrease the required storage volume of a TSF, and improve the shear strength of the deposited tailings.

12.2.1 Waste Rock

The generation and management of waste rock is primarily a challenge encountered in open-pit mining because of the overall mass of uneconomical rock that must be removed to exhume ore. In contrast, underground mining is more effective at targeting and exhuming the ore body, which can considerably decrease the amount of waste rock generated. The process of producing and managing waste rock, which includes blasting, excavating, hauling, and dumping for end-state management in piles, combined with the geologic makeup of the parent rock, can yield a broad range of coarse-grained materials (e.g., Figure 12-1a, b). Two key challenges encountered in waste rock management are the strength of the material, quantified by an effective stress friction angle (ϕ'), and the presence of sulfide minerals that can result in ARD.

A relationship between ϕ' and effective normal stress (σ_v') for rockfill materials is shown in Figure 12-2 on the basis of Blight (2010). Waste rock is typically deposited in piles via dumping, and the density of the waste rock pile and σ_n' at a given location in the pile increase with subsequent deposition and trafficking from surface equipment (i.e., mechanical

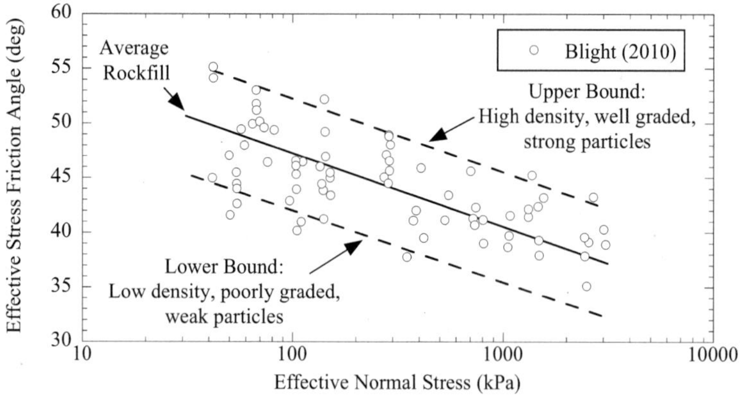

Figure 12-2. Relationship between effective stress friction angle (ϕ') and effective normal stress (σ_n') for rockfill materials.
Source: Blight (2010).

compaction). Adequately evaluating the ϕ' of waste rock is difficult because of the presence of large particles (e.g., > 300 mm); however, ϕ' can be estimated on the basis of rockfill properties, measured in the laboratory via scalping large particles to replicate parallel gradations (Marschi et al. 1972), or measured in the laboratory via large-scale laboratory equipment (Marsal 1967). The ϕ' in Figure 12-2 shows a range of approximately 10 degrees for a given σ_n', whereby higher ϕ' are representative of high-density, well-graded rockfill with strong particles and lower ϕ' are representative of lower-density, poorly graded rockfill with weak particles. The decrease in ϕ' with increasing σ_n' is attributed to the suppression of dilation and increased particle breakage at a higher stress. The range of ϕ' for a given material that may be anticipated over the broad range of σ_n' encountered in a waste rock pile can be simulated with a nonlinear shear strength envelope (e.g., Frossard et al. 2012).

The development of ARD is attributed to the presence of (1) sulfide minerals, (2) water, and (3) oxygen. An excellent source of information pertaining to ARD is presented in the Global Acid Rock Drainage Guide (INAP 2009) that is available to all practicing engineers. The presence of sulfide minerals in rock is natural and the development of acidic waters containing high metal contents can occur naturally as well as from mining activities. Conditions that lead to ARD are exacerbated in mining because of the exposure of large volumes of material that can contain sulfide minerals (rock or tailings) to atmospheric air and water, combined with an increased surface area of materials that enhances chemical reactions. These factors are encountered in all mines that target rock bodies containing sulfide minerals. Sources of ARD include waste rock piles, ore and low-grade ore stockpiles, tailings, heap leach materials, pit walls, and underground workings. The main reasons that ARD is typically linked to waste rock more so than the other sources are the large volumes of waste rock generated in open-pit mines combined with exposure to atmospheric air and precipitation. The range of potential prevention and mitigation strategies for ARD are extensive and typically focus on eliminating one of the three key ingredients (i.e., sulfides, oxygen, or water).

12.2.2 Mine Tailings

Mine tailings are the by-product of ore processing and the physical and chemical characteristics are a function of the ore mineralogy and physical–chemical extraction processes. Curves representing lower-bound (small particles), upper-bound (large particles), and average particle-size distributions (PSDs) for mine tailings compiled from the literature are shown in Figure 12-3 (Gorakhki and Bareither 2017). The upper-bound PSD represents a silty sand, whereas the lower-bound PSD represents a nearly 100% fine-grained material. The plasticity of mine

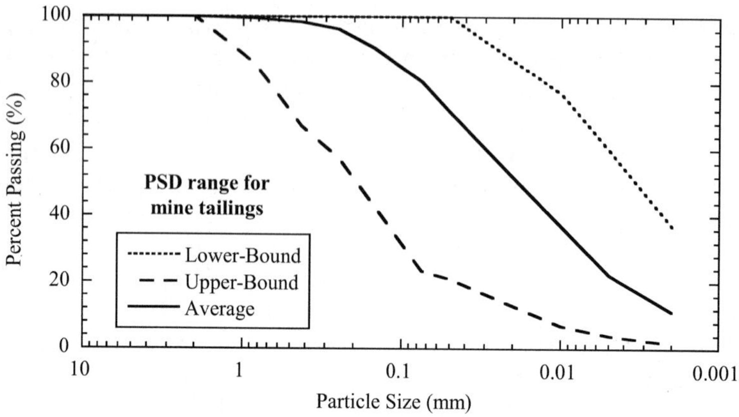

Figure 12-3. Average, upper-bound, and lower-bound particle-size distributions for mine tailings based on a review of literature.
Source: Gorakhki and Bareither (2017).

tailings is typically low such that the fine fraction of most mine tailings classifies as low plasticity silt. However, the presence of clay minerals in the ore can lead to mine tailings that classify and behave more as high-plasticity clays (e.g., phosphatic clays and oil sand tailings) (Carrier et al. 1983).

Mine tailings discharged from the milling and ore extraction process typically contain high water contents and exist as slurry. The consistency of mine tailings is more appropriately characterized by the concentration of solids, *SC* (i.e., solid mass/total mass), which can be modified on the basis of dewatering techniques. The nomenclature for four general types of mine tailings as a function of *SC* is identified in the tailing continuum shown in Figure 12-4. The yield stress (τ_y), which is analogous to undrained shear strength (S_u) of soils, is included in Figure 12-4 to capture the general strength–based break points between conventional slurry, thickened, paste, and filtered tailings. Specific values of τ_y or S_u to separate the

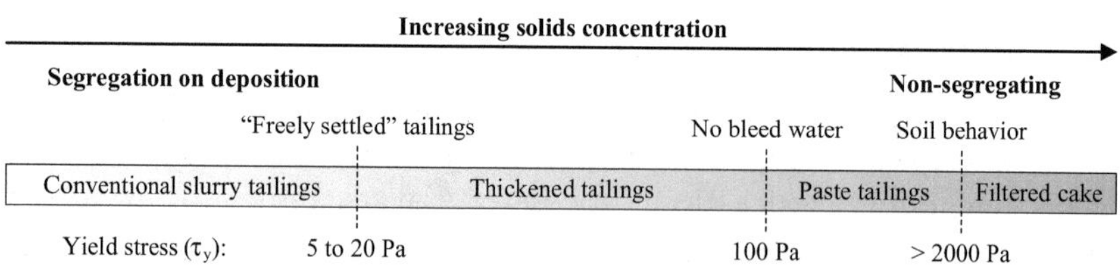

Figure 12-4. Schematic of the tailing continuum as a function of yield stress and solid concentration. Material behavior ranges from a "fluid-like" behavior of traditional slurry tailings to a "soil-like" behavior of filtered cake.

consistency states of mine tailings have not been broadly accepted (e.g., Jewell and Fourie 2006, Boger 2009), and the ranges identified in Figure 12-4 represent general ones used in practice.

A relationship between S_u (or τ_y) and water content for gold mine tailings presented in Blight (2010) is shown in Figure 12-5 along with the ranges of τ_y identified in Figure 12-4. Undrained shear strength in Figure 12-5 was measured via a laboratory vane shear and viscometer at two different rates of shear strain. The range of water content (or *SC*) over which the consistency of mine tailings can vary from conventional slurry tailings to filtered (dewatered) cake is dependent on the tailing composition. Reducing the water content, or increasing the solid content, such that tailings exhibit strength representative of filtered tailings transitions tailings from a rheological material that behaves as a fluid to a solid material that behaves as a soil.

Slurry-deposited tailings present challenges such as particle segregation, low shear strength, susceptibility to liquefaction, high settlement potential, and reduced TSF capacity. Dewatering via thickening and/or filtering can minimize some challenges encountered with slurry tailings. However, aspects such as tailing production rate, capital and operational costs, location of the TSF, and regulations, among others, can constrain opportunities to dewater tailings and control the type of tailing disposal and management practice at a given mine.

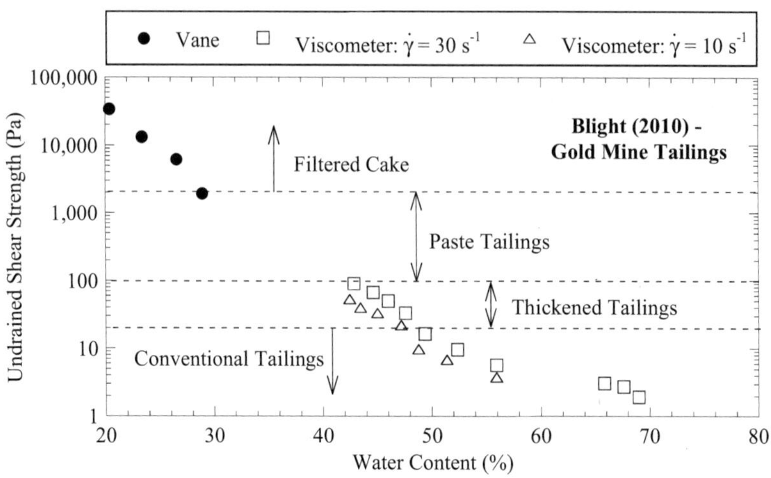

Figure 12-5. Relationships between undrained shear strength and water content for gold mine tailings. Undrained shear strength measured with laboratory vane shear and viscometer at two different rates of shear strain ($\dot{\gamma}$).
Source: Gold mine tailings, adapted from Blight (2010).

12.3 MINE WASTE MANAGEMENT

Common sources of mine waste generated at a given mine site that require management can include waste rock, tailings, spent ore on heap leach pads, process water, and contact water. Each mine waste source presents unique challenges to practicing engineers that, when combined with geographic location and climatic conditions of a given mine site, should be considered unique for any project. However, the general practices of mine waste management, particularly for waste rock and tailings, are similar based on years of development, consistency in proven techniques, and regulations that govern acceptable methodologies. The extent of this section is limited to waste rock and tailings, and this content is presented in concise, generalized discussions.

Barrier systems for containment of mine waste, which include liners and covers, may be required for any given project depending on local regulations, potential contaminants of concern, location, climate, and/or best practice, among numerous other factors. The four sections of containment systems (Chapters 8 to 11) in this manual are all relevant to barrier systems for mine waste. Considering the extent of coverage within this manual, as well as other relevant publications (e.g., Rowe et al. 2004, Albright et al. 2010), barrier systems are not elaborated herein for mine waste.

12.3.1 Waste Rock Piles

Waste rock is conventionally managed in piles that can reach heights exceeding 400 m and volumes exceeding 1 billion-m^3 (Eaton and Broughton 2010). Waste rock is transported via truck or conveyor and is mechanically placed in a dry to moist state, with water added primarily for dust control. The size of waste rock piles can lead to the development of high stress that can cause stability concerns within the foundation soils. Key challenges of waste rock piles are related to geotechnical and geochemical stability. Geotechnical stability is evaluated via the shear strength of the waste rock (e.g., Frossard et al. 2012, Duncan et al. 2014) and geochemical stability is evaluated via the assessment of potential acid-generating materials (INAP 2009).

12.3.2 Tailing Storage Facilities

Tailing storage facilities include two fundamental components: (1) dam/embankment to retain the tailings, and (2) impounded tailings. Schematics of upstream, downstream, and centerline tailing dam construction methods are shown in Figure 12-6. All three construction methods rely on a starter dam that allows tailings to be discharged behind

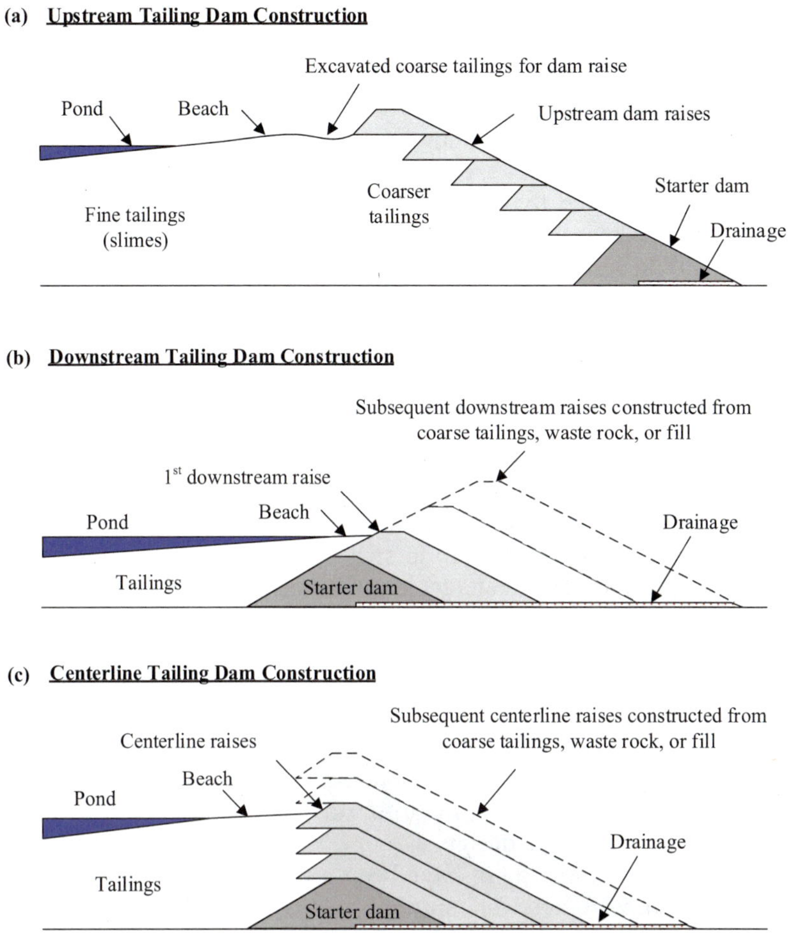

Figure 12-6. Schematics of (a) upstream, (b) downstream, and (c) centerline tailing dam construction methods used for mine tailing management in tailing storage facilities.
Source: IIED and WBSCD (2002).

the dam to form a tailing impoundment. Starter dams are commonly constructed of earth and/or rock fill material and include internal drainage to manage seepage. The impounded tailings can exist at any *SC* along the tailing continuum (Figure 12-4). However, the three tailing dam construction methods have developed from the management of conventional slurry tailings. A common approach is to deposit tailings via spigotting along the crest of a dam to allow coarse fraction settling and beach formation close to the crest. Alternatively, coarse tailings (i.e., sands) can be removed via cyclone and used for subsequent increases of the tailing dam above the starter dam. The overflow material from the cyclone is then discharged into the facility to form the main body of the impounded tailing deposit.

Two common types of TSFs are cross-valley impoundments and ring-dyke impoundments. Cross-valley impoundments are used in hilly and mountainous regions where topographic relief supports dam construction across the valley such that the impounded volume is bordered by the valley walls (natural containment) and tailing dam (constructed containment). In contrast, ring-dyke impoundments are used in areas with flat topography and include dams constructed on all sides to create a storage basin in the middle. Tailings can also be stored in the open pits where ore was exhumed; however, this practice requires an available open pit at a given mine site, which is not common during active mining operations.

Geotechnical challenges related to TSFs include stability of the dam slopes, foundation soils, and impounded tailings as well as settlement of the impounded tailings (i.e., impoundment capacity). Environmental challenges related to TSFs include surface erosion, dust control, and surface and groundwater management. Geotechnical stability of TSFs is evaluated on the basis of conventional limit equilibrium stability practices used in geotechnical engineering (e.g., Duncan et al. 2014), which should appropriately employ liquefied shear strength in such analyses where potentially liquefiable materials are present (e.g., Robertson et al. 2019, Fourie et al. 2022). The construction of larger tailing dams, particularly in high seismic regions, has led practitioners to employ more sophisticated approaches to appropriately incorporate stress–strain behavior and dynamic analysis into TSF design.

Over the last decade, there have been a number of high-profile TSF failures that have been well studied and led to changes in tailing management. These failures include Mount Polley (Morgenstern et al. 2015), Fundao (Morgenstern et al. 2016), Cadia (Jefferies et al. 2019), and Feijão (Robertson et al. 2019, CIMNE 2021). The reader is encouraged to read the forensic reports on each case history as well as the numerous publications associated with the failures. In each of the failures, tailing liquefaction was triggered, and although there were similarities between sites, each case is unique and should be evaluated in detail for a comprehensive understanding of the site conditions, engineering design, and operations that led to failure.

An important learning from the Mount Polley and Cadia failures was the need for an extensive subsurface investigation, particularly in geologic environments that include multiple and discontinuous layers within the foundation soils. In both cases, a unique layer in the foundation soils transitioned to a contractive and strain-softening layer when sheared undrained as the height of the embankment above increased. The potential for undrained failure of these unique, localized layers within the embankment foundations was not captured during initial site investigations.

The Fundão failure occurred in 2015 and the Feijão in 2019, both in the state of Minas Gerais, Brazil. Although each site and failure was unique and complex, a few key factors were shared. First, both tailing facilities had upstream-raised embankments that included stepbacks in the embankment as the raising progressed. These stepbacks caused subsequent embankment raises to be constructed over layers of previously deposited finer-grained tailings (i.e., slimes). The presence of slimes decreased permeability within the impounded tailings, resulting in a higher-than-desired phreatic surface. Finally, both tailing facilities included sandy tailings that were brittle, contractive, and saturated. Liquefaction was triggered in these sandy tailings, which resulted in catastrophic flow failures. An important takeaway from engineering practice is that if liquefaction-susceptible materials are present, it can be simply assumed that liquefaction will occur in the materials, in which case low-strength estimates for evaluation need to be used (Silvis and de Groot 1995, Morgenstern 2018).

The capacity of a TSF is critical to appropriately plan for storage of generated tailings such that ore processing can continue unimpeded. Mass balance of the tailing production stream is based on the solid tailing mass and process water, whereby methods to reduce, reuse, or recycle either the solid or liquid fraction decrease the required storage volume. Slurry tailings discharged into the impoundment area of a TSF will transition from slurry to soil, as shown schematically in Figure 12-7. Flocculated tailing particles will settle out of suspension to aggrade into a soil deposit. As tailing particles settle out of suspension, the clear, supernatant water forms a pond within the TSF that can be collected for subsequent reuse in mining operations.

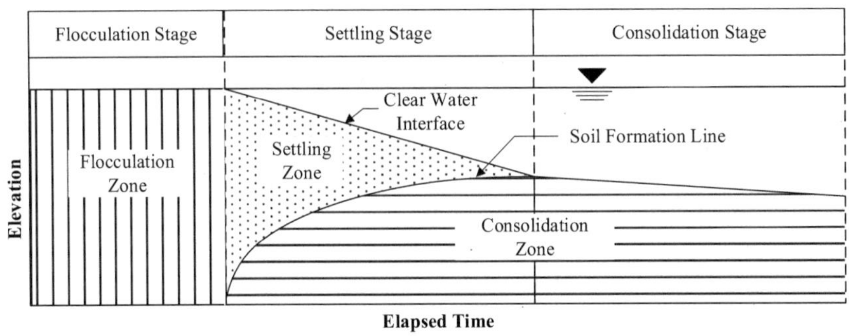

Figure 12-7. Conceptual schematic of the temporal transition of slurry to soil in a tailing impoundment.
Source: Imai (1981).

REFERENCES

Albright, W., C. Benson, and W. Waugh. 2010. *Water balance covers for waste containment: Principles and practice*. Reston, VA: ASCE.

Blight, G. 2010. *Geotechnical engineering for mine waste storage facilities*. Boca Raton, FL: CRC Press.

Boger, D. V. 2009. "Rheology and the resource industries." *Chem. Eng. Sci.* 64 (22): 4525–4536.

Bussière, B. 2007. "Colloquium 2004: Hydrogeotechnical properties of hard rock tailings from metal mines and emerging geoenvironmental disposal approaches." *Can. Geotech. J.* 44 (9): 1019–1052.

Carrier, D., L. Bromwell, and F. Somogyi. 1983. "Design capacity of slurried mineral waste ponds." *J. Geotech. Geoenviron. Eng.* 109 (5): 699–716.

CIMNE (International Centre for Numerical Methods in Engineering). 2021. *Computational analyses of dam I failure at the Corrego de Feijão mine in Brumadinho*. Final Report. Barcelona: CIMNE.

Duncan, J. M., S. G. Wright, and T. L. Brandon. 2014. *Soil strength and slope stability*. 2nd ed. Hoboken, NJ: Wiley.

Eaton, T., and S. Broughton. 2010. "Design and operation of large waste dumps." *EduMine*. Vancouver. www.edumine.edu.

Fourie, A., R. Verdugo, A. Bjelkevik, L. A. Torres-Cruz, et al. 2022. "Geotechnics of mine tailings: A 2022 state of the art." In *Proc. 20th ICSMGE–State of the Art and Invited Lectures*. Crawley, WA, Australia: Australian Geomechanics Society. Accessed March 6, 2024. https://www.researchgate.net/profile/Andy-Fourie/publication/360639583_20ICSMGE_State_of_the_Art_-_Tailings/links/628300df7a08f263d551ec84/20ICSMGE-State-of-the-Art-Tailings.pdf.

Frossard, E., W. Hu, C. Dano, and P. Y. Hicher. 2012. "Rockfill shear strength evaluation: A rational method based on size effects." *Géotechnique* 62 (5): 415–427.

Gorakhki, M. H., and C. A. Bareither. 2017. "Unconfined compressive strength of synthetic and natural mine tailings amended with fly ash and cement." *J. Geotech. Geoenviron. Eng.* 143 (7): 1–14.

Herrington, R. 2021. "Mining our green future." *Nat. Rev. Mater.* 6 (6): 456–458.

IIED (International Institute for Environment and Development) and WBSCD (World Business Council for Sustainable Development). 2002. "Breaking new ground: Mining, minerals, and sustainable development (MMSD)." Accessed March 6, 2024. https://www.iied.org/9084iied.

Imai, G. 1981. "Experimental studies on sedimentation mechanism and sediment formation of clay materials." *Soil Found.* 21 (1): 7–20.

INAP (International Network for Acid Prevention). 2009. "Global acid rock drainage guide (GARD guide)." Accessed March 6, 2024. www.gardguide.com.

Jefferies, M., N. Morgenstern, D. Van Zyl, and J. Wates. 2019. *Report on NTSF embankment failure Cadia Valley operations for Ashurst Australia.* Subiaco, Australia: Newcrest Mining Limited.

Jewell, R., and A. Fourie. 2006. *Paste and thickened tailings—A guide.* 2nd ed. Crawley, WA, Australia: Australian Centre for Geomechanics.

Luong, J. H. T., C. Tran, and D. Ton-That. 2022. "A paradox over electric vehicles, mining of lithium for car batteries." *Energies* 15 (21): 7997.

Marsal, R. J. 1967. "Large scale testing of rockfill materials." *J. Soil Mech. Found. Div.* 93 (2): 27–43.

Marschi, N. D., C. K. Chan, and H. B. Seed. 1972. "Evaluation of properties of rockfill materials." *J. Soil Mech. Found. Div.* 98 (SM1): 95–114.

Morgenstern, N. R. 2018. "Geotechnical risk, regulation and public policy." *Soil Rock* 41 (2): 107–129.

Morgenstern, N. R., S. G. Vick, and D. Van Zyl. 2015. "Report on Mount Polley tailings storage facility breach. Report of independent expert engineering investigation and review panel." Accessed March 6, 2024. ReportonMountPolleyTailingsStorageFacilityBreach.pdf (mount polleyreviewpanel.ca).

Morgenstern, N. R., S. G. Vick, C. B. Viotti, and B. D. Watts. 2016. "Fundão tailings dam review panel report on the immediate causes of the failure of the Fundão dam." Accessed March 6, 2024. fundao-2016.pdf (resolutionmineeis.us).

Qiu, Y., and D. C. Sego. 2001. "Laboratory properties of mine tailings." *Can. Geotech. J.* 38 (1): 183–190.

Robertson, P. K., L. D. Melo, D. J. Williams, and W. Wilson. 2019. "Report of the expert panel on the technical causes of the failure of Feijão Dam I." Accessed March 6, 2024. https://www.resolutionmineeis.us/sites/default/files/references/robertson-et-al-2019.pdf.

Rowe, R. K., R. M. Quigley, R. W. I. Brachman, and J. R. Booker. 2004. *Barrier systems for waste disposal.* 2nd ed. New York: E & FN Spon.

Silvis, F., and M. B. de Groot. 1995. "Flow slides in the Netherlands: Experience and engineering practice." *Can. Geotech. J.* 32 (6): 1086–1092.

Song, Y., Z. Zhang, Y. Zhang, and J. Cheng. 2022. "Technological innovation and supply of critical metals: A perspective of industrial chains." *Res. Policy* 79 (December): 103144.

Vick, S. G. 1990. *Planning, design, and analysis of tailings dams.* Vancouver: BiTech Publishers.

PART 6
CONTAMINATED GROUND INTERVENTIONS

CHAPTER 13
EXCAVATION AND REPLACEMENT

Carsten H. Floess, Ph.D., P.E., D.GE, F.ASCE

13.1 INTRODUCTION

Excavation and replacement (commonly called *dig and haul, hog and haul,* and others) is frequently used to remediate contaminated land. This option is practicable wherever impacted soil is readily accessible, the excavated material is suitable for disposal in a municipal waste [Resource Conservation and Recovery Act (RCRA) Subtitle D], hazardous waste (RCRA Subtitle C0), or polychlorinated biphenyl waste (Toxic Substance Control Act) landfill, and the cost for excavation and disposal makes this option competitive with other potential remediation options. It is often the solution of choice for heavily contaminated zones where source removal is crucial to reducing the potential for long-term impacts to groundwater.

Excavated soil is often landfilled, although it is sometimes feasible to treat this material ex situ (aeration, thermal treatment, etc.) and subsequently reuse it as clean backfill. As an example, ex situ thermal treatment is commonly used for soils impacted with manufactured gas plant (MGP) waste, typically transporting the soils to a nearby thermal treatment facility. Manufactured gas plant-impacted soils have also been treated by solidifying them with cementitious reagents using in situ auger mixing or other mixing techniques.

The primary advantages of excavation and replacement include the following:

- Low-tech operation that is familiar to environmental contractors is possible;
- Source material is removed;

- Visual inspection/verification is feasible; and
- Clean backfill, if properly compacted, facilitates future site development.

Excavation and replacement become more difficult or even impracticable where contaminated soil extends close to or under structures, where contamination is deep, where groundwater is shallow or where artesian pressures occur, and in urban locations where odors, vibrations, and truck traffic may make the process challenging. Alternate options, for example, in situ treatment or solidification, may also be more cost-effective than excavation and replacement.

Although the focus herein is land-based remediation, excavation (i.e., dredging) is also used for remediation in open water settings. This is briefly discussed at the end of this section.

13.2 EXCAVATION SUPPORT

Excavation shoring is required if an open cut is not feasible. This may occur if the excavation is deep, if the excavation extends below groundwater, or if site constraints such as adjacent structures or property lines limit the use of side slopes. A typical sheet pile excavation cell is illustrated in Figure 13-1. In this example, a single row of internal cross bracing was used, and the impacted soil was removed to a depth of about 6 m using a long-stick excavator working from the top of excavation. Anticipated surcharge loads resulting from heavy equipment positioned immediately behind the sheeting need to be considered in the shoring design.

Figure 13-1. Steel sheet pile cell with internal bracing.

In Figure 13-1, the sheet piles extended down into a low-permeability soil layer, creating a perimeter cutoff in the upper, more pervious soil. With a cutoff in place, infiltration from below and minor leakage through sheet pile interlocks were small and could easily be managed using sumps and pumping.

Excavations, unless very small, are commonly subdivided into a series of subcells, as shown in Figure 13-2. This has the following advantages:

- Groundwater is more manageable in smaller subcells;
- Shoring (typically sheet piles) can be reused, minimizing material costs; and
- Excavation and backfill can proceed in one subcell, while shoring is simultaneously installed in an adjacent subcell(s), leading to efficient operations.

The shape of contaminant plumes can be highly irregular, requiring excavations with complex shapes. Shoring for such irregular shapes can be intricate and costly to construct, especially if each excavation cell requires a unique shape. Often, it is beneficial to simplify the shape of excavation cells and to use consistent geometry. Figure 13-3 provides an example of a simplified cell layout for an irregularly shaped site.

In this example, remediation excavation was originally planned to use a series of 12 internally braced sheet pile cofferdams. These excavation cells were approximately equidimensional but had varying corner angles and dimensions. Each shape was unique. The contractor proposed a simpler, more economical excavation concept, smoothing out the perimeter boundary and using fewer internally braced sheet pile cells with consistent

Figure 13-2. Concurrent excavation/backfill and subcell construction.

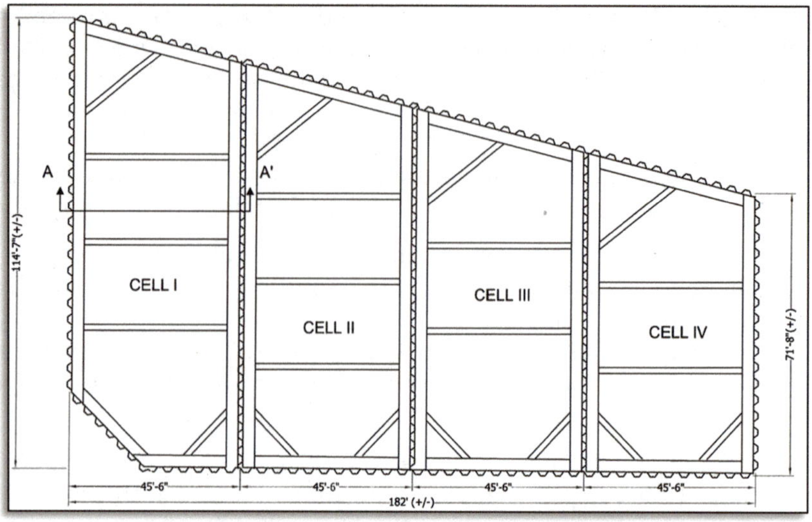

Figure 13-3. Excavation cell layout for an irregularly shaped site.

geometry. The cell widths were kept constant, whereas the cell lengths became progressively shorter, allowing an efficient reuse of internal bracing with minimal field fabrication.

Deadman anchors or drilled-in tiebacks can also be used in lieu of cross bracing. Figure 13-4 provides an example where sheet piles with drilled-in tiebacks were used to support a railroad embankment, allowing removal of impacted soils at a former junk yard site. Because excavation did not

Figure 13-4. Drilled-in tieback anchors to support railroad embankment.

Figure 13-5. Circular excavation cell with steel ring beams.

extend below groundwater and there were no other site constraints, the remaining sides of the excavation cell were open cut.

Deadmen or drilled-in tiebacks are also appropriate for large excavation cells that would otherwise require heavy cross bracing for the long spans. Alternatively, it is sometimes practical to use a circular excavation cell with steel or reinforced concrete ring beams, also providing an excavation area unencumbered by cross bracing, as illustrated in Figure 13-5. This example shows removal of a remnant circular gas holder at a former MGP site.

The examples presented herein focus on sheet pile cofferdams because this is the most common method of excavation support, and many remediation contractors are familiar with sheet pile construction. Soldier piles and lagging may also be viable for excavations above groundwater. Other shoring methods can also be considered [e.g., secant piles or soil mixing (in situ solidification or jet grouting) to create gravity walls]. These methods generate minimal construction vibrations and can be used to advantage at vibration-sensitive sites. However, they all require specialty contractors and they will often be more expensive to implement compared to sheet pile installations.

13.3 SHEET PILE INSTALLATION

Many remediation projects are located at former industrial or developed sites. As such, remnant underground foundations and structures are common, forming obstructions for sheet pile installation. If buried foundations and structures are shallow, they can usually be removed by preexcavating along the sheet pile alignment.

Underground utilities are always an important consideration, particularly at urban sites. Underground utilities can be located using available mapping, surface features (e.g., manholes), geophysics, and test holes. This effort is referred to as subsurface utility engineering. Where practical, interfering utilities are removed and relocated to allow safe excavation. If relocation is not feasible, the interfering utility must be protected while excavating around it to the extent practical. Overhead utilities may also interfere, requiring shielding or relocation to facilitate safe excavation.

Dense soils or soils containing cobbles and boulders can also be problematic, for example, glacial till. It is sometimes difficult to judge whether sheet piles can successfully be driven into such dense soils. Several options can be considered to evaluate sheet pile drivability:

1. Obtain input from local sheet pile contractors,
2. Analyze sheet pile drivability using wave equation analysis, and
3. Drive representative test piles during the design phase.

It is common practice to predrill with an auger along the sheet pile alignment to loosen the soil and probe for potential obstructions. If necessary, a heavy spud can be driven to break or "push" small boulders and obstructions to the side. Predrilling is often performed at a frequency of one to two auger holes per sheet pile pair. Intermediate auger holes can then be drilled on a contingency basis where boulders/obstructions are frequent or where the soil is found to be exceptionally dense. Predrilling can also be performed continuously, with each auger hole tangent to adjacent holes. Predrilling below excavation level should be carefully evaluated to ensure that undesirable vertical hydraulic pathways are not created.

Contingency plans should be developed in case deep obstructions are encountered that cannot readily be removed. For example, the sheet pile alignment could be adjusted in the field to bypass such an obstruction, or it may be feasible to simply leave a few sheet piles "short" where an isolated obstruction is encountered.

13.4 CONSTRUCTION VIBRATIONS AND PROTECTION OF ADJACENT STRUCTURES

Sheet piles are normally driven with a vibratory hammer, although an impact hammer may also be used in dense soils. Both methods create vibrations (and noise), and the following issues should be considered:

1. *Will vibrations cause cosmetic (e.g., plaster cracks) or structural damage to adjacent structures?* The US Bureau of Mines (Siskind et al. 1980)

criteria are commonly used to protect structures from vibration damage, with acceptable vibration levels on the order of 25 mm/s. More stringent criteria are often specified to protect structures in poor condition or historic structures, as low as 2.5 mm/s.

2. *Will vibrations cause densification of granular soils, leading to settlement of adjacent structures?* Criteria for determining vibration levels that will cause densification of loose granular soils are less well established than those that exist for cosmetic or structural damage. On the basis of experience, soil densification and building settlement has occurred at vibration levels below those established for cosmetic/structural damage (Lacy and Gould 1995). One concept that could be used to develop vibration criteria to limit soil densification is *volumetric cyclic threshold shear strain*. Below the *volumetric cyclic threshold shear strain*, soil particles are not displaced with respect to one another, and the soil's volume remains practically unchanged during cyclic loading (Hsu and Vucetic 2004, Dobry and Abdoun 2011). Above the *volumetric cyclic threshold shear strain*, the soil particles are displaced irreversibly with respect to one another, resulting in a permanent change in volume and ground settlement. In any case, if adjacent buildings are founded on loose granular soils, a vibration criterion more stringent than that used for preventing cosmetic/structural damage should be considered to minimize the risk of settlement damage.
   Essentially, all values for *volumetric cyclic threshold shear strain* for normally consolidated sand lie in the range of $0.5 \times 10^{-2}\%$ and $2.0 \times 10^{-2}\%$, with the majority grouping at about $1 \times 10^{-2}\%$ (0.0001) (Dobry and Abdoun 2011). This threshold value is notably independent of sand type, depositional environment, density, and confining pressure. The *volumetric cyclic threshold shear strain*, however, is sensitive to overconsolidation, increasing with increasing overconsolidation ratio (Dobry and Abdoun 2011). In essence, the *volumetric cyclic threshold shear strain* is a fundamental property of soils related to the minimum level of strain needed to start gross sliding and rearrangement of individual soil particles.
3. *Will vibrations and noise be disturbing to people in occupied buildings?* People are sensitive to vibration and vibrations as low as 2.5 mm/s can be troublesome. Therefore, it is important to understand how neighboring structures are being used and to evaluate whether vibrations may be problematic to their occupants.

The following procedures are suggested to minimize the risk of vibration issues during construction:

- Inform neighboring property owners and other interested parties of planned construction activities and anticipated vibration/noise levels.

Figure 13-6. Sheet pile installation with an ABI hydraulic press adjacent to a historic brick building.

- Mitigate vibration/noise impacts to the extent practicable. Vibration impacts can be minimized to some degree by using a high-frequency variable moment vibratory hammer. The basic principle of a variable moment hammer is that it can be brought up to its operating frequency before the vibration mode is activated, avoiding resonance during startup and shutdown. A hydraulic press can also be considered when sheet piles need to be installed in very close proximity (e.g., within about 5 m) to adjacent structures, as illustrated in Figure 13-6. The use of a press essentially eliminates vibrations. Shrouds have also been used around pile hammers with some success to reduce noise impacts.
- Perform preconstruction condition surveys of adjacent structures and monitor vibration/noise levels during construction. Preconstruction surveys are intended to document the condition of existing structures and provide a baseline for subsequent damage assessment. Such surveys are typically performed on buildings within about 30 m from sheet pile installation. Continuous vibration monitoring should also be performed at the closest structure(s) while pile driving. These measurements document vibration levels and provide warning of potential damage if vibration levels approach project threshold values. In addition, survey points should be established on nearby structures to monitor potential settlement during construction activities. These can be monitored continuously using a total station or surveyed on a periodic basis. Continuous monitoring is generally preferred.

It is also noted that sheet piles will deflect under load. Therefore, sheet piles in close proximity to structures must have adequate bracing and stiffness to minimize deflection and risk of associated settlement damage to the structure. Lateral deflection of sheeting can be monitored using survey points, inclinometers, or similar means.

13.5 GROUNDWATER CONTROL

Collection, treatment, and disposal of contaminated water can be costly, and it is usually beneficial to control and limit the quantity of water entering the excavation. Sometimes, it can be cost-effective to construct sheeted excavation cells to cut off water inflow, even if an open cut were technically feasible. In deep, highly permeable soils, it may also be cost-effective to construct a bottom seal using jet grouting or other means to control upwelling.

13.6 SUBSURFACE INVESTIGATION AND DESIGN

Environmental investigations typically include borings and test pits to evaluate and delineate the extent of contamination and type of contaminants. In general, on the one hand, because of cost and schedule, environmental borings are performed using direct push technology or similar methods, and geotechnical borings with standard penetration test (SPT) blow counts are lacking or extend to inadequate depth. On the other hand, high-quality groundwater data are usually available.

The first step in designing excavation shoring is to plan and execute a geotechnical investigation using test borings with SPT blow counts and/or cone penetration tests. The number and distribution of borings should be adequate to develop a realistic soil profile(s) suitable for shoring analysis. Borings should extend to a depth of two to three times the planned depth of remedial excavation. It is also important to identify subsurface conditions that may impact the design, such as the presence of remnant foundations or other obstructions, low-permeability confining layers, cobbles/boulders, shallow rock, and artesian groundwater. If clay soils are encountered, representative undisturbed samples of these soils should be recovered for laboratory strength testing, particularly if the clays are soft to medium stiff.

The importance of geotechnical borings and laboratory testing is often overlooked in environmental projects. Therefore, an experienced geotechnical engineer should be included in the design team.

13.7 OPEN WATER EXCAVATION (DREDGING)

Contaminated sediments can also be remediated by removal using mechanical or hydraulic dredging techniques. Mechanical dredging is generally accomplished using an environmental clamshell that is designed to cut a flat bottom surface and incorporating tight fitting gaskets to control leakage and spillage as the clamshell is raised. Dredge excavations are often capped with a layer of clean sand fill; alternatively, they can be allowed to backfill by natural sedimentation.

Similar to remediation on land, the work area can be isolated by building a sheet pile cofferdam, as illustrated in Figure 13-7. In this example, because water was shallow, the cofferdam was dewatered allowing excavation "in-the-dry" using conventional excavation equipment. Alternatively, excavation can be performed as a dredging operation in a flooded cofferdam.

A cofferdam is useful in containing turbidity and controlling offsite migration of contaminants. Such a cofferdam, however, may not be practical or feasible in deep water because of large lateral loads created by water currents, waves, ice, tidal fluctuations, and collisions with floating debris, work boats, or other vessels. In this case, the work area is best contained using turbidity curtains, air bubble curtains, or similar means. Such curtains, however, provide less reliable containment than sheet piles and are subject to damage during storm events.

Figure 13-7. Sheet pile cofferdam extending out into open water; winter conditions.

Remediation of contaminated sediments in small creeks and rivers can be accomplished by constructing a temporary bypass channel or by building upstream and downstream cofferdams and pumping water around the work area. This allows excavation "in-the-dry" using conventional excavation equipment.

REFERENCES

Dobry, R., and T. Abdoun. 2011. "3rd Ishihara lecture. An investigation into why liquefaction charts work: A necessary step toward integrating the states of art and practice." In *Proc., 5th Int. Conf. on Geotechnical Earthquake Engineering*. Santiago, Chile: International Society for Soil Mechanics and Geotechnical Engineering.

Hsu, C., and M. Vucetic. 2004. "Volumetric threshold shear strain for cyclic settlement." *J. Geotech. Geoenviron. Eng.* 130 (1): 58–70.

Lacy, H. S., and J. P. Gould. 1985. "Settlement from pile driving in sands." In *Proc., Vibration Problems in Geotechnical Engineering,* Geotechnical Engineering Division, Detroit. Reston, VA: ASCE.

Siskind, D. E., M. S. Stagg, J. W. Kopp, and C. H. Dowding. 1980. *Structure response and damage produced by ground vibration from surface mine blasting*. Report of Investigations 8507. Washington, DC: Bureau of Mines.

CHAPTER 14
ELECTROKINETIC REMEDIATION

Arvin Farid

14.1 INTRODUCTION

Groundwater contamination has become a widespread problem in need of urgent attention from various federal and state agencies. According to the Environmental Protection Agency (USEPA 2021), "groundwater contamination is a common issue at Superfund sites. Of the National Priorities List sites where EPA has selected remedies, EPA addressed groundwater contamination at approximately 85 percent of them." Several remediation techniques have been developed to either physically remove or contain contaminants, or chemically or biologically transform contaminants into less toxic forms. Examples of ex situ physical techniques are soil excavation-and-disposal/treatment, soil washing, and groundwater pump-and-treat. Two examples of in situ physical removal of volatile organic compounds are air sparging (AS) and soil vapor extraction. A few examples of in situ chemical treatment techniques are in situ chemical oxidation (ISCO) and in situ chemical reduction (ISCR). In addition to natural biodegradation, engineered biological treatment techniques such as enhanced in situ bioremediation and phytoremediation have been developed (Farid et al. 2019a, Lima et al. 2017). Each of these in situ remediation techniques faces limitations imposed by the hydrophysical and chemical and biological properties of environments (e.g., pH, temperature, and hydraulic conductivity). Most remediation techniques rely on either removing contaminants out of, or moving material that helps with remediation (including microbial species or the needed nutrients) into, the contaminated area. Hence, with most of these in situ remediation techniques, the hydraulic conductivity of the subsurface materials is a controlling/limiting factor. Consequently, most of these in situ remediation

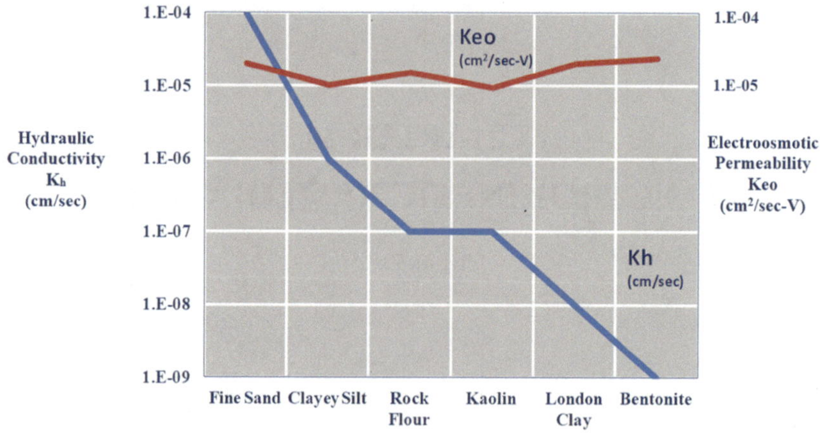

Figure 14-1. Permeability of various soils subjected to hydraulic head (hydraulic conductivity, K_h) in seepage flow and subjected to electric potential (electroosmotic permeability, K_{eo}) in EK.
Source: Cox (2018).

methods are more successful in soils with higher hydraulic conductivity, such as sand, while facing challenges and limitations in soils with lower hydraulic conductivity, such as clay. Researchers have attempted to find solutions to these limitations. Electrokinetics (EK) is an in situ remediation technique applicable to low-hydraulic-conductivity soils. EK leverages electric fields for remediation. The reason for this is that the transport using EK is controlled by electrical and not hydraulic conductivity, and the electrical properties of water-saturated soil media do not vary much between coarser-grained and fine-grained soils, as illustrated in Figure 14-1.

Using electric fields for various geotechnical and geoenvironmental applications, including sediment dewatering and stabilization (Casagrande 1949, Raats et al. 2002, Yang et al. 2011, Ayodele et al. 2018), in situ remediation of soil contaminated with hazardous waste (Alshawabkeh and Acar 1992, Pamukcu and Wittle 1992, Lageman 1993, Acar et al. 1995, Pamukcu et al. 1997, Hansen et al. 1997, Ottosen et al. 1997, Reddy and Cameselle 2009, Alshawabkeh 2009, Pamukcu 2009, ESTCP 2018), and more recently, resource recovery [L. Ottosen and I. Rorig-Dalgaard "Method and device for removing an ionic impurity from building structures." Patent WO2009124890, 2009; Wittle et al. 2009, 2011; Ghazanfari et al. 2012; Pamukcu et al. 2016; Couto et al. 2020], has, after decades of research and development, become viable for field-scale applications. Many of the works cited previously support the potential effectiveness of using EK methods for soil and sediment remediation employing direct electric fields and the resulting direct current (DC). Recent research has also shown promise in advanced applications of EK for triggering induced polarization and oxidation/reduction that transform residing contaminants to their benign reaction products in clay soils (Wittle et al. 2009; Brosky and Pamukcu 2013; Sun et al. 2015; Gomes et al. 2014, 2015; Zanko et al. 2020), for driving reagents to enhance desorption or bioremediation (Lima et al. 2012, Riis et al. 2012, Cameselle et al. 2013, Gill et al. 2014), for capturing transported contaminants in strategically placed reactive permeable barriers (Vieira dos Santos et al. 2017, Zhou et al. 2016, Ramírez et al. 2015), and for using the electromagnetic field to transport water in soil (Azad et al. 2013, 2015; Farid et al. 2015). EK fundamentals and design factors are detailed in the following section.

14.2 HOW ELECTROKINETICS WORKS

The focus of this section is on the fundamentals of the use of a direct electric field and the resulting DC in soil remediation. In such a method, a direct electric potential is applied across electrodes inserted into the soil, resulting in an induced direct electric current in the subsurface across the electrodes (Acar and Alshawabkeh 1993). In wet soils, DC can induce the migration of ionic and surface-charged constituents as well as the flow of pore water to a directed location within the field. Commonly referred to as *electrokinetics* or EK, this process is a proven, sustainable technology that can transport water and substances residing in or near the aqueous phase in clayey soil at a significantly higher rate than what can be practically achieved by hydraulic methods. Using EK, contaminants can be transported to extraction wells by electroosmosis and ion migration. A DC electric field can directly move charged particles and, therefore, can be used to transport and remove many ionized materials (e.g., inorganics, heavy metals, and ionic radioactive compounds) in the soil matrix. A direct electric field and the resulting DC can also transport noncharged (polar or nonpolar) materials (e.g., petroleum hydrocarbons) by the advective force of the electroosmotic flow of water (Pamukcu 1994; Pamukcu et al. 1995, 2016; Lima et al. 2011; Ghazanfari et al. 2012, 2014).

In addition to the induced transport of contaminants to an electrode site where they can be electrodeposited or pumped out of extraction wells, a number of methods such as ion exchange, redox, and sorption can also be used to enhance electrokinetic remediation as cited.

The major flow mechanisms induced by EK are ionic migration (also known as *electromigration*), electrophoresis, electroosmotic advection (or electroosmosis), and, to a lesser extent, diffusion (Alshawabkeh 2001). Table 14-1 presents the methods used in conjunction with EK and relevant properties controlling flow mechanisms or processes impacted by EK.

Table 14-1. Methods to which Electrokinesis Can Be Coupled, Controlling Properties, Mechanisms, and Processes.

Method	EK-BIO	EK-ISCO	EK-ISCR	EK-SALT	EK-TAP
Process	Bioremediation	In situ chemical oxidation	In situ chemical reduction	Electrical migration	Chemical oxidation and thermal activation
Factors	Electron donors, acceptors, and/or microbes	Perman-ganate (MnO_4^-)	Nano zero-valent iron	pH buffers	Persulfate ($S_2O_8^{2-}$)

Source: Fan et al. (2016).

Ionic migration is the induced movement of charged particles (ions) present in the saturated medium toward the oppositely charged electrode. When an electric field/potential is applied across soils, a movement is induced on ions, causing the transport of charged ions in the pore fluid toward the electrode opposite in polarity (i.e., metal cations move toward the cathode and anions move toward the anode) (Ajar et al. 1997). This creates an electric current caused by ion migration. *Ionic mobility* is a term describing the rate of migration of any specific ion under a unit electric field. A more complete term is used in soils to account for ion mobility, soil porosity, and tortuosity, referred to as *effective ionic mobility*. Rates of contaminant extraction and removal from soils using EK depend on the effective ionic mobilities of contaminants.

Electrophoresis, conversely, is the induced mobilization of colloids (i.e., suspended particles) through sufficiently large soil macropores of the soil resulting from the applied electric potential. Electrophoresis then drives fluids through soils within the medium in the opposite direction. This flow of fluids in the opposite direction is called *electroosmosis*.

Electroosmosis is the fluid transport with respect to a solid wall because of an applied electric potential across the wall. When an electric potential is applied across saturated soils or a membrane, the fluid will move from one side to the other side of the soil or membrane because of the electromotive force. Electroosmotic advection is the advection of contaminants by the induced flow of pore water caused by electroosmosis when soil porosity and zeta potential are conducive to the flow (Lima et al. 2017).

Electrolysis is a side process occurring during the application of EK, which can interact or even interfere with the effectiveness of EK. Electrolysis, the breakdown (ionization) of water into H^+ and OH^-, causes H^+ generation at the anode location reducing the pH and OH^- generation at the cathode increasing the pH (Lima et al. 2017, Alshawabkeh 2001, Ottosen et al. 2000, Acar and Alshawabkeh 1993, Pamukcu and Wittle 1992).

More detailed descriptions of equations governing these fundamental flow mechanisms are beyond the scope of this chapter (for more details, see Acar et al. 1995, Alshawabkeh 2001, Ghazanfari et al. 2012).

The increase of the acidity at the anode and alkalinity at the cathode can have twofold negative impacts, first the self-impeding effect caused by the change in environmental conditions (i.e., the electric resistivity and redox potential of the medium) near electrodes. In addition, this pronounced change in the pH can negatively impact microbial life, including indigenous and exogenous microorganisms that facilitate natural biodegradation and engineered bioremediation, respectively. Adjusting pH at the electrode sites or blocking the entrance of the electrolysis products into the treated medium using exchange membranes may be the solution to this hindering effect. However, the installation of exchange membranes may prove difficult and costly. Another problematic EK side effect, in addition to the change in the soil pH, is the temperature increase induced by the DC, also negatively impacting the aforementioned microbial activities that facilitate biodegradation and bioremediation. However, numerous laboratory and field experiments have shown that, even for long-term applications, the thermal effects may remain minimal when a low current density is maintained—to alleviate temperature increase as well as electrochemical effects that stem from pH and redox changes.

To address the several shortcomings of EK and other traditional techniques and enhance remediation effectiveness, researchers have attempted to combine EK with more conventional techniques such as ISCO, ISCR, bioremediation, and phytoremediation (Meinel et al. 2020, Lima et al. 2017). Meinel et al. (2020) studied EK-bio, a combination of EK and bioremediation. As mentioned in a review by Lima et al. (2017), other researchers have also combined EK with the use of permeable reactive barriers (Zhou et al. 2016, Vieira dos Santos et al. 2017, Huang and Cheng 2012, Weng et al. 2007, Chung and Lee 2007, Moon et al. 2005), which has attracted revitalized recent interest in EK and excitement in the research community.

14.3 ELECTROKINETICS APPLICATION IN THE FIELD

Figure 14-2 shows a schematic of the field implementation of EK. The figure shows electrodes (a cathode array and an anode array) inserted into the ground and connected via an alternating current (AC)/DC converter to the AC electric power source. The AC/DC converter converts the AC

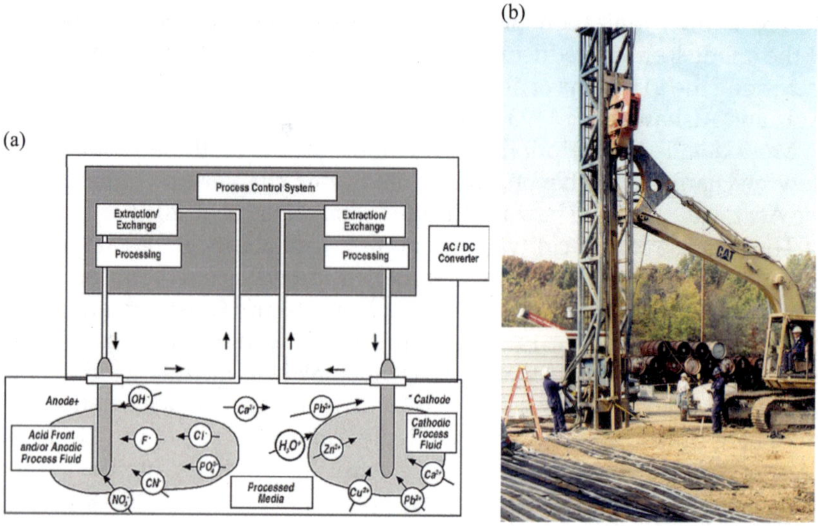

Figure 14-2. (a) Schematic of electrokinetic remediation, (b) electrode installation in the field.
Source: (a) FRTR (2017), (b) TC (2017).

supply to DC to then be input into electrode arrays. The DC then induces the flow of subsurface ions and water toward electrodes, with cations (i.e., positively charged particles, such as, metal and ammonium ions, and positively charged organic compounds) migrating toward the cathode and anions (i.e., negatively charged particles, such as chloride, cyanide, fluoride, and nitrate ions, as well as negatively charged organic compounds) migrating toward the anode. As previously described, the DC also electrolyzes water molecules into H^+ and OH^-, creating acid and base fronts at the anode and the cathode, respectively.

Remediation may include "enhanced removal" and/or "in situ treatment" (i.e., without removal). In other words, EK can induce the flow of contaminants toward one of the electrodes (later to be extracted using pumping systems) or ionize inorganic contaminants. The two are labeled as *extraction/exchange* and *processing*, which are achieved using a "process control system" above ground.

Enhanced removal is used for the treatment of groundwater contaminated with heavy metals. The enhanced removal system extract contaminants collected at each polarized electrode for subsequent ex situ treatment. The contaminant removal at the electrode can be achieved using methods such as electroplating at the electrodes; precipitation or coprecipitation at the electrodes; pumping of water near the electrodes; or complexing with ion exchange resins (FRTR 2017).

In situ treatment is more commonly used for the treatment of organic contaminants. In situ treatment is achieved using the electroosmotic transport of contaminants through a treatment zone located between the electrode arrays. To enhance in situ treatment, the polarity of the electrodes is also reversed periodically to alternate the direction of contaminant movement through the treatment zone. The frequency of polarity reversal is decided on the basis of contaminant transport rate through the treatment zone.

14.4 ELECTROKINETICS GOVERNING FACTORS

This section provides a brief explanation of factors to be considered during the design of an EK remediation system.

14.4.1 Electric Voltage and Current

In most EK applications, soil electric resistivity/conductivity and electrochemical properties control the electric current, typically in the order of a few amperes per square meter orthogonal to the direction of the voltage gradient between two electrodes. One should be aware that, although higher current densities may generate a preferential acidic environment enhancing the electrokinetic process, it can also increase the total ionic concentration, resulting in a decrease in the overall electroosmotic flow (Alshawabkeh 2001). Hence, for remediation cases where most of the contaminants are in their ionic forms or may be surface-charged, this would not be a problem as the majority of the contaminant transport would take place by electromigration. The problems with high-voltage gradients and high-current densities lie with the ensuing electrode reactions and/or other unpredicted redox reactions such as competing pH levels developed at the electrodes, potential weathering of the porous media during long-term applications, unresolved gas generation at the electrodes and within pores, alternating pore-water pressures, and polarization effects, all of which can impede transport of the ionic constituents by causing them to accumulate in zones between the electrodes (Muraoka et al. 2011, Loch et al. 2005). The electroosmotic flow and its enhancement are of particular importance when dealing with nonpolar organic or other contaminants that are transported in the small micellar or colloidal form with the advective action of electroosmotic flow (Pamukcu 1994; Ghazanfari et al. 2012, 2014). Therefore, after characterizing the site soils and the contaminant types, an appropriate current level can be selected depending on the dominant transport mechanism anticipated (i.e., electroosmosis and electromigration). After selecting the appropriate electric current, the electric resistivity/conductivity needs to be measured to decide on the electric-voltage gradient across electrodes for the desired electrode spacing, time requirements, and optimization of energy/cost consumption. A voltage gradient on the order of 100 V/m and a current density of 35 mA/cm^2 can be used as an initial estimate of system requirements (Alshawabkeh 2001).

14.4.2 Electrode Placement

Because of the three-dimensional (3D) electric field and current-flow patterns, there will always be areas/volumes of soil where the lower-than-applied average current is reached, referred to as inactive areas. To reduce inactive areas, electrodes need to be placed closer to each other (lower spacing between same-polarity electrodes), which results in a higher number of electrodes and, in turn, higher implementation cost. Thus, the required number of electrodes to address a target treatment area depends on the spacing between electrodes of the same polarity (e.g., cathode–cathode spacing) to minimize inactive zones.

A popular current pattern is an axisymmetric (i.e., radial) current toward a center electrode (e.g., locating a cathode as the center electrode with a circle of outer electrodes as anodes), allowing accumulations of the cationic contaminants in a focused zone around the cathode. In such a configuration, outer electrodes (anodes) are placed at a desired spacing around the center cathode. The electrodes can also be placed in a square, triangular, or hexagonal (honeycomb) configuration (Alshawabkeh et al. 1999). The engineer in charge of the design should be aware that different configurations result in various patterns of inactive zones and should strategically locate the electrodes to properly address the area targeted for remediation.

14.4.3 Soil Type

Both hydraulic and electrokinetic flow can be induced in granular soils. However, in fine-grained soils where the hydraulic flow is slow and may be difficult to induce, a much faster electroosmotic flow can be established using EK designed with proper engineering measures. This makes EK remediation highly suitable for fine-grained materials (clay and silt). Clay mineralogy and pore-fluid properties can, however, affect the EK flow and remediation effectiveness. Pamukcu and Wittle (1992) and Wittle and Pamukcu (1993) demonstrated that higher water content, a high degree of saturation, low ionic strength, and low activity in soils directly correlate with a higher transport of contaminants by electroosmotic advection and ionic migration. High plasticity (e.g., of montmorillonite) and high acid/base buffer capacity may require excess acid and/or enhancement reagents to desorb and solubilize contaminants before they can be transported through the subsurface to be removed (Alshawabkeh et al. 1997,

Alshawabkeh 2001). Reddy et al. (1997) demonstrated that the presence of iron oxides in soils such as glacial till impedes hexavalent chromium [Cr(VI)] electrokinetic transport, whereas it shows no negative impact in kaolinite and sodium montmorillonite on Cr(VI) electrokinetic extraction. The results of that study indicated better remediation efficiency in kaolinite followed by sand with 10% sodium montmorillonite, whereas sodium montmorillonite showed the lowest Cr(VI) removal efficiency.

14.4.4 Contamination Level

Contamination level is another factor controlling the efficiency of electrokinetic remediation (Alshawabkeh 2001). In general, remediation at higher initial contamination is more efficient, and as remediation progresses and the contamination level decreases, the contaminant removal rate can potentially reach a limit, depending on the sorption and exchange capacity and the age of the contamination (Hamed 1990, Hamed et al. 1991, Pamukcu and Wittle 1992, Wittle and Pamukcu 1993). For example, Hamed et al. (1991) studied the electrokinetic remediation of kaolinite clay at various lead (Pb^{2+}) concentrations below and above the soil cation-exchange capacity.

Numerous researchers (Runnells and Larson 1986; Runnells and Wahli 1993; Lageman et al. 1989; Lageman 1993; Eykholt 1992; Eykholt and Daniel 1994; Acar et al. 1992, 1993, 1994, 1996; Pamukcu et al. 1997, 2004) demonstrated remediation of soils at a wide range of heavy metal contamination.

Removal of radionuclides from clayey soil has also been demonstrated to be feasible (Roh et al.,). Acar et al. (1992, 1993, 1994, 1996) and Ugaz et al. (1994) demonstrated that thorium and uranium can be removed from kaolinite. However, precipitation of radionuclides, in the form of nonconductive hydroxide gel (e.g., thorium hydroxide gel and uranium hydroxide gel) at hydroxide solubility limits surrounding the cathode, impeded further remediation.

A higher concentration of contaminants may result in similar effects hindering complete remediation. In such cases, it was recommended to enhance the process (e.g., use of acetic acid) to neutralize such precipitation and increase the hydroxide solubility levels. Pamukcu and Wittle (1993) showed near-complete removal of surrogate radionuclides such as Sr, Cs, and U at the vicinity of the positive electrode (anode) in clay upon 1 pore volume of water transported toward the negative electrode (cathode). Transient pH gradients through the clay matrix affected the metal movement, as the metal accumulated at the discharge location attributed to precipitation and reduced mobility in their path of migration. Radionuclides' control—in soils/clays/liquid organic wastes—using direct electric fields was also documented by others (Maes et al. 1999, Kim et al. 2003, Valdovinos et al. 2016). Valdovinos et al. (2016) showed more than 50% removal of 24Na and 99mTc, whereas in uranium, there was less than 50% removal (Kim et al. 2003).

When applied without chemical or biological aids for process controls, EK has faced some challenges mainly because of the technique's side effects stemming from electrode reactions (Reddy 2010). A number of techniques, including the use of anolyte/catholyte pH control, chelating, or complexing agents, as well as using ion-selective membranes, have also been employed to effectively alleviate some of these side effects on the basis of site-specific conditions (Ottosen et al. 2000).

14.4.5 Cost Estimation

The total costs for electrokinetic remediation field implementation can be divided into four main components: (1) electric energy consumption, (2) fabrication of electrodes, (3) installation of electrodes, and (4) operation costs. As discussed, these are controlled by electrode spacing and configuration, which are themselves controlled by the soil resistivity/ conductivity and geochemical properties. Two other potential costs are the cost of reagents for remediation or operation, if necessary, and post-treatment monitoring and decommissioning, as necessary.

14.5 EXAMPLES OF ELECTROKINETIC REMEDIATION

The following are a few examples of EK remediation processes applied at the field scale. Figures 14-3 and 14-4 show sequential photographs of the evolution of EK treatment demonstrated over a period of 1 year at the Erie Pier Confined Disposal Facility (CDF), Duluth, Minnesota. The CDF was constructed in the mid-1970s by the US Army Corps of Engineers (USACE) to contain dredged material from the Duluth-Superior Harbor. The EK treatment (referred to here as electrochemical geooxidation) was conducted to remove or degrade polycyclic aromatic hydrocarbons (PAHs) and polychlorinated biphenyls (PCBs) in the dredged mud. A controlled pilot-scale treatment was designed using the fresh sediment dredged from the Minnesota Slip site in Duluth Harbor. Details of this demonstration, including test cell preparation, material and electrode placement, system startup and operation, monitoring, and sampling, can be found in Wittle et al. (2009). The year-end analysis showed an overall 20% reduction in PAHs in the treatment cell compared to that of a control cell, which may still not be sufficient and require further treatments. EK has also been attempted for PCB remediation, but this needs further research and evaluation (Fan et al. 2016).

Figure 14-3. Preparation for electrokinetic treatment; 2009 field electrokinetic cell construction at the Erie Pier site (USACE).
Source: Wittle et al. (2011).

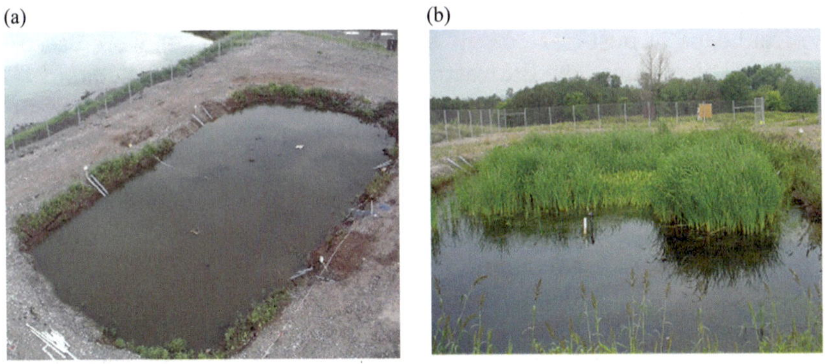

Figure 14-4. Sites treated with electrokinetics: (a) 2002 aerial view of test cell conditions, Erie Pier site (USACE); (b) 2003 ground-level view of test cell conditions, Erie Pier site (USACE).
Source: Wittle et al. (2009).

Figure 14-5. Electrokinetic treatment heaps with vertical steel electrodes, Copenhagen, Denmark.
Source: Wittle et al. (2009).

In 2006, the Soil Recycling Company in Denmark started examining the performance of EK treatment of 16,200 t of petroleum-hydrocarbon-contaminated sandy silt soil, piled up in 11 heaps. Figure 14-5 shows typical heaps of post-treated soil where hollow steel pipes were used as sacrificial electrodes. The EK treatment of all heaps was completed in 177 days by which the total petroleum hydrocarbons were brought down to substantially below regulatory clean-up levels. The details of this work can be found in Wittle et al. (2009).

As another field application example, Figure 14-6 shows a pilot-scale EK test setup operated to evaluate the effectiveness of induced polarization–type EK treatment for reducing PCB levels under anaerobic conditions typical of saturated and submerged sediments. Polychlorinated biphenyl-contaminated sediments from New Bedford Harbor (NBH), Massachusetts, were tested in these treatment tests. The treatment of both aerated and unaerated (anaerobic) samples of the NBH sediments began in November 2003 and lasted more than 4 years until early July 2008. At the end of the first 30-month period, the sample analyses indicated about a 30% reduction in total PCB levels in unaerated test cells compared to the control. The final sampling in March 2010 showed that total PCBs were reduced in the

Figure 14-6. Electrokinetic treatment facility for New Bedford Harbors sediments, showing the sediment-treatment area and tests cells with 0.9 m × 0.9 m lids (for scale).
Source: Zanko et al. (2020).

unaerated test cell by more than 90% relative to the 2003 starting levels and the control sample average. Conversely, PCB levels in sediments in the aerated test cell remained relatively unchanged. The ultimate results indicated that induced polarization EK was effective at degrading PCBs in a reducing environment, as detailed by Zanko et al. (2020).

14.6 FUTURE PROSPECT

As mentioned, the fundamental flow mechanisms induced by the EK method are electromigration, electroosmotic advection, electrophoresis, and electrolysis. Despite decades of research, these phenomena are not fully understood (Yeung 2011). According to Lima et al. (2017), example variables requiring a more in-depth understanding are the electric-current flow through porous soil; the microscale interactions at the liquid–liquid and soil–liquid interfaces within porous media; diffuse-double layer compression caused by EK; and the variability in osmotic flow rate, electroosmotic drag, and electroosmotic induction of displacement.

In addition, more research is needed to find ways to be able to effectively complement conventional remediation methods using EK, resulting in new complemented methods like bioremediation (EK-Bio) (Meinel et al. 2020), and microbially induced calcite precipitation (Keykha et al. 2014). Full-scale remediation projects using EK-Bio have been conducted at NAS and US Navy facilities and reports have been published about these (e.g., Cox 2018).

According to Lima et al. (2017), EK, given its versatility, is a method with potential and undiscovered capabilities for more sustainable remediation. In addition to complementing other traditional geoenvironmental techniques with EK, there are several potential new applications for EK. According to Lima et al. (2017), researchers have begun to study the applications of EK for an in situ manipulation of the redox state for redox-sensitive metals such as chromium (Lima et al. 2017, Sun et al. 2015, Brosky and Pamukcu 2013, Pamukcu et al. 2004); remediation of emerging contaminants such as per- and polyfluoroalkyl substances, radionuclides, polychlorinated biphenyl (PCB), and pesticides (Vieira dos Santos et al. 2017, López-Vizcaíno et al. 2017); and oil extraction and remediation of oil-contaminated media (Pamukcu et al. 2016; Ghazanfari et al. 2014, 2012; Ghazanfari et al. 2014; Haroun et al. 2013; Shalabi et al. 2012; Al Shalabi et al. 2012; Wittle et al. 2011). Another example of a potential application is the use of EK for element recycling in the management of waste (e.g., phosphorus) (Guedes et al. 2014).

Furthermore, to enhance the effectiveness of EK applications, the use of new catalytic materials has also been studied (Linley et al. 2014). More recently, EK has also been shown to be effective in recovering rare earth elements (REE) from coal ashes (Couto et al. 2020), while prospecting the use of the method to recover critical materials, including REE, from municipal solid waste ashes, which is also in need of further research.

REFERENCES

Acar, Y. B., and A. N. Alshawabkeh. 1993. "The principles of electrokinetics." *Environ. Sci. Technol.* 27 (1): 1–5.

Acar, Y. B., A. Alshawabkeh, and R. J. Gale. 1993. "Fundamentals of extracting species from soils by electrokinetics." *Waste Manage.* 13 (2): 141–151.

Acar, Y. B., R. J. Gale, A. N. Alshawabkeh, and R. E. Marks. 1995. "Electrokinetic remediation: Basics and technology status." *J. Hazard. Mater.* 40 (2): 117–137.

Acar, Y. B., J. T. Hamed, A. Alshawabkeh, and R. J. Gale. 1994. "Cd(II) removal of cadmium (II) from saturated kaolinite by the application of electrical current." *Géotechnique* 44 (2): 239–254.

Acar, Y. B., H. Li, and R. J. Gale. 1992. "Phenol removal from kaolinite by electrokinetics." *J. Geotech. Geoenviron. Eng.* 118 (11): 1837–1852.

Acar, Y. B., M. F. Rabbi, R. J. Gale, E. E. Ozsu, and A. N. Alshawabkeh 1996. "Enhanced soil bioremediation with electric fields." *Chemtech* 26 (4): 40–44.

Acar, Y. B., M. F. Rabbi, and E. E. Ozsu. 1997. "Electrokinetic injection of ammonium and sulfate ions into sand and kaolinite beds." *J. Geotech. Geoenviron. Eng.* 123 (3): 239–249.

Al Shalabi, E. W., B. Ghosh, M. Haroun, and S. Pamukcu. 2012. "The application of direct current potential to enhancing waterflood recovery efficiency." *Pet. Sci. Technol.* 30 (20): 2160–2168.

Alshawabkeh, A. N. 2001. *Basics and application of electrokinetic remediation.* Boston: Northeastern University.

Alshawabkeh, A. N. 2009. "Electrokinetic soil remediation: Challenges and opportunities." *Sep. Sci. Technol.* 44 (10): 2171–2187.

Alshawabkeh, A. N., and Y. B. Acar. 1992. "Removal of contaminants from soils by electrokinetics: A theoretical treatise." *J. Environ. Sci. Health A* 27 (7): 1835–1861.

Alshawabkeh, A. N., S. K. Puppala, Y. B. Acar, R. J. Gale, and M. Bricka. 1997. "Effect of solubility on enhanced electrokinetic extraction of metals." In *Proc., In Situ Remediation of the Geoenvironment (In Situ Remediation '97)*, 532–544. CONF-971032, Reston, VA: ASCE.

Alshawabkeh, A. N., A. T. Yeung, and M. R. Bricka. 1999. "Practical aspects of in-situ electrokinetic extraction." *J. Environ. Eng.* 125 (1): 27–35.

Ayodele, A. L., S. Pamukcu, R. A. Shrestha, and O. A. Agbede. 2018. Electrochemical soil stabilization and verification." *Geotech. Geol. Eng.* 36 (2): 1283–1293.

Azad, M., H. Sangrey, A. Farid, J. Browning, and E. Barney-Smith 2013. "Electromagnetic stimulation of two-phase transport in water-saturated media for geoenvironmental applications." *Geotech. Test. J.* 13 (1): 97–106.

Azad, S., A. Farid, and J. Browning. 2015. "Consequence of EM stimulation on hydraulic conductivity of a bentonite clayey sample." *Environ. Geotech. J.* 2 (4): 211–223.

Brosky, R. T., and S. Pamukcu. 2013. "Role of DDL processes during electrolytic reduction of Cu(II) in a low oxygen environment." *J. Hazard. Mater.* 262: 878–882.

Cameselle, C., R. A. Chirakkara, and K. R. Reddy. 2013. "Electrokinetic-enhanced phytoremediation of soils: Status and opportunities." *Chemosphere* 93 (4): 626–636.

Casagrande, L. 1949. "Electro-osmosis in soils." *Géotechnique* 1 (3): 159–177.

Chung, H. I., and M. Lee. 2007. "A new method for remedial treatment of contaminated clayey soils by electrokinetics coupled with permeable reactive barriers." *Electrochim. Acta* 52 (10): 3427–3431.

Couto, N., A. R. Ferreira, V. L. M. Lopes, S. C. Peters, et al. 2020. "Electrodialytic recovery of rare earth elements from coal ashes." *Electrochim. Acta* 359: 136934. https://doi.org/10.1016/j.electacta.2020.136934.

Cox, E. 2018. "Electrokinetic-enhanced amendment delivery for remediation of low-permeability and heterogenous material: Effective cleanup of chlorinated solvents in clay and silt using electrokinetic techniques." ESTCP Project ER-201325. Accessed May 2023. https://serdp-estcp-storage.s3.us-gov-west-1.amazonaws.com/s3fs-public/project_documents/ER-201325_Final_Report.pdf.

ESTCP (Environmental Security Technology Certification Program). 2018. "Electrokinetic-enhanced (EK-enhanced) amendment delivery for remediation of low permeability and heterogeneous materials." ER-201325. Accessed March 7, 2024. https://serdp-estcp.mil/projects/details/2a2e46b1-60c4-4469-8b9e-92526cf2bc86.

Eykholt, G. R. 1992. "Driving and complicating features of electrokinetic treatment of soils." Ph.D. thesis. University of Texas at Austin, Dept. of Civil, Architectural, and Environmental Engineering.

Eykholt, G. R., and D. E. Daniel. 1994. "Impact of system chemistry on electroosmosis in contaminated soil." *J. Geotech. Eng.* 120 (5): 797–815.

Fan, G., Y. Wang, G. Fang, X. Zhu, et al. 2016. "Review of chemical and electrokinetic remediation of PCBs contaminated soils and sediments." *Environ. Sci. Processes Impacts* 9 (18): 1114–1128.

Farid, A., M. Azad, J. Browning, and E. Barney-Smith. 2015. "Electromagnetically induced transport for soil/groundwater remediation." *J. Geotech. Geoenviron. Eng.* 141 (3): 04014115.

Farid, A., A. Najafi, J. Browning, and E. Barney-Smith. 2019a. "Effects of air injection pressure on airflow pattern of air sparging." *Environ. Geotech.* 8 (7): 1–13. https:// doi.org/10/.1680/jenge.18.00096.

Farid, A., A. Najafi, J. Browning, and E. Barney-Smith. 2019b. "Electromagnetic waves' effect on airflow during air sparging." *J. Contam. Hydrol.* 220: 49–58. https://doi.org/10.1016/j.jconhyd.2018.11.004.

FRTR (Federal Remediation Technologies Roundtable). 2017. "Remediation technologies screening matrix and guide." Accessed May 2017. https://frtr.gov/matrix2/top_page.html.

Ghazanfari, E., S. Pamukcu, M. Pervizpour, and Z. Karpyn. 2014. "Investigation of generalized relative permeability coefficients for electrically assisted oil recovery in oil formations." *Transp. Porous Media* 105 (1): 235–253.

Ghazanfari, E., R. A. Shrestha, A. Miroshnik, and S. Pamukcu. 2012. "Electrically assisted liquid hydrocarbon transport in porous media." *Electrochim. Acta* 86: 185–191.

Gill, R. T., M. J. Harbottle, J. W. N. Smith, and S. F. Thornton. 2014. "Electrokinetic-enhanced bioremediation of organic contaminants:

A review of processes and environmental applications." *Chemosphere* 107: 31–42. https://doi.org/10.1016/j.chemosphere.2014.03.019.

Gomes, H. I., C. Dias-Ferreira, A. B. Ribeiro, and S. Pamukcu. 2014. "Influence of electrolyte and voltage on the direct current enhanced transport of iron nanoparticles in clay." *Chemosphere* 99: 171–179. https://doi.org/10.1016/j.scitotenv.2014.05.112.

Gomes, H. I., J. M. Rodríguez-Maroto, C. Dias-Ferreira, S. Pamukcu, et al. 2015. "Numerical prediction of diffusion and electric field-induced iron nanoparticle transport." *Electrochim. Acta* 181: 5–12. https://doi.org/10.1016/j.electacta.2014.11.157.

Guedes, P., N. Couto, L. M. Ottosen, and A. B. Ribeiro. 2014. "Phosphorus recovery from sewage sludge ash through an electrodialytic process." *Waste Manage.* 34 (5): 886–892.

Hamed, J. 1990. "Decontamination of soil using electro-osmosis." Ph.D. diss. Louisiana State University, Dept. of Civil and Environmental Engineering.

Hamed, J., Y. B. Acar, and R. J. Gale. 1991. "Pb(II) removal from kaolinite by electrokinetics." *J. Geotech. Eng.* 117 (2): 241–271.

Hansen, H. K., L. M. Ottosen, B. K. Kliem, and A. Villumsen. 1997. "Electrodialytic remediation of soils polluted with Cu, Cr, Hg, Pb, and Zn." *J. Chem. Technol. Biotechnol.* 70 (1): 67–73.

Haroun, M. R., G. V. Chilingar, S. Pamukcu, J. K. Wittle, et al. 2013. "Optimizing electroosmotic flow potential for electrically enhanced oil recovery (EEORTM) in carbonate rock formations of Abu Dhabi based on rock properties and composition." Proceedings of International Petroleum Technology Conference, IPTC-13812-MS, Doha, Qatar. https://doi.org/10.2523/IPTC-13812-MS.

Huang, Y.-C., and Y.-W. Cheng. 2012. "Electrokinetic-enhanced nanoscale iron reactive barrier of trichloroethylene solubilized by Triton X-100 from groundwater." *Electrochim. Acta* 86: 177–184. https://doi.org/10.1016/j.electacta.2012.03.048.

Keykha, H. A., B. B. K. Huat, and A. Asadi. 2014. "Electrokinetic stabilization of soft soil using carbonate-producing bacteria." *Geotech. Geol. Eng.* 32 (4): 739–747.

Kim, G. N., O. Wansuk, H. J. Won, and W. K. Choi. 2003. "Removal of cesium and cobalt from soil around TRIGA reactor using electrokinetic method." *J. Industr. Eng. Chem.* 9: 306–313.

Lageman, R. 1993. "Electroreclamation. Applications in the Netherlands." *Environ. Sci. Technol.* 27 (13): 2648–2650.

Lageman, R., P. Wieberen, and G. Seffinga. 1989. "Electro-reclamation: Theory and practice." *Chem. Ind. (London)* 9: 585–590. https://doi.org/10.1007/s00226-004-0267-z.

Lima, A. T., A. Hofmann, D. Reynolds, C. J. Ptacek, et al. 2017. "Environmental electrokinetics for a sustainable subsurface." *Chemosphere* 181: 122–133. https://doi.org/10.1016/j.chemosphere.2017.03.143.

Lima, A. T., P. J. Kleingeld, K. Heister, and J. P. G. Loch. 2011. "Removal of PAHs from contaminated clayey soil by means of electro-osmosis." *Sep. Purif. Technol.* 79 (2): 221–229. https://doi.org/10.1016/j.seppur.2011.02.021

Lima, A. T., L. M. Ottosen, K. Heister, and J. P. G. Loch. 2012. "Assessing PAH removal from clayey soil by means of electro-osmosis and electrodialysis." *Sci. Total Environ.* 435–436: 1–6. https://doi.org/10.1016/j.scitotenv.2012.07.010.

Linley, S., Y. Liu, C. J. Ptacek, D. W. Blowes, et al. 2014. "Recyclable graphene oxide-supported titanium dioxide photocatalysts with tunable properties." *ACS Appl. Mater. Interfaces* 6 (7): 4658–4668.

Loch, J. P. G., K. Richter, and T. J. S. Keijzer. 2005. "Coupling between chemical and electrical osmosis in clays." In Vol. 125 of *IUTAM Symp. on Physicochemical and Electromechanical Interactions in Porous Media. Solid Mechanics and Its Applications*, edited by G. M. L. Gladwell, J. Huyghe, P. A. Raats, and S. C. Cowin, 283–288. Dordrecht, Netherlands: Springer.

López-Vizcaíno, R., C. Risco, J. Isidro, S. Rodrigo, et al. 2017. "Scale-up of the electrokinetic fence technology for the removal of pesticides. Part II: Does size matter for removal of herbicides?" *Chemosphere* 166: 549–555. https://doi.org/10.1016/j.chemosphere.2016.09.114.

Maes, N., H. Moors, A. Dierckx, P. De Cannière, et al. 1999. "The assessment of electromigration as a new technique to study diffusion of radionuclides in clayey soils." *J. Contam. Hydrol.* 36 (3–4): 231–247.

Meinel, M., R. Krajmalnik-Brown, and C. I. Torres. 2020. "Coupled electrokinetic and biological remediation method leads to improved treatment of chlorinated solvents at high sulfate, transport limited sites." *Environ. Sci.: Water Res. Technol.* 6 (10): 2926–2937.

Moon, J.-W., H.-S. Moon, H. Kim, and Y. Roh. 2005. "Remediation of TCE-contaminated groundwater using zero valent iron and direct current: Experimental results and electron competition model." *Environ. Geol.* 48 (6): 805–817.

Muraoka, A., K. Boukheddaden, J. Linarès, and F. Varret. 2011. "Two-dimensional Ising-like model with specific edge effects for spin-crossover nanoparticles: A Monte Carlo study." *Phys. Rev.* B 84: 054119.

Ottosen, L. M., H. K. Hansen, and C. B. Hansen. 2000. "Water splitting at ion-exchange membranes and potential differences in soil during electrodialytic soil remediation." *J. Appl. Electrochem.* 30 (11): 1199–1207.

Ottosen, L. M., H. K. Hansen, S. Laursen, and A. Villumsen. 1997. "Electrodialytic remediation of soil polluted with copper from wood preservation industry." *Environ. Sci. Technol.* 31 (6): 1711–1715.

Pamukcu, S. 1994. *Electrokinetic removal of coal tar constituents from contaminated soils.* Washington, DC: Electric Power Research Institute.

Pamukcu, S. 2009. "Electrochemical transport and transformations." In Chap. 2 in *Electrochemical remediation technologies for polluted soils, sediments, and groundwater*, edited by K. R. Reddy and C. Camaselle, 29–65. New York: Wiley.

Pamukcu, S., I. Filipova, and J. K. Wittle. 1995. "The role of electroosmosis in transporting PAH compounds in contaminated soils." In *Proc., Symp. on Electrochemical Technology Applied to Environmental Problems*, 252–266. Pennington, NJ: Electrochemical Society.

Pamukcu, S., R. A. Shrestha, A. B. Ribeiro, and E. P. Mateus. 2016. "Electrically induced displacement transport of immiscible oil in saline sediments." *J. Hazard. Mater.* 313: 185–192. https://api.semanticscholar.org/CorpusID:23717306.

Pamukcu, S., A. Weeks, and J. K. Wittle. 1997. "Electrochemical extraction and stabilization of selected inorganic species in porous media." *J. Hazard. Mater.* 55 (1–3): 305–318.

Pamukcu, S., A. Weeks, and J. K. Wittle. 2004. "Enhanced reduction of Cr(VI) by direct electric current in a contaminated clay." *Environ. Sci. Technol.* 38 (4): 1236–1241.

Pamukcu, S., and J. K. Wittle. 1992. "Electrokinetic removal of selected heavy metals from soil." *Environ. Prog.* 11 (3): 241–250.

Pamukcu, S., and J. K. Wittle. 1993. "Electrokinetics for removal of low-level radioactivity from soil." In *Proc., Annual USDOE Low-Level Radioactive Waste Management Conf.*, 256–278. https://www.osti.gov/biblio/343725.

Raats, M. H. M., A. J. G. van Diemen, J. Lavèn, and H. N. Stein. 2002. "Full scale electrokinetic dewatering of waste sludge," *Colloids Surfaces A* 210 (2-3): 231–241. https://doi.org/10.1016/S0927-7757(02)00380-1.

Ramírez, E. M., C. S. Jimenez, J. V. Camacho, M. A. R. Rodrigo, et al. 2015. "Feasibility of coupling permeable bio-barriers and electrokinetics for the treatment of diesel hydrocarbons polluted soils." *Electrochim. Acta* 181: 192–199. https://doi.org/10.1016/j.electacta.2015.02.201.

Reddy, K. R. 2010. "Technical challenges to in-situ remediation of polluted sites." *Geotech. Geol. Eng.* 28 (3): 211–221.

Reddy, K. R., and C. Cameselle, eds. 2009. *Electrochemical remediation technologies for polluted soils, sediments, and groundwater*. New York: Wiley.

Reddy, K. R., U. S. Parupudi, S. N. Devulapalli, and C. Y. Xu. 1997. "Effects of soil composition on the removal of chromium by electrokinetics." *J. Hazard. Mater.* 55 (1–3): 135–158.

Riis, C., M. Bymose, E. Cox, J. Wang, et al. 2012. "Successful pilot test of electrokinetic enhanced bioremediation (EK-BIO) as an innovative remedial approach for PCE DNAPL source area. In *Proc., NORDROCS 2012: 4th Int. Conf. on Nordic Joint Meeting on Remediation of Contaminated Sites.*

Roh, Y., S. Y. Lee, M. P. Elless, and K. S. Cho. 2000. "Electro-enhanced remediation of radionuclide-contaminated groundwater using zero-valent iron." *J. Environ. Sci. Health, Part A Toxic/Hazard. Subst. Environ. Eng.* 35: 1043–1059. https://api.semanticscholar.org/CorpusID:98126021.

Runnells, D. D., and J. L. Larson. 1986. "A laboratory study of electromigration as a possible field technique for the removal of contaminants from ground water." *Groundwater Monit. Rem.* 6 (3): 85–91.

Runnells, D. D., and C. Wahli. 1993. "In situ electromigration as a method for removing sulfate, metals, and other contaminants from ground water." *Groundwater Monit. Rem.* 13 (1): 121–129.

Shalabi, E. W., M. Al Haroun, B. Ghosh, and S. Pamukcu. 2012. "The stimulation of sandstone reservoirs using DC potential." *Pet. Sci. Technol.* 30 (20): 2137–2147.

Sun, T. R., S. Pamukcu, L. M. Ottosen, and F. Wang. 2015. "Electrochemically enhanced reduction of hexavalent chromium in contaminated clay: Kinetics, energy consumption, and application of pulse current." *Chem. Eng. J.* 262: 1099–1107. https://api.semanticscholar.org/CorpusID:95755397.

TC (Terran Corporation). 2017. *Protecting, restoring, and enhancing the environment*. Beavercreek, OH: Terran Corp.

Ugaz, A., S. Puppala, R. J. Gale, and Y. B. Acar. 1994. "Electrokinetic soil processing complicating features of electrokinetic remediation of soils and slurries: Saturation effects and the role of the cathode electrolysis." *Chem. Eng. Commun.* 129 (1): 183–200.

USEPA (US Environmental Protection Agency). 2021. "How superfund addresses groundwater contamination." Accessed March 2021. https://www.epa.gov/superfund/how-superfund-addresses-groundwater-contamination#:~:text=Groundwater%20contamination%20is%20a%20common%20issue%20at%20Superfund,sites%20protect%20human%20health%20and%20the%20environment%20by%3A.

Valdovinos, V., F. Monroy-Guzmán, and E. Bustos. 2016. "Electrokinetic removal of radionuclides contained in scintillation liquids absorbed in soil type Phaeozem." *J. Environ. Radioact.* 162–163: 80–86. https://doi.org/10.1016/j.jenvrad.2016.05.017.

Vieira dos Santos, E., C. Sáez, P. Canizares, C. A. Martínez-Huitle, et al. 2017. "Reversible electrokinetic adsorption barriers for the removal of atrazine and oxyfluorfen from spiked soils." *J. Hazard. Mater.* 322: 413–420. https://doi.org/10.1016/j.jhazmat.2016.10.032.

Weng, C.-H., Y.-T. Lin, T. Y. Lin, and C. M. Kao. 2007. "Enhancement of electrokinetic remediation of hyper-Cr(VI) contaminated clay by zero-valent iron." *J. Hazard. Mater.* 149 (2): 292–302.

Wittle, J. K., D. G. Hill, and G. V. Chilingar. 2011. "Direct electric current oil recovery (EEOR)d a new approach to enhancing oil production. Energy sources, Part A recover." *Util. Environ. Eff.* 33 (9): 805–822.

Wittle, J. K., and S. Pamukcu. 1993. *Electrokinetic treatment of contaminated soils, sludges, and lagoons*. DOE/CH-9206. Washington, DC: US Department of Energy.

Wittle, J. K., S. Pamukcu, D. Bowman, L. M. Zanko, et al. 2009. "Field studies on sediment remediation." In Chap. 32 in *Electrochemical remediation technologies for polluted soils, sediments, and groundwater*, edited by K. R. Reddy and C. Camaselle, 661–696. New York: Wiley.

Yang, G. C. C., M.-C. Chen, and C.-F. Yeh. 2011. "Dewatering of a biological industrial sludge by electrokinetics-assisted filter press." *Sep. Purif. Technol.* 79 (2): 177–182.

Yeung, A. T. 2011. "Milestone developments, myths, and future directions of electrokinetic remediation." *Sep. Purif. Technol.* 79 (2): 124–132.

Zanko, L. M., J. K. Wittle, and S. Pamukcu. 2020. "Case study: Electrochemical geo-oxidation (ECGO) treatment of Massachusetts New Bedford harbor sediment PCBs." *Electrochim. Acta* 354: 136690. https://doi.org/10.1016/j.electacta.2020.136690.

Zhou, M., S. Zhu, Y. Yi, and T. Zhang. 2016. "An electrokinetic/activated alumina permeable reactive barrier-system for the treatment of fluorine-contaminated soil." *Clean Technol. Environ. Policy* 18 (8): 2691–2699.

CHAPTER 15

GROUND IMPROVEMENT FOR GEOENVIRONMENTAL ENGINEERING PRACTICE

Anirban De, Ph.D., P.E.

15.1 INTRODUCTION

Sometimes it is necessary to perform ground improvement at sites with environmental challenges to modify the engineering characteristics of the subsurface material such that they are suitable for the project. At traditional waste disposal sites (landfills), ground improvement may be used to increase the strength properties and/or reduce the compressibility characteristics inherent of many waste materials. At sites with other types of environmental issues, ground improvement methods may be used to reduce the in situ hydraulic conductivity, to immobilize contaminants, or to solidify liquids. The goal of ground improvement associated with waste disposal sites is typically one of the following:

1. Retrofit active or closed landfill to improve long-term global stability,
2. Increase active landfill capacity through the compression of in-place waste,
3. Reduce potential for a differential settlement of the landfill cover system,
4. Enhance bearing capacity and reduce the settlement of structures or surface construction planned as part of postclosure development over a closed landfill, and
5. Provide subgrade stability for liner construction to minimize settlement issues related to cell construction and leachate collection and removal system performance.

For the most part, ground improvement methods commonly used to solve geoenvironmental problems are a subset of the broader array of geotechnical ground improvement methods available to geotechnical engineers, with some special considerations to account for the nature of the site and the specific materials involved.

This chapter includes discussions of methods currently used in the industry and reported in the literature. Although the primary focus of the section is on waste disposal sites (landfills), other types of geoenvironmental applications are also discussed.

15.2 COMMON GROUND IMPROVEMENT METHODS

15.2.1 Introduction

The more common methods of ground improvement for geoenvironmental applications are

- Preloading,
- Roller compaction,
- Deep dynamic compaction (DDC),
- Columns of improved material,
- Grouting,
- Soil mixing, and
- In situ stabilization/solidification.

The main features of each of these improvement methods, along with brief descriptions of relevant case histories, are presented in the following sections. More extensive discussions and case studies are provided by Bouzza and Kavazanjian (2001), Evans et al. (2021), Gifford et al. (1990), Van Impe and Bouazza (1996), and Welsh (1983).

15.2.2 Preloading

A discussion on the compressibility and settlement of municipal solid waste (MSW) is provided in Chapter 5. The mechanism of waste settlement can be attributed to a combination of physical and mechanical processes (such as compression, particle reorientation, voids collapse, and fines migration), chemical processes, dissolution, and biological degradation. Primary settlement occurs mainly because of the physical and mechanical processes and is usually completed during the first 4 months of load placement (Sowers 1973, Sharma and De 2007). Secondary compression, which mainly occurs because of chemical and biological degradation in MSW landfills, is a time-dependent phenomenon and may be characterized by way of Equation (15-1) from Sharma and De (2007):

$$\Delta H_s = \Delta H_{(\mathrm{EL})} = C_{\alpha(\mathrm{EL})} H_1 \log \frac{t_2}{t_1} \quad (15\text{-}1)$$

where

$\Delta H_{(EL)}$ = Settlement at time t_2 after external load application,
t_1 = Time for primary settlement (usually 1 to 4 months),
t_2 = Time elapsed between external load application and measurement,
H_1 = Thickness of refuse fill at the end of the primary settlement, and
$C_{\alpha(EL)}$ = Coefficient of secondary compression caused by external loading.

On the one hand, secondary settlement is the most important component of settlement in the case of MSW landfills because of the relatively large proportion of degradable materials, which undergo significant volume change over time. On the other hand, construction debris are primarily inert and demonstrate negligible secondary settlement. Secondary settlement of hazardous waste depends on the nature of the material.

Values of C_α depend on many factors, including waste composition, method of placement, waste age, site-specific environmental conditions (such as leachate recirculation), and waste thickness. On the basis of a review of a series of case histories, Sharma and De (2007) recommend an overall range of values of C_α to be between 0.01 and 0.07. The values tend to be smaller when the waste material has been subjected to pretreatment (such as dynamic compaction or roller compaction), because of the reduction in void space and, therefore, air to contribute to further decomposition.

Bareither and Kwak (2015) have evaluated a number of MSW settlement models and found that the geomechanical empirical models do not capture the true mechanism of biodegradation. Model parameters that take into account MSW characteristics and biodegradation conditions at the landfill are more appropriate. More details on this are provided in Chapter 5.

de Melo et al. (2012) presented a case history where preloading was used at the Fresh Kills Landfill in New York, as part of treatment of in-place waste material, to prepare the material as foundation for a proposed roadway. At Fresh Kills, MSW was placed in the area between the mid-1960s and the late 1990s (i.e., between one and four decades prior to preloading). External load was applied by means of stockpile soil, placed to a height of about 6.1 m, with the top deck 33.5 m wide and 192 m long. Settlement was monitored following stockpile placement. On the basis of available literature, de Melo et al. (2012) assumed that primary settlement of the waste was completed within the first one month after load placement and any settlement occurring past the first month was considered secondary settlement. A coefficient of secondary compression, $C_{\alpha(EL)}$, was back-calculated using Equation (15-1). The calculated values of $C_{\alpha(EL)}$ were between 0.016 and 0.026, with an average value of 0.022. This falls on the lower side of the range recommended by Sharma and De (2007). According to de Melo et al. (2012), the preloading at Fresh Kills Landfill was placed in an area that had about 6.1 m of waste placed during the late 1990s, which itself was underlain by waste from the 1960s, which was more than 12.2 m thick in places. Therefore, the value of $C_{\alpha(EL)} = 0.022$ obtained from this case study was characteristic of older waste that had undergone some compression under self-weight prior to being subjected to external preloading. The case study from Fresh Kills by de Melo et al. (2012) demonstrates that surcharge loading is a viable method for ground improvement over existing MSW and that reasonable methods to estimate secondary compression exist on the basis of previous studies.

The modified secondary compression index expresses the rate of long-term settlement of MSW, which is assumed to be constant, although as discussed in Chapter 5, this is not generally the case. Other empirical models have also been recommended, such as the rheological model, the power creep model (Edil et al. 1990), and the hyperbolic model (Ling et al. 1998). Overall, none of the models account for the true biochemical mechanisms that are responsible for the majority of long-term settlement; instead, they have been used to match field settlement data and guide extrapolations to the future.

15.2.2.1 Roller Compaction. Roller compaction is a common method for densification and compression of soil under load. Avalle and McKenzie (2005) described a case where waste material at a disposal site proposed for residential development was compacted using a noncircular (square, with rounded corners) impact roller. The 8 t × (71 kN) roller was towed by a tractor and settlement at the surface was monitored. Because of its noncircular shape, the impact roller applies an impact on the ground as it turns, while also applying compression. The site is underlain by approximately 3 to 4 m of MSW placed in a former basalt quarry. Compaction induced an average settlement of about 120 mm after about 30 passes of the impact roller.

The effectiveness of improvement from the roller compaction was measured by employing a noninvasive geophysical method known as continuous surface wave system (CSWS), which indicated an increase of 30% to 40% in the shear wave velocity up to a depth of 2 m below the ground surface. Vibration monitoring was undertaken to measure the impact of roller compaction on residential areas adjacent to the site and the results indicated that the peak particle velocity expected at the structure closest to the site boundary was about 3 mm/s (Bouzza and Avalle 2006).

Roller compaction, or other similar surface compaction methods, are suitable for stabilizing loose, uncompacted material to a relatively shallow depth, typically less than 2 m (Hausmann 1990). Because these improvement methods impact the properties of soil near the ground surface, they are suitable for cases where the load transfer occurs at relatively shallow depths, such as caused by a shallow footing.

15.2.2.2 Deep Dynamic Compaction. In deep dynamic compaction (DDC), improvement is achieved by repeatedly dropping a heavy tamper, typically 44.5 to 222 kN, from a predetermined height, typically 6 to 24 m. The action is conducted in several passes or phases, usually in a grid pattern The craters formed by this process are subsequently filled with granular soil and leveled mechanically, following which a final surface compaction effort is completed either by a low-energy ironing pass or by several passes of conventional roller compaction.

The depth of improvement (D) is typically estimated using the following empirical equation:

$$D = n \cdot \sqrt{W \cdot H} \tag{15-2}$$

where

n = Empirical factor, which depends on the subsurface conditions at the site;
W = Weight of the tamper; and
H = Drop height of the tamper.

The product of W and H is also known as the drop energy.

Zekkos et al. (2013) compiled an extensive database of published technical information on DDC projects at 56 MSW disposal sites. On the basis of this information, Zekkos et al. (2013) concluded that DDC is a viable ground improvement method at MSW sites. However, the depth of improvement in MSW is smaller than in cohesionless soil materials under comparable conditions because of the significant dampening of the impact energy as it travels through MSW (compared to the rate of dampening in most inorganic soil materials). Zekkos et al. (2013) estimated the value of parameter n to be 0.4 ± 0.05 for MSW sites that DDC has been implemented successfully at dozens of landfill sites in an effort to provide foundation support to lightly loaded structures, pavements, and other surface features.

The depths of ground improvement at MSW sites have been successfully evaluated using a range of methods such as settlement measurement load tests employing soil embankments and in situ testing such as SPT, CPT, and seismic geophysics. As described by Zekkos et al. (2013), settlement monitoring before and especially after improvement using embankment and large-scale tests has been used to confirm the reduced compressibility of the waste mass. For seismic geophysics, surface wave methods have been used at solid waste landfills as they have the advantage of being noninvasive, and thus, they do not require the performance of boreholes or drilling. Kavazanjian et al. (1996) and Zekkos et al. (2014) described projects at which ground improvement using dynamic compaction was verified by using geophysical methods such as measurements of surface wave properties before and after improvement.

15.2.2.3 Columns of Improved Material. When site development on top of existing landfilled waste involves the construction of settlement-sensitive structures (e.g., buildings), compressibility and bearing capacity of the underlying material are critical design considerations. In-place waste material typically has low bearing capacity and high compressibility, which are undesirable characteristics for supporting shallow foundations. If surface ground improvement methods (as described previously) cannot achieve adequate improvement in waste compressibility and bearing capacity to support the structures to meet design criteria, deep foundation support elements may need to be included in the design. The installation of conventional deep foundations, such as piles, is often not an acceptable option at modern landfills because of concerns for penetrating bottom liners of the landfill cell or damaging leachate collection and removal systems (when present). Downdrag on piles caused by MSW settlement over time is also a concern.

Jedele and Buschmeier (2013) presented a case study of redevelopment at an old industrial site at a brownfield location on the coast of Lake Erie in Michigan. Because of the nature of contaminated soil and groundwater immediately underlying the site, the regulatory agency did not permit the foundation systems for the structures at the site to extend to the underlying rock formation, as it could compromise the groundwater quality. Instead, shallow foundations supported on columns of improved material were selected as an alternative.

Options considered at this site included controlled modulus columns (CMCs) and rammed aggregate piers (RAPs). The advantage of these two options was that each of them could be terminated in a competent bearing soil, without the need to penetrate the underlying rock formation.

The final option selected at the site was rigid inclusion columns with a diameter of 305 mm and approximately 6 m long, terminating in a stiff clay bearing stratum. The rigid inclusions were composed of grout columns installed in groups. The rigid inclusions were not intended to directly carry the load from the overlying foundations, but rather serve as stiffening elements in the soil mass to reduce the overall settlement. The CMCs were located 150 mm below the bottom of the spread footings. In areas supported on slab on grade, the inclusions were installed 0.6 to 0.9 m below the bottom of the slab on grade. A layer of compacted granular material, known as the *load transfer platform* (LTP), was installed between the top of the rigid inclusions and the bottom of the spread footings or the slab on grade. The use of the LTP eliminated the need to use a separate reinforced structural slab, which would be necessary with a pile foundation. The soil-inclusion composite mass had stiffness that was significantly greater than that of the local untreated ground. Settlements measured during field load tests met the project specifications of 25 mm total and 12.5 mm differential settlement under long-term design loading conditions.

A second case summary of use of rigid inclusion columns at a superfund landfill site is presented by the Environmental Protection Agency (USEPA 2017). The use of rigid inclusion columns eliminated the need for deep foundations. The gravity loads from a low-rise distribution center are transferred to the rigid inclusion columns (formed by injecting grout into vertical holes augered into the ground) through an LTP made of compacted granular soil.

15.2.2.4 Grouting. The process of grouting generally involves the injection of an additive into soil to alter its engineering properties. Different types of grouting methods are available and they are selected on the basis of the goals of a project.

Compaction grouting is generally used when it is necessary to reduce the voids and increase the overall density of the soil. This, in turn, increases the stiffness and reduces the compressibility. Permeation grouting is preferred when it is necessary to reduce the hydraulic conductivity of the soil by filling the voids with a grout that has high mobility (i.e., low viscosity), without altering the overall structure of the soil matrix. Jet grouting can be used to create columns of relatively stiff material such as a soil-cement mixture (often called *soilcrete*). These columns can be used to form relatively impermeable vertical barriers when they are located side by side, with a slight overlap between neighboring columns. More details on this are provided in Chapter 16. Jet grouting can also be used to create a horizontal barrier of limited thickness at a certain depth below the ground surface, without any need for an excavation.

D'Antonio et al. (2014) presented a case history on ground improvement at an old landfill site in Los Angeles. This site, where waste placement occurred during the period between 1934 and 1953, has since been closed and a hydraulic fill placed over the site, as part of subsequent development. A roadway located over this fill has experienced large settlement, primarily attributed to the degradation and compression of waste over time, leading to pavement distress and a need for repeated and expensive maintenance.

Options for ground improvement at this site were severely restricted because of various compelling concerns. Potential contamination at the site eliminated any option that required excavation and/or dewatering because of concerns for worker exposure. The chosen improvement method had to minimize impact in terms of schedule, noise, traffic, and other types of disruption because of the site's proximity to sensitive residential development. Concerns about the nature of the underlying natural soil meant that the vertical extent of any ground improvement had to be restricted within the fill zone (hydraulic fill and waste), which extended to a depth of 4.6 m. Finally, any chosen method had to accommodate the existence of numerous utility lines.

Compaction grouting was selected for ground improvement, as this method was able to meet the many stringent restrictions at the site. The need for dewatering or in situ soil removal was eliminated and the operation could be executed within a limited footprint, without causing unacceptable disruptions. Compaction grouting was applied in two stages—the primary grouting consisted of a 2.4 m square pattern, with secondary holes located between the primary holes. Compaction grouting is considered a "self-adaptive" technique, where larger grout volumes are initially consumed to treat areas of large voids and, as the soil is improved and void volumes reduced, higher soil resistance pressure is encountered and less grout volume is consumed.

Cone penetration tests (CPTs) were conducted pre- and post construction to estimate the extent of ground improvement. The volume of grout injected at each depth was also monitored. A significant increase in the cone tip resistance (q_c) was observed after the grouting. An analysis of the soil behavior type index of the improved soil indicated that relatively sensitive fine grain soils, which were identified prior to grouting, were eliminated post improvement.

The application of jet grouting to construct vertical barriers, intended to impede the transport of contaminants from a pollution source, is discussed in Chapter 16.

An emerging area of grouting with potential applications in the field of geoenvironmental engineering is the use of biobased grouts, which utilize biogeochemical processes to form precipitates in granular soil. The precipitates may be formed either by micro-organisms, in a process known as microbially induced calcite precipitation (MICP), or by enzyme actions, in a process known as enzyme-induced calcite precipitation (EICP). The presence of precipitates, such as calcium carbonate ($CaCO_3$), reduces hydraulic conductivity and increases stiffness and strength by filling the voids and providing interparticle bonding. This is an emerging field, with limited field applications, but holds significant promise. Detailed discussions regarding the current state of the art and potential applications of biogrouts are provided in Al Qabany et al. (2012), DeJong et al. (2013), and Khodadi et al. (2017). Guo et al. (2018) discuss a potential application of MICP to repair desiccation cracking in bentonite on a laboratory scale.

15.2.2.5 Soil Mixing. The process of in situ soil mixing introduces stabilizing agents that bind to the existing contaminated materials to improve physical properties such as increased compressive strength, decreased compressibility, and decreased hydraulic conductivity. The stabilizing agents may also improve chemical properties by immobilizing the contaminants and rendering them less mobile. Typical inorganic binders used in geoenvironmental applications include cement, fly ash, and lime. USEPA (2009) provided a comprehensive review of commonly used soil mixing technologies for geoenvironmental applications.

Andromalos and Carr (2016) included a summary of soil-mixing projects in environmental applications. Fatahi and Khabbaz (2013) described the injection of pozzolanic mixtures, specifically fly ash and quicklime, to improve the engineering properties of MSW, in an attempt to make it suitable for redevelopment. Laboratory-scale tests were conducted, in which samples with MSW, mixed with fly ash and quick lime, were tested. A total of five different mixes were tested, each containing both fly ash and quick lime in different proportions. Specifically, a mechanical mixer was used to mix fly ash (5% to 20%) and quick lime (1.7% to 6.7%)—both as a percentage of dry weight of MSW. The results indicated that the introduction of pozzolanic materials enhanced the shear strength properties of the MSW mixtures (increase in friction angle by as much as 10 degrees) and made them less compressible (30% and 36% reduction in the compression index and secondary compression index, respectively), thus making them better suited for proposed development over the closed landfill.

15.2.2.6 In situ stabilization/solidification. The process of stabilization uses chemical reactions to reduce the leachability of toxic contaminated materials and convert them into stable insoluble forms. It is a technique that has been used extensively for many decades and particularly at superfund sites with a rich publication and documentation record. For example, EPA (USEPA 1992) has published a resource guide as well as a technology review. The process of stabilization chemically binds free liquids, immobilizes contaminated materials, and/or reduces solubility. Solidification converts waste materials into a solid state through processes that encapsulate the contaminated material, giving them structural rigidity. The migration of contaminants is restricted because of a decrease in the surface area exposed to leaching or by coating the contaminated material with relatively low-permeability materials. (Bates and Hills 2015, Sharma and Lewis 1994). Among reagents commonly utilized for in situ stabilization/solidification (ISS) are portland cement, quick lime, fly ash, cement kiln dust, and others.

A recent application of large-scale ISS was used in the cleanup of the Gowanus Canal in Brooklyn, New York. The canal received waste from industries such as manufactured gas plants, paper mills, tanneries, and chemical plants, in addition to combined sewer overflows, since the mid-1800s. The canal sediments contain high levels of coal tar, creosote, and other organic nonaqueous phase liquid (NAPL) contaminants. There are two primary sediment layers of concern: a soft, unconsolidated overlying layer (average thickness 3 m) and a deeper underlying native sediment. The upper sediment contains organic and inorganic contaminants such as coal tar and contaminated solids, whereas the deeper native sediment is impacted by residuals from manufactured gas plants and NAPL.

The upper sediment layer could be removed by dredging, treated ex situ, and used as a landfill cover component. However, the deeper native sediment could not be removed because of concerns on the potential impact on the foundations of bulkheads along the shoreline, as well as the potential vertical migration of the underlying contaminants. Therefore, the most suitable approach to remediation was an ISS of the native sediments, followed by the placement of an active cap layer across the length of the canal.

The implementation of ISS involved bench-scale testing, followed by field pilot studies, before the actual remediation was implemented on-site. The objectives of the testing were to determine the final mix designs, demonstrate the feasibility of applying ISS through the canal water body, and the scalability of the entire operation in the field. The reagents used in the application included portland cement Type I/II and Type V, as well as ground granulated blast furnace slag (GGBFS). The mix also included an antiwashout fluid and sodium bentonite. (Olean, et al. 2016). This is an ongoing project at the time this section was written, and reports of postconstruction observation and performance are not yet available in the technical literature.

15.3 CONCLUSIONS

As described in this chapter, ground improvement methods are widely used for the treatment and remediation of contaminated sites. The methods range from traditional geotechnical approaches like preloading and compaction to the more specialized ones such as soil mixing and in situ stabilization/solidification. The design and implementation steps of the more conventional methods are generally well-documented and ample case histories exist to examine their feasibility for specific applications. The more specialized methods have been less commonly used and require careful planning and feasibility studies, prior to implementation. It is important for detailed case studies to be published so that potential users are informed of the steps involved in design and construction, as well as long-term performance of the ground improvement methods at contaminated sites.

REFERENCES

Al Qabany, A., K. Soga, and C. Santamarina. 2012. "Factors affecting efficiency of microbially induced calcite precipitation." *J. Geotech. Geoenviron. Eng.* 138 (8): 992–1001.

Andromalos, K. B., and J. A. Carr. 2016. "Environmental soil mixing: A 30-year update on means, methods, applications, and advances." In *Geo-Chicago 2016*, Geotechnical Special Publication 273. Reston, VA: ASCE.

Avalle, D. L., and R. W. McKenzie. 2005. "Ground improvement of landfill site using the square impact roller." *Aust. Geomech.* 40 (4): 15–21.

Bareither, C. A., and S. Kwak. 2015. "Assessment of municipal solid waste settlement models based on field-scale data analysis." *Waste Manage.* 42: 101–117.

Bates, E., and C. Hills. 2015. *Stabilization and solidification of contaminated soil and waste: A manual of practice*. Cluj-Napoca, Romania: HyggeMedia.

Bouzza, A., and D. L. Avalle. 2006. "Verification of the effects of rolling dynamic compaction using a continuous surface wave system." *Aust. Geomech.* 41 (2): 101–108.

Bouzza, A., and E. Kavazanjian Jr. 2001. "Construction on former landfills." In *Proc. 2nd ANZ Conf. on Environmental Geotechnics*, 467–482. Sydney: Australian Geomechanics Society.

D'Antonio, R., G. Lay, J. Wan, N. Su, et al. 2014. "Admiralty Way ground improvement case study." In *Geo-Congress 2014*, 2491–2502. Reston, VA: ASCE.

DeJong, J. T., K. Soga, E. Kavazanjian Jr., S. E. Burns, et al. 2013. "Biogeochemical processes and geotechnical applications: Progress, opportunities and challenges." *Géotechnique* 63 (4): 287–301.

de Melo, L., R. El-Sherbiny, W. Steier, and M. Salem. 2012. "Evaluation of waste compressibility due to preloading at the Fresh Kills landfill." In *GeoCongress 2012*, 4184–4193. Reston, VA: ASCE.

Edil, T. B., V. J. Ranguette, and W. W. Wuellner. 1990. "Settlement of municipal refuse." In *Geotechnics of waste fills—Theory and practice*, ASTM STP 1070, edited by A. O. Landva and G. D. Knowles. West Conshohocken, PA: ASTM.

Evans, J., D. Ruffing, and D. Elton. 2021. *Fundamentals of ground improvement engineering*. Boca Raton, FL: CRC Press.

Fatahi, B., and H. Khabbaz. 2013. "Influence of fly ash and quicklime addition on behaviour of municipal solid wastes." *J. Soils Sediments* 13: 1201–1212.

Gifford, G. P., A. O. Landva, and V. C. Hoffman. 1990. "Geotechnical considerations when planning construction on a landfill." In *Geotechnics of waste fills—Theory and practice*, ASTM STP 1070, edited by A. O. Landva and G. D. Knowles, 41–56. West Conshohocken, PA: ASTM.

Guo, Y., M. Loria, K. Rhoades, and X. Yu. 2018. "Effects of microbial induced calcite precipitation on bentonite cracking remediation." In *Proc., IFCEE 2018*, Geotechnical Special Publication 296. Reston, VA: ASCE.

Hausmann, M. R. 1990. *Engineering principles of ground modification*. New York: McGraw-Hill.

Jedele, L. P., and B. Buschmeier. 2013. "Ground improvement for redevelopment of former landfill." In *Proc., 7th Int. Conf. on Case Histories in Geotechnical Engineering*. Paper No. 6.09b. Rolla, MS: Missouri University of Science and Technology.

Kavazanjian, E. Jr., N. Matasovic, K. H. Stokoe II, and J. D. Bray. 1996. "In-situ shear wave velocity of solid waste from surface wave measurements." In *Environmental Geotechnics*, edited by M. Kamon, 97–102. Rotterdam, Netherlands: Balkema.

Khodadi, T. H., E. Kavazanjian Jr., L. van Paassen, and J. DeJong. 2017. "Bio-grout materials: A review." In *Grouting 2017*, Geotechnical Special Publication 288. Reston, VA: ASCE.

Ling, H. I., D. Leshchinsky, Y. Mohri, and T. Kawabata. 1998. "Estimation of municipal solid waste landfill settlement." *J. Geotech. Geoenviron. Eng.* 124 (1): 21–28.

Olean, T. J., J. L. Gentry, A. J. Salter-Blanc, T. M. Himmer, et al. 2016. "In-canal stabilization/solidification of NAPL-impacted sediments." *Rem. J.* 26 (3): 9–25.

Sharma, H. D., and A. De. 2007. "Municipal solid waste landfill settlement: Postclosure perspectives." *J. Geotech. Geoenviron. Eng.* 133 (6): 619–629.

Sharma, H. D., and S. P. Lewis. 1994. *Waste containment systems, Waste stabilization, and landfills*. New York: Wiley.

Sowers, G. F. 1973. "Settlement of waste disposal fills." In *Proc., 8th Int. Conf. on Soil Mechanics and Foundation Engineering*, 207–210. Moscow.

USEPA (US Environmental Protection Agency). 1992. *Solidification/stabilization resource guide*. EPA/542-B-99-002. Washington, DC: USEPA.

USEPA. 2009. *Technology performance review: Selecting and using solidification/stabilization treatment for site remediation*. EPA/600/R-09/148. Washington, DC: USEPA.

USEPA. 2017. *Rising opportunities at capped superfund sites*. Washington, DC: USEPA.

Van Impe, W. F., and A. Bouazza. 1996. "Densification of domestic waste fills by dynamic compaction." *Can. Geotech. J.* 33 (6): 879–887.

Welsh, J. P. 1983. "Dynamic deep compaction of sanitary landfill to support superhighway." In Vol. 3 of *Proc., 8th European Conf. on Soil Mechanics and Foundation Engineering*, 319–321. Rotterdam, Netherlands: A.A. Balkema.

Zekkos, D., M. Kabalan, and M. Flanagan. 2013. "Lessons learned from case histories of dynamic compaction at municipal solid waste sites." *J. Geotech. Geoenviron. Eng.* 139 (5): 738–751.

Zekkos, D., A. Sahadewa, R. D. Woods, and K. H. Stokoe II. 2014. "Development of model for shear-wave velocity of municipal solid waste." *J. Geotech. Geoenviron. Eng.* 140 (3): 04013030.

PART 7
BARRIERS

CHAPTER 16

VERTICAL BARRIER WALLS FOR GEOENVIRONMENTAL APPLICATIONS

Jeffrey C. Evans, Ph.D., P.E., D.GE, F.ASCE and
Kristin M. Sample-Lord, Ph.D., P.E., M.ASCE

16.1 INTRODUCTION

Barrier walls have long been used to control vertical and horizontal migrations of contaminants in the subsurface. Historically, the principal design parameters have been hydraulic conductivity and compatibility of the barrier material with the contaminated environment in which the barrier is placed. Vertical barriers to resist horizontal movement of groundwater and contaminants have been widely deployed and are most frequently keyed into a low-permeability aquitard. Horizontal barriers to resist vertical movement have been rarely deployed in geoenvironmental applications and therefore receive only brief mention in this chapter. The primary focus of this chapter is therefore the means and methods to construct in situ a vertical barrier of low hydraulic conductivity for geoenvironmental containment. Note that Chapters 8 and 9 address surface-built containment such as landfill liner and cover systems, respectively.

16.2 VERTICAL BARRIER DEPLOYMENT OPTIONS

Vertical barriers are most commonly built by surrounding the contaminant source and, where possible, around the entire zone of contamination (Figure 16-1a). In this scenario, the containment system consists of the vertical barrier, an underlying low-permeability layer into which the barrier is keyed, a cover to minimize infiltration into the contaminant source zone, and limited inboard pumping to develop an inward gradient. In this way, the advection component of contaminant transport is eliminated, and diffusive transport is counteracted by the inward groundwater flow.

Although the most effective configuration for the vertical barrier is to be keyed 0.6 to 0.9 m (2 to 3 ft) into a low-permeability layer (Van Deuren et al. 2002), this is not always possible. Where no low-permeability layer to key into is present, two options are available. First, and most commonly, the wall can be left "hanging" and the inboard pumping rates increased to establish inward flow both through and beneath the wall. Second, a horizontal barrier beneath the site could be constructed, most commonly with jet grouting. However, the latter option is rare.

Where site and subsurface conditions permit, vertical barriers are deployed on the upgradient side of the contaminated site and coupled with pumping wells on the downgradient portion of the site (Figure 16-1b). This system, a combination of containment and pump-and-treat remediation, allows for the capture of contaminants in the plume downstream of the source. By installing an upgradient vertical barrier, uncontaminated groundwater does not flow into the site, reducing the quantity of water that needs to be handled by the pump-and-treat component of the system.

Occasionally, such as at Rocky Mountain Arsenal in Colorado (Konikow and Thompson 1984), a vertical barrier wall is installed downgradient of the contamination along with pumping wells immediately upstream of the vertical barrier (Figure 16-1c). In these situations, it is deemed beneficial to allow groundwater to enter the zone of contamination to "flush" contaminants from the zone for capture by the pump-and-treat system, while at the same time minimizing the quantity of downgradient water that needs to be treated. The vertical barrier prevents downgradient flow

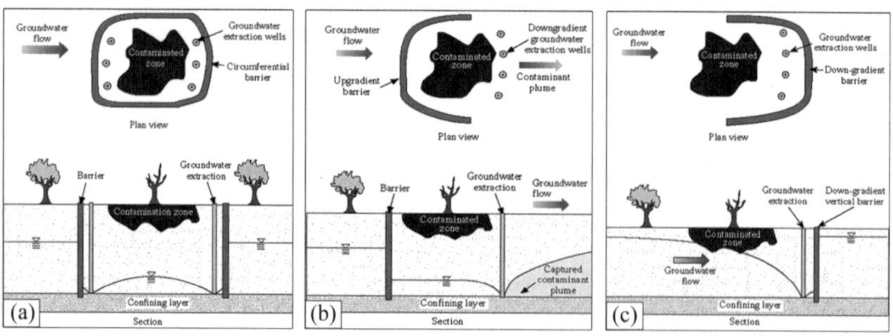

Figure 16-1. (a) Circumferential cutoff wall with inboard pumping, (b) upgradient vertical barrier to prevent the inflow of clean groundwater and help with plume capture, (c) downgradient vertical barrier to assist with flushing contaminants from the site.
Source: LaGrega et al. (2010).

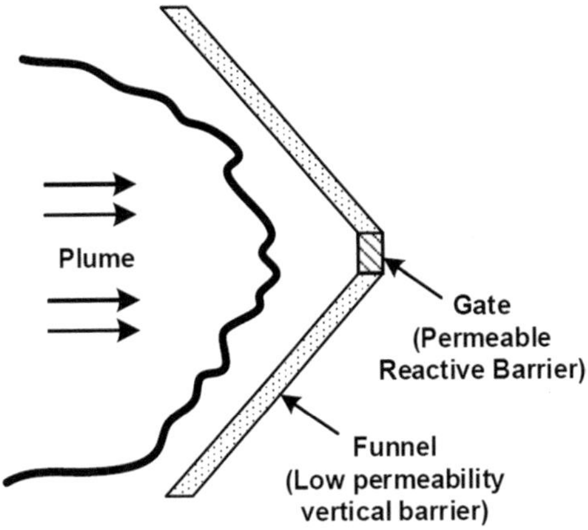

Figure 16-2. Schematic of the "funnel and gate" remediation system utilizing vertical barriers.

into the pumping wells, reducing the flow quantities to the pump-and-treat system.

Finally, the efficacy of a "funnel and gate" remediation system is dependent on vertical barriers to serve as the funnel (Figure 16-2). Permeable reactive barriers serve as the gate. Permeable reactive barriers are discussed in Chapter 17 of this book.

16.3 VERTICAL BARRIER TYPES

There are several means and methods by which a vertical barrier can be installed in the subsurface. Understanding how each is constructed is key to choosing the correct vertical barrier for any given contaminated site. The most common types of vertical barriers currently used for geoenvironmental applications are described subsequently. Discussion regarding future directions and specialty amendments to improve sorption (e.g., activated carbon) and other performance parameters of vertical barriers is not included in this chapter, but it can be found in the literature (e.g., Bierck and Chang 1994, Evans and Prince 1997, Malusis et al. 2009).

16.3.1 Soil–Bentonite Slurry Trench Cutoff Walls

Soil–bentonite (SB) slurry trench cutoff walls are by far the most frequently used vertical barriers for geoenvironmental applications in the United States (Evans 1993, USEPA 1998). SB walls have proven effective for a wide range of site, subsurface, and contamination conditions, while generally being the most economical. The technique derives its name from

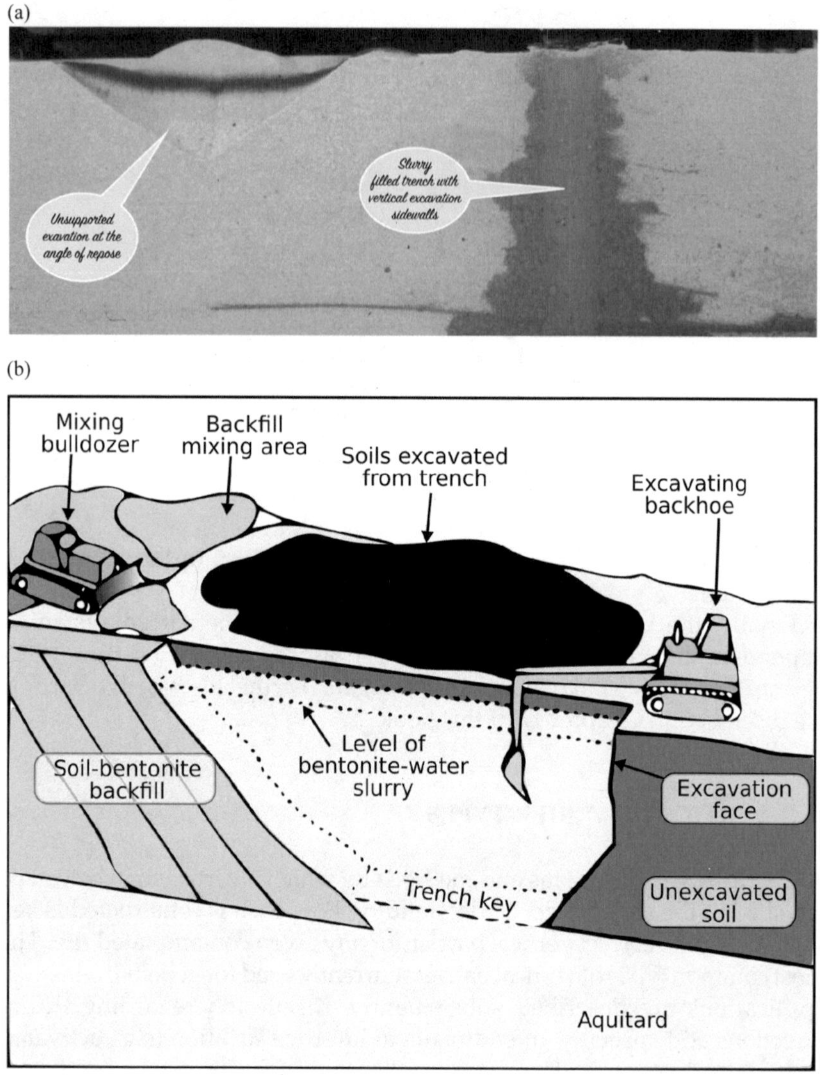

Figure 16-3. (a) Trench support with slurry, (b) schematic of an SB cutoff wall construction.
Source: (a) J. Evans, (b) After LaGrega et al. (2010).

the construction method; that is, a trench (typically 0.6 to 1.2 m wide), filled with bentonite–water slurry to provide trench stability, is excavated and the slurry is then replaced with SB backfill. As shown in the left side of the photograph in Figure 16-3a, excavation in sand results in a stable slope at the angle of repose. In comparison, the right side of Figure 16-3a shows a vertical cut in the same sand that is stable. Also shown in Figure 16-3a is the low-permeability filter cake on the trench walls necessary for trench stability. The slurry is made by blending approximately 5% to 6% bentonite to water (by mass). Slurry is typically stored on-site for a minimum of 24 h before use for the bentonite to fully hydrate. Desirable properties of the slurry include a target Marsh viscosity of 40 s, a filtrate loss of < 25 mL, and $6.5 < \text{pH} < 10$ (e.g., Evans 1993). Backfill is made by blending excavated soils (or imported soils if excavated soils are unsuitable) with the bentonite–water slurry to form a homogeneous backfill that has the consistency of high slump concrete. If no additional dry bentonite is added, the final bentonite content of the mixture is approximately 1%. The fines content of the backfill prior to the addition of slurry or additional dry bentonite is ideally greater than 20% to 25% to produce a hydraulic conductivity of $< 1 \times 10^{-9}$ m/s. For backfill made from coarser-grained soils, additional dry bentonite is blended with the backfill to produce the necessary low hydraulic conductivity. The bentonite content of SB backfills typically ranges from 1% to 5% (Sharma and Reddy 2004). Although adding additional dry bentonite is useful in reducing hydraulic conductivity, it may be better to add additional fines for better contaminant resistance (see the section on longevity). Once mixed, the backfill is placed in the trench, displacing the slurry, to form the permanent subsurface barrier. To act as a thick viscous liquid and displace the slurry, a slump of 100 to 150 mm is normally required. The complete process is shown schematically in Figure 16-3b, and construction site photographs are shown in Figure 16-4.

The design phase for an SB barrier begins with the establishment of performance criteria. The SB backfills are primarily designed to provide low hydraulic conductivity (typically $< 10^{-6}$ to 10^{-9} m/s) to minimize advective transport of contaminants through the wall. The design also includes the evaluation of constructability. During the design phase, test borings along the proposed alignment are drilled to evaluate stratigraphy and the depth to a low-permeability material for a key (if available). Samples of each stratigraphic unit are analyzed for grain size distribution and plasticity. With these results, the expected grain size distribution including fines content can be computed by using weighted averages based on stratigraphic layer thicknesses. This exercise is done for multiple locations along the alignment to define the range of expected backfill formulations. With these results, composite samples can be mixed for laboratory permeability testing as well as compatibility testing if appropriate. During these bench-scale studies, bentonite is added via slurry to form a sample of the appropriate slump, and the need to test samples with added dry bentonite is assessed. Once the candidate samples are determined, these composite samples are tested at the appropriate state of consolidating stress.

When unconfined compressive strength is also required, a variation of the SB backfill is soil–cement–bentonite (SCB). SCB is similar to SB, except that dry portland cement (5%) is blended in the backfill. The unconfined

Figure 16-4. (a) SB slurry wall excavation, (b) SB backfill.
Source: J. Evans.

compression strength of SB is zero, whereas the addition of cement in SCB results in unconfined strength values typically between 100 and 700 kPa (Ryan and Day 2002). It is essential to note that the hydraulic conductivity of the SB backfill is highly stress dependent and the state of stress of the backfill in the trench is far less than geostatic (Evans et al. 1995, Filz 1996, Ruffing et al. 2010). The high compressibility of the SB backfill results in the backfill moving downward with respect to the sidewalls of the trench, resulting in load transfer from the SB within the trench to the material in the formation adjacent to the trench.

16.3.2 Cement–Bentonite Slurry Trench Cutoff Walls

Cement–bentonite (CB) slurry trench cutoff walls are commonly used in Europe but less so in the United States. These vertical barriers are constructed using a one-step process whereby the trench is filled with a slurry composed of bentonite, water, and cementitious material consisting of either portland cement or a combination of this cement and granulated ground blast furnace slag (Jefferis 1997). Rather than displace the slurry with a backfill as in the SB technique, the slurry is left to harden in place and forms the permanent barrier (Figure 16-5). As a result, CB walls do not require a source of suitable materials to prepare backfill and the completed wall is therefore independent of site geology. However, these types of wall use substantial quantities of portland cement and granulated ground blast furnace slag (hereafter called slag) and are thus frequently more expensive than SB cutoff walls.

Mix designs vary and results are somewhat dependent on the mixing water chemistry and the source of the slag. Typical mixtures have 10% to 20% cementitious materials mixed with a bentonite–water slurry. Where the cementitious material is 100% portland cement, the resulting hydraulic conductivity is in the range of 1×10^{-7} m/s to 1×10^{-8} m/s which, for geoenvironmental applications, is usually too high. Fortunately, blending portland cement with slag at a ratio of about 20/80 (portland/slag) can lower the hydraulic conductivity by about two orders of magnitude. The dependence of hydraulic conductivity on the ratio between slag and portland cement is illustrated with data in Figure 16-6a. As shown in Figure 16-6a, there is no substantial improvement in hydraulic conductivity for up to 70% slag replacement (i.e., up to 70% slag substituted for portland cement in the mixture on a dry weight basis). However, hydraulic conductivity decreases to $< 10^{-8}$ m/s for slag replacements of 80% and

(a) (b)

Figure 16-5. (a) Excavation of a slag–CB wall through the core of a dam, (b) hardened slag–CB.
Source: J. Evans.

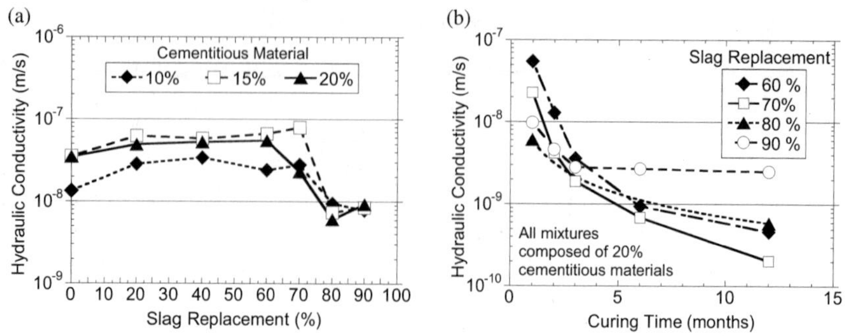

Figure 16-6. (a) Influence of slag content on the hydraulic conductivity of slag–CB mixtures cured for one month, (b) influence of curing time on the hydraulic conductivity of 20% cementitious slag–CB mixtures.
Source: Opdyke and Evans (2005).

higher. Some quantity of portland cement is required in the mix because 100% slag will not set, unless the hydration reaction is sparked by another additive such as hydrogen peroxide. Proprietary mixtures that do not use portland cement to initiate the curing process are also available.

An added complication to the choice of slag–CB is that the measured values of hydraulic conductivity are rather time dependent. For example, shown in Figure 16-6b is the relationship between hydraulic conductivity and time for 20% cementitious mixtures having 60%, 70%, 80%, and 90% slag. For these mixtures, the hydraulic conductivity declined as much as two orders of magnitude over a time span of one year. Although decreasing hydraulic conductivity with time is an excellent characteristic of a geoenvironmental barrier, complications arise regarding acceptance criteria at the end of construction. Specification writers and owners need to accommodate the considerable decline in hydraulic conductivity over time in contract documents.

16.3.3 Mix-in-Place Methods (Panel Mixing)

There are advantages to mixing in-place soils with slurry using in situ mixing equipment, rather than excavating soils, blending slurry, and placing backfill in a slurry-filled trench as done with SB wall construction. Using the panel method for in situ mixing of a vertical barrier requires a specialized mixing tool such as a multiple auger tool or a counter-rotating soil excavating and mixing tool such as a hydrofraise from Soletanche Bachy or a cutter soil mixer from Bauer. These tools have several features in common, including the ability to excavate soils vertically downward and to mix soil and injected slurry to produce a homogeneous barrier material. Normally, deployment is done by first excavating primary panels followed by secondary panels, as shown in Figure 16-7a. Figures 16-7b and 16-7c illustrate the multiple auger tool and cutter soil mixer equipment, respectively.

Slurry formulation for mix-in-place panel methods is normally similar to that used for slag–CB slurry trench cutoff walls. The completed barrier wall is a mixture of in situ soil and slurry, the solid content is high, and the material usage (slag, portland cement, and bentonite) is low compared to slag–CB slurry trench walls. The principal limitation is that the mixing tools can blend soils only over a limited vertical range of the soil profile and not the entire depth. For example, in one project where there was a peat layer, it was first necessary to remove the peat prior to blending the soil profile (Arnold et al. 2011). Nonetheless, for many sites, the subsurface

Figure 16-7. Mix-in-place method using augers: (a) schematic (b) multiauger equipment, (c) cutter soil mixer equipment.
Source: (a) Ruffing and Evans (2023), (b, c) Courtesy of Malcom Drilling.

materials are suitable for blending and can be readily penetrated with mix-in-place tools. Further, even with variability in the in situ soils, the resulting hydraulic conductivity is largely dependent on the slag–CB mixture filling the soil voids.

16.3.4 Mix-in-Place Methods (Continuous Trench Mixing)

The trench remixing and deep wall (TRD) method employs a multisection cutter post and revolving chain to excavate and blend the loosened soil in an ascending current of slurry in a manner that ultimately produces a highly homogeneous blend of slurry and soil (Evans and Garbin 2009). A schematic of the TRD method and a photograph of the equipment are shown in Figure 16-8. Similarly, the DeWind One-Pass trencher employs a one-piece cutter bar and revolving chain to both excavate and blend the loosened soil with slurry (Figure 16-9).

As with panel mixing, a wide variety of slurry formulations are possible depending on the needs of the project. Most commonly, the slurry is a slag–CB mixture, but bentonite–water slurry and other blends have been used. The principal advantage of continuous trench mixing is that the entire vertical profile is blended with the slurry, resulting in a more uniform mixture from top to bottom. As a result, small layers of unsuitable soils can often be blended with the remaining soils in the stratigraphy to produce a suitable cutoff wall material. An example of the uniformity of the resulting barrier using continuous trench mixing is shown in Figure 16-10. This downhole photograph from the TRD wall at Herbert Hoover Dike in Florida shows the blending of sand, shells, limestone fragments, and peat (in discontinuous pockets). As noted previously, the use of mix-in-place panel methods for this stratigraphy would have first required removal of the peat layer, as the entire vertical profile is not blended with these tools.

(a) (b)

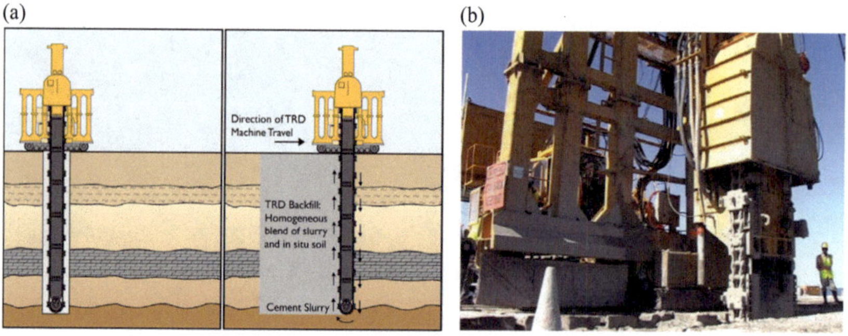

Figure 16-8. (a) Schematic of the trench remixing and deep wall method, (b) TRD equipment.
Source: (a) Garbin et al. (2009), (b) Evans (2007).

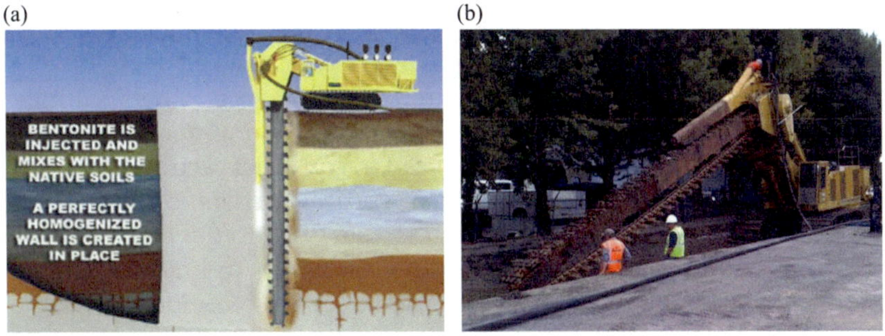

Figure 16-9. (a) Schematic of a one-pass trencher, (b) one-pass trencher equipment. Source: (b) Courtesy of Dewind.

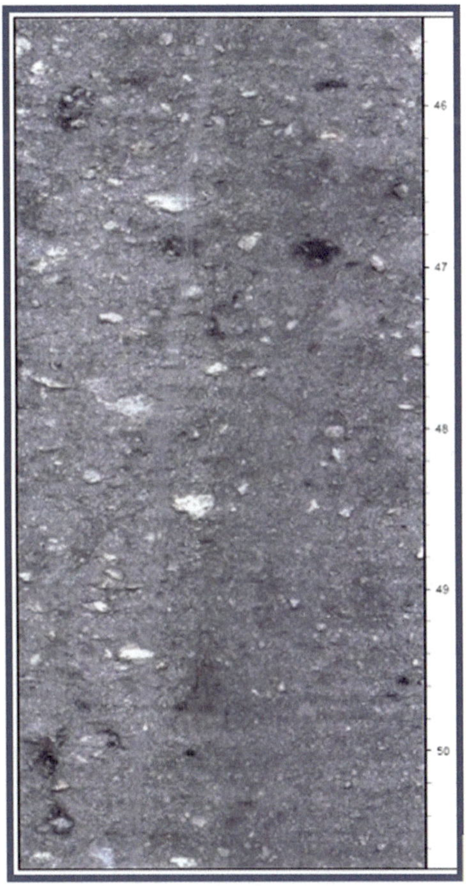

Figure 16-10. Downhole photograph of a TRD mixed backfill. Source: Courtesy of Keller-NA.

16.3.5 Other Vertical Barrier Methods

Although the foregoing methods are the most used techniques to construct a vertical barrier in geoenvironmental projects, other techniques have been used on occasion. Vibrating beam, jet grouting, and conventional (permeation) grouting have all been used to construct vertical barriers for environmental projects. However, these vertical barriers can have higher permeability and cost more than the slurry cutoff walls described previously (Sharma and Reddy 2004). As these techniques are less commonly used, only a limited discussion of each is provided.

Grouted barrier walls are typically constructed with overlapping grout columns in rows (Pearlman 1999). The constructability of vertical grout walls depends on both the permeability of the soil and the viscosity of the grout. The vibrating beam thin cutoff wall (Figure 16-11) employs a wide-flange steel beam equipped with a trailing fin to guide subsequent beam injections into the previous insertion and with grout injection nozzles to pump self-hardening slurry into the void created by the beam as the beam is withdrawn. The advantage of the method for geoenvironmental sites relates to the fact that no subsurface soils (which are often contaminated) are brought to the surface, reducing health and safety risks. The principal disadvantage is the relative thickness of the wall (100 to 150 mm) compared to typical slurry walls of 750 mm. Of course, a slag–CB slurry wall of 750 mm uses considerably more material than a 150 mm vibrating beam wall, resulting in considerably greater material costs.

Methods for jet grouting include the injection of only grout (single rod), injection of grout and air (double rod), or injection of grout, air, and water (triple rod) (Pearlman 1999). The grout mixture is injected via jets in the drill rod at high pressures of 35 to 40 MPa (approximately 5,000 to 6,000 psi) into the surrounding soil/rock (Figure 16-12). Overlapping, homogenous columns are constructed to create a continuous wall (curtain). The

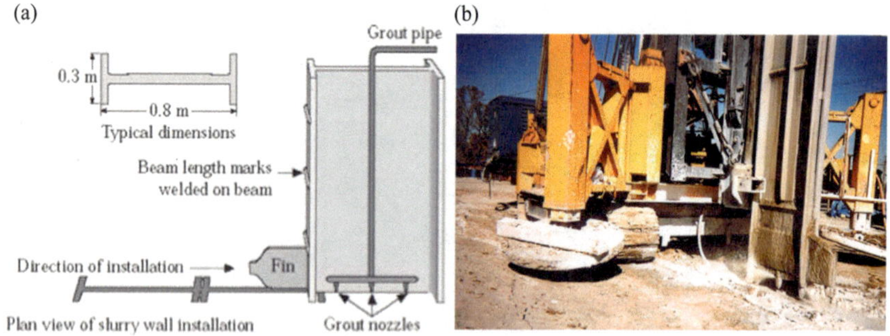

Figure 16-11. (a) Vibrating beam cutoff wall schematic, (b) vibrating beam equipment.
Source: Courtesy of Slurry Systems Inc.

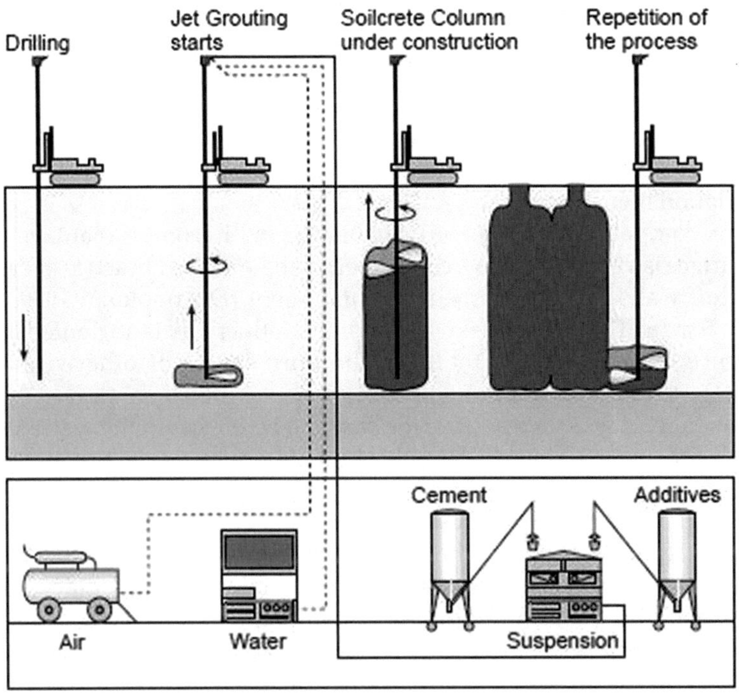

Figure 16-12. Construction of a vertical barrier wall with the jet grouting method. Source: Filz and Mitchell (1995).

advantage of jet grouting is its versatility; jet-grouted barriers can be constructed in a wide range of soil types and at angles allowing for both vertical and horizontal barriers. However, it can be difficult to ensure that the barrier is continuous and free of gaps. Also, excess pressures may develop during grouting, which need to be monitored closely, and large volumes of spoils may be produced (Rumer and Mitchell 1995, Sharma and Reddy 2004).

Conventional (permeation) grouting techniques inject low-viscosity grout into the soil at lower pressures than jet grouting. The two primary methods of permeation grouting are point injection or the tube-a-manchette method using a sleeve pipe. Permeation grouting, however, is typically limited to soil formations with moderate to high permeability ($> 10^{-3}$ cm/s).

16.4 LONGEVITY CONSIDERATIONS

Regardless of the method chosen to construct a vertical barrier for geoenvironmental conditions, consideration must be given to the longevity of the barrier. The long-term performance of a vertical barrier can be impacted by incompatibility with subsurface contamination, shrinkage and cracking of cementitious barrier materials, wetting and drying stresses in the zone of the fluctuating water table, desiccation, and hydraulic fracturing. The impacts of environmental stresses are dependent on the nature and magnitude of the stresses and the barrier material type and installation method.

The compatibility of a vertical barrier with contaminants in the subsurface is virtually always considered in the standard practice of design of cutoff walls for environmental containment (D'Appolonia 1980, Day 1994). For an SB backfill, the low hydraulic conductivity is attributed to the presence of hydrated bentonite in the pore spaces of otherwise inert gravels, sands, and silts and interclod spaces between clods of low-permeability clays that comprise the base soil before bentonite water slurry is added to make the backfill flowable into the slurry trench. The bentonite, a highly swelling smectite clay, is highly effective in producing a backfill with hydraulic conductivity less than 1×10^{-7} cm/s, even at low bentonite contents (e.g., 1% bentonite in the backfill, by mass). Unfortunately, this swelling is reversible in the presence of inorganic salts and organic fluids of low dielectric constant. With respect to inorganics, as the cation valence and salt concentration increase, the detrimental impact on the hydrated bentonite in the backfill increases. For organics, the principal concern is nonaqueous phase liquids. Given the relatively low solubility of organics, dissolved organics have little effect on the dielectric constant of the permeating fluid and thus little effect on the bentonite (Evans et al. 1985). The general practice for the laboratory evaluation of the compatibility of an SB backfill is to permeate design mixtures until two to three pore volumes of permeant have passed through the sample. It is important that the tests be conducted at effective stresses expected in the field, as high effective stresses have been shown to mute the negative impacts of chemical incompatibility (Acar et al. 1985).

Cementitious mixtures are more complicated than those with bentonite only. First, during the hydration process, the bentonite in its characteristic crystalline form, morphs into the crystalline structure of the hydrated slag–CB or SCB barrier mixtures. Hence, lessons from studies on bentonite or bentonite–soil mixtures cannot be extrapolated to slag–CB or SCB barriers. Limited studies to date show that organics have little detrimental effect on slag–CB mixtures (Soga et al. 2013). It is hypothesized that some inorganics such as sulfates may cause degradation (Fratalocchi et al. 2006) but only after long-term permeation (more than the two to three pore volumes used for SB). Immersion tests are often used to evaluate the potential for degradation in response to exposure to site chemicals (Garvin and Hayes 1999). These tests may be considered conservative, as they do not receive the benefit of confinement.

Volume change (shrinkage) in cementitious barriers can be expected (Evans and Jefferis 2014). Bleed-, filtration-, and hydration-induced (autogenous) shrinkage, thermal shrinkage, and creep are all important for cementitious barriers. All types of wall materials may also undergo chemically induced volume changes from interactions with the permeating fluid. In regions of the barrier above the zone of saturation, there also may be drying shrinkage. For cementitious barriers, tensile stresses induced by shrinkage may exceed the tensile strength of the material, resulting in cracking (Cermak et al. 2012). For an SB backfill, shrinkage of the bentonite within the pores of the base soil matrix is largely irreversible, and therefore, care must be taken to avoid drying suction pressures and desiccation.

16.5 SUMMARY

Vertical barriers for geoenvironmental applications have a long and successful record going back nearly 40 years. Several lengthy publications provide additional detail on the techniques discussed (e.g., Paul et al. 1992, Evans 1993, Xanthakos 1994) along with innumerable technical papers on the subject. The goal of this chapter was to highlight the current practice in geoenvironmental engineering with regard to vertical barriers and to provide a select few references as a start for the more interested reader to delve further into any aspect of the information presented.

REFERENCES

Acar, Y. B., A. B. Hamidon, S. D. Field, and L. Scott. 1985. "The effect of organic fluids on hydraulic conductivity of compacted kaolinite." In *STP874, Hydraulic barriers in soil and rock*, edited by A. Johnson, R. Frobel, N. Cavalli, and C. Pettersson, 171–187. West Conshocken, PA: ASTM.

Arnold, M., K. Beckhaus, and U. Wiedenmann. 2011. "Cut-off wall construction using cutter soil mixing: A case study." *Geotechnik* 34 (1): 11–21.

Bierck, B. R., and W. C. Chang. 1994. "Contaminant transport through soil–bentonite slurry walls: Attenuation by activated carbon." In *Proc., Water Environmental Federation Specialty Conf. on Innovative Solutions for Contaminated Site Management*, 461–472. Reston, VA: ASCE.

Cermak, J., J. C. Evans, and G. J. Tamaro. 2012. "Evaluation of soil–cement–bentonite wall performance—Effects of backfill shrinkage." In *Proc., DFI 4th Int. Conf. on Grouting and Deep Mixing*. Hawthorne, NJ: International Conference Organization for Grouting, Inc. (ICOG) and the Deep Foundations Institute.

D'Appolonia, D. J. 1980. "Soil–bentonite slurry trench cutoffs." *J. Geotech. Eng.* 106 (4): 399–417.

Day, S. R. 1994. "The compatibility of slurry cutoff wall materials with contaminated groundwater." In *Hydraulic conductivity and waste contaminant transport in soils*, ASTM STP 1142, edited by D. E. Daniel and S. J. Trautwein, 284–299. West Conshohocken, PA: ASTM.

Evans, J. C. 1993. "Vertical cutoff walls." In Chap. 17 in *Geotechnical practice for waste disposal*, edited by D. E. Daniel, 430–454. London: Chapman & Hall.

Evans, J. C. 1995. "Soil- and cement-based vertical barriers with focus on materials." In *Section 2, Assessment of barrier containment technologies—A comprehensive treatment for environmental applications*, 5–43. https://clu-in.org/download/contaminantfocus/dnapl/treatment_technologies/containment-1995.pdf.

Evans, J. C. 2007. "The TRD method: Slag-cement materials for in situ mixed vertical barriers." In *Geo-Denver 2007: New Peaks in Geotechnics*, Geotechnical Special Publication 172, 1–11. Reston, VA: ASCE.

Evans, J. C., M. Costa, and B. Cooley. 1995. "The state of stress in soil-bentonite slurry trench cutoff walls." In *Proc., Geoenvironment 2000: Characterization, Containment, Remediation and Performance in Environmental Geotechnics*, Geotechnical Special Publication 46, 1173–1191. Reston, VA: ASCE.

Evans, J. C., and E. J. Garbin. 2009. "The TRD method for in situ mixed vertical barriers." In *Proc., US-China Workshop on Ground Improvement Technologies*, Geotechnical Special Publication 188, 271–280. Reston, VA: ASCE.

Evans, J. C., and S. A. Jefferis. 2014. "Volume change characteristics of cutoff wall materials." In *Proc., 7th Int. Congress on Environmental Geotechnics*,10–14. Melbourne, Australia. London: International Society of Soil Mechanics and Geotechnical Engineering.

Evans, J. C., I. J. Kugelman, and H. Y. Fang. 1985. "Organic fluid effects on the strength, deformation, and permeability of soil–bentonite slurry walls." In *Proc., 17th Mid-Atlantic Industrial Waste Conf.*, 275–291. Bethlehem, PA: Technomic Publishing Co.

Evans, J. C., and M. J. Prince. 1997. "Additive effectiveness in minerally-enhanced slurry walls." In *Proc., In Situ Remediation of the Geoenvironment*, Geotechnical Special Publication 71, 181–196. Reston, VA: ASCE.

Filz, G. M., and J. K. Mitchell. 1995. "Design, construction and performance of soil- and cement-based vertical barriers." In *Section 3, Assessment of barrier containment technologies—A comprehensive treatment for environmental applications*, 45–75. https://clu-in.org/download/contaminantfocus/dnapl/treatment_technologies/containment-1995.pdf.

Filz, G. M. 1996. "Consolidation stresses in soil–bentonite back-filled trenches." In *Proc., 2nd Int. Congress on Environmental Geotechnics*, edited by M. Kamon, 497–502. Osaka, Japan: International Congress on Environmental Geotechnics.

Fratalocchi, E., E. Pasqualini, and P. Balboni. 2006. "Performance of a cement–bentonite cut-off wall in an acidic sulphate environment." In *Proc., 5th Int. Conf. Environmental Geotechnics*, 133–139. London: International Society for Soil Mechanics and Geotechnical Engineering.

Garbin, E. J., J. C. Evans, and J. D. Hussin. 2009. "The TRD method for in situ mixed vertical barriers." In *Proc., US–China Workshop on Ground Improvement Technologies*, Geotechnical Special Publication 188, 271–280. Reston, VA: ASCE.

Garvin, S. L., and C. S. Hayes. 1999. "The chemical compatibility of cement–bentonite cut-off wall material." *Constr. Build. Mater.* 13 (6): 329–341.

Jefferis, S. A. 1997. "The origins of the slurry trench cut-off and a review of cement–bentonite cutoff walls in the UK." In *Proc., Int. Containment Technology Conf. and Exhibition*, 52–61. Tallahassee, FL: Florida State University.

Konikow, L., and D. Thompson. 1984. "Groundwater contamination and aquifer reclamation at the Rocky Mountain Arsenal, Colorado." In Chap. 6 in *Groundwater contamination*, 93–103. Washington, DC: National Academies Press.

LaGrega, M. L., P. L. Buckingham, and J. C. Evans. 2010. *Hazardous waste management*. 2nd ed. Long Grove, IL: Waveland Press.

Malusis, M. A., E. J. Barben, and J. C. Evans. 2009. "Hydraulic conductivity and compressibility of a model soil–bentonite backfill amended with activated carbon." *J. Geotech. Geoenviron. Eng.* 135 (5): 664–672.

Opdyke, S. M., and J. C. Evans. 2005. "Slag–cement–bentonite slurry walls." *J. Geotech. Geoenviron. Eng.* 131 (6): 673–681.

Paul, D. B., R. R. Davidson, and N. J. Cavalli. 1992. *Slurry walls: Design, construction and quality control*. STP1129-EB. West Conshohocken, PA: ASTM.

Pearlman, L. 1999. *Subsurface containment and monitoring systems: Barriers and beyond*. Washington, DC: USEPA, National Network of Environmental Management Studies.

Ruffing, D. G., J. C. Evans, and M. A. Malusis. 2010. "Prediction of earth pressures in soil–bentonite cutoff walls." In *ASCE GeoFlorida 2010 Advances in Analysis, Modeling and Design*, Geotechnical Special Publication 199, 2416–2425. Reston, VA: ASCE.

Rumer, R. R., and J. K. Mitchell. 1995. *Assessment of barrier containment technologies—A comprehensive treatment for environmental applications*. Baltimore: US Department of Energy, US Environmental Protection Agency, Dupont Company.

Ryan, C. R., and S. R. Day. 2002. "Soil–cement–bentonite slurry walls." In *Deep foundations 2002: An international perspective on theory, design, construction, and performance*, 713–727. Reston, VA: ASCE.

Sharma, H. D., and K. R. Reddy. 2004. *Geoenvironmental engineering*. Hoboken, NJ: Wiley.

Soga, K., K. Joshi, and J. C. Evans. 2013. "Cement bentonite cutoff walls for polluted sites." In *Coupled phenomena in environmental geotechnics*, edited by M. Manassero and D. Dominijanni, 149–168. Boca Raton, FL/Netherlands: CRC Press/Balkema.

USEPA (US Environmental Protection Agency). 1998. *Evaluation of subsurface engineered barriers at waste sites*. Washington, DC: USEPA, Office of Solid Waste and Emergency Response.

Van Deuren, J., T. Lloyd, S. Chhetry, R. Lious, R., et al. 2002. *Remediation technologies screening matrix and reference guide*. 4th ed. Washington, DC: US Army Environmental Center.

Xanthakos, P. P. 1994. *Slurry walls as structural systems*. Hoboken, NJ: Wiley.

CHAPTER 17

ZERO-VALENT IRON PERMEABLE REACTIVE BARRIERS FOR IN SITU TREATMENT OF ORGANICS AND METALS IN GROUNDWATER

Andrzej Przepiora, P.Geo; Claire F. Wildman, Ph.D., P.E.; and Megan Hart, Ph.D.

17.1 INTRODUCTION

A permeable reactive barrier (PRB) is defined as a continuous, in situ hydraulically permeable treatment zone emplaced in the subsurface to intercept and then destroy or immobilize contaminants in groundwater (USEPA 1998, ITRC 2005). As groundwater flows through a PRB under the natural hydraulic gradient, dissolved contaminants interact with the reactive media, leading to either their transformation to less harmful compounds or their fixation to the reactive materials (Figure 17-1). Permeable reactive barriers are designed to be more permeable than the surrounding aquifer matrix, passively treating groundwater readily flowing through the material under the natural gradient, without significantly altering groundwater hydrogeology or plume morphology. At the same time, a correctly implemented PRB creates a "barrier" to migration of the targeted dissolved contaminants (Figure 17-1). A variety of materials can be utilized as permeable barrier material, including

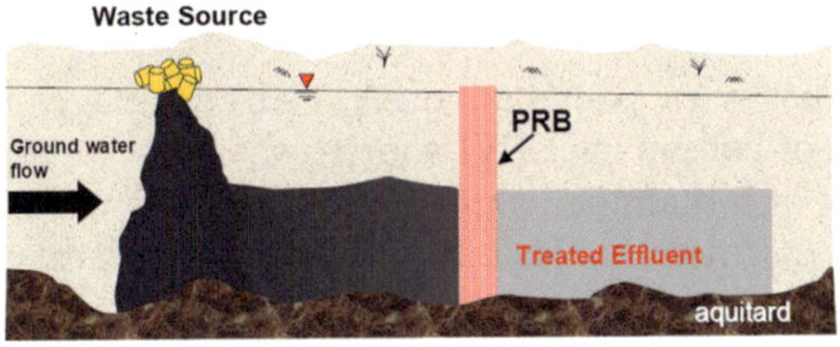

Figure 17-1. Conceptual schematic of a PRB for the treatment of contaminated groundwater.

zero-valent iron (ZVI), carbonaceous sources, alkaline materials, granular activated carbon, zeolites, and new materials. Depending on the reactive material used, contaminant treatment may occur through physical, chemical, or biological processes via destructive and/or nondestructive reactions, as discussed in the following.

Zero-valent iron is the most common PRB material. It is applicable for the treatment of various chlorinated organics, various dissolved metals, and radionuclides. During treatment, ZVI oxidizes to form ferrous and ferric iron (Fe[II] and Fe[III]) as contaminants are reduced. Other reactive media, including carbonaceous sources (compost, mulch), alkaline materials (limestone, magnesium hydroxide), granular activated carbon, zeolites, and so on, have also been deployed in PRBs to treat metals and some inorganic and organic compounds (ITRC 2011).

The oldest full-scale PRB, which is composed of ZVI, was installed in 1995 for the treatment of chlorinated ethenes (Warner et al. 2005). Since that first commercial application, PRBs have become an established technology in the field of groundwater remediation, with a published record of more than 200 field installations since its inception, predominantly using ZVI materials (ITRC 2011). Although there are many published review articles, monographs, and guidance documents regarding ZVI PRB applications (e.g., ITRC 2005, 2011; USEPA 1998; Gillham et al. 2010; Guan et al. 2015; Fu et al. 2014), this chapter provides a concise and practical guide to the implementation of ZVI PRBs for the treatment of chlorinated aliphatic hydrocarbons (CAHs) and certain redox-reactive metals such as chromium and arsenic. This chapter details the application stages for this groundwater remediation method, including the initial feasibility assessment, site data collection, laboratory treatability studies, development of design parameters, PRB construction techniques, and long-term performance monitoring.

17.2 ZERO-VALENT IRON PERMEABLE REACTIVE BARRIER APPLICABILITY

This section provides a summary of ZVI PRB advantages and limitations, lists the types of commercial ZVI sources that can be applied in PRBs, details the type of contaminants treated by ZVI along with related process chemistry, and discusses factors influencing the long-term performance of ZVI PRBs.

17.2.1 Feasibility Evaluation

The main advantage of a PRB is that it provides passive treatment after installation, and no active pumping or aboveground treatment is required.

Because no aboveground installed structures are needed, the aboveground portions of the affected property can be used productively while groundwater below the surface is being cleaned up. Also, when applied in the right site conditions, ZVI material is typically used up slowly and, therefore, PRBs have the potential to passively treat a plume over several years or decades, with operation and maintenance limited to performance monitoring. Upon treatment with a ZVI PRB, groundwater geochemistry is typically altered to both a higher, more alkaline pH and a lower, more reducing oxidation–reduction potential. Potential downgradient impacts of the geochemical changes, particularly for naturally occurring elements in site soil, should be considered during the feasibility evaluation. Depending on the design of the PRB and the longevity of the specific type of ZVI used (granular or microscale ZVI), the PRB may have to be replaced periodically; however, it is expected that such maintenance would be required infrequently, if at all, within the remediation timeframe.

Other advantages of ZVI PRBs used in groundwater remediation include the following:

- Zero-valent iron is compatible with the subsurface environment; it does not cause adverse chemical reactions or create by-products when reacting with constituents in the contaminant plume and does not act as a possible source of contaminants itself.
- Degradation reactions and ZVI's chemical behavior are well understood.
- Various commercial sources of ZVI are readily available, which allows for selecting the most suitable material considering the site conditions to minimize constraints on groundwater flow and to ensure contaminant treatment with the selected installation method.
- Zero-valent iron materials are available at a low to moderate cost and the materials' reactivity persists over long periods of time.

Limitations to the application of the ZVI PRB technology include the following:

- Incomplete plume capture or treatment resulting from insufficient site characterization may result in contaminant plume migration underneath or around the PRB. An underestimated groundwater velocity value may result in inadequate groundwater residence time in the PRB.
- Certain CAHs are not treated by ZVI [e.g., 1,2-dichloroethane (1,2-DCA), dichloromethane (DCM)], and the presence of these compounds, either in the target groundwater or created from the reaction of ZVI with parent compounds, may require additional treatment.
- Reactivity of ZVI decreases over time because of surface passivation caused by corrosion products (i.e., the precipitation of iron oxy-

hydroxides, carbonates, and other secondary minerals and/or surface films). The expected site-specific rate of ZVI material passivation is a key design parameter, and appropriate safety factors should be applied to ensure effective long-term performance.

- Capital cost of ZVI PRB application is typically higher than that of other remedial technologies. However, because of greater longevity, the lifetime cost is typically lower over remedial timeframes of a decade or more.
- Currently, there are no technically and economically viable rejuvenation methods for a mature ZVI PRB where ZVI has become deactivated and is no longer reactive toward target contaminants.

17.2.2 Zero-Valent Iron Materials

The ZVI used for PRB applications falls into two general categories on the basis of particle size and manufacturing process (Table 17-1). The most commonly used ZVI in a PRB configuration is granular, millimeter-size material, which is manufactured by recycling scrap gray cast iron from the automotive industry. This material is typically applied in PRBs constructed using excavation-and-backfill installation. It has a grain size range of 2.0 to 0.25 mm, which provides an average hydraulic conductivity of about 5×10^{-4} m/s and a specific surface area in a range of 0.5 to 1.5 m^2/g (ITRC 2005). Microscale ZVI (mZVI) is typically emplaced using injection-based techniques where trenching cannot be performed because of site access or installation depth limitations. These microscale ZVI materials have a higher specific surface area and typically a higher reactivity compared to granular ZVI; however, their longevity is typically less than the granular sources. Nanoscale ZVI materials have an even higher specific surface area and initial reactivity, but a relatively short life, and are typically applied

Table 17-1. Commercial ZVI Media Used in PRB Applications.

Scale	Size range	SSA (m^2/g)	Vendors
Millimeter	0.1–2 mm	0.5–1.5	Connelly, Peerless, RioTinto, GMA, ISPAT, iPutec
Micrometer	20–200 μm 1–20 μm	3–5	Hepure, RioTinto, Högenäs, CERES, ISP, BASF
Nanometer*	20–200 nm	1–58	Nano Iron s.r.o., Hepure

Source: ITRC (2005).

*Nanoscale ZVI is used in direct applications in source zones and not as a PRB.

for source zone treatment rather than in PRBs for long-term plume migration control (ITRC 2005).

17.2.3 Contaminants Treated

The majority of the previously applied ZVI PRBs targeted the treatment of CAHs, redox-sensitive metals, such as chromium and arsenic, and cationic metals, as enumerated in Table 17-2.

Table 17-2. Overview of Constituents That Can Be Treated by ZVI PRBs.

Type	Specific constituents
Chlorinated ethenes and ethanes*	Tetrachloroethene (PCE); trichloroethene (TCE); cis 1,2-dichloroethene (cDCE); trans 1,2-dichloroethene (tDCE); 1,1-dichloroethene (1,1-DCE); vinyl chloride (VC); hexachloroethane; 1,1,2,2-tetrachloroethane; 1,1,1,2-tetrachloroethane, 1,1,1-trichloroethane (1,1,1-TCA); 1,1,2-trichloroethane (1,1,2-TCA); 1,1-dichloroethane (1,1-DCA)
Chlorinated methanes* and propanes	Tetrachloromethane (PCM); trichloromethane (TCM); 1,2,3-trichloropropane; 1,2-dichloropropane
Explosives	Hexahydro-1,3,5-trinitro-1,3,5-triazine (RDX); 1,4,6-trinitrotoluene (TNT); octahydro-1,3,5,7-tetranitro-1,3,5,7-tetrazocine (HMX)
Other organics	Tribromomethane (TBrM); 1,2-dibromomethane; 1,2-dibromo-3-chloropropane; trichlorotrifluoroethane (Freon 113); trichlorofluoromethane (Freon 11); hexachlorocyclohexane (HCHs); carbon disulfide; chloropicrin
Metals (immobilization as solid phase via reduction and/or precipitation)	Chromium; arsenic; uranium; selenium; cationic metals (copper, nickel, and zinc)

Source: ITRC (2005).

*1,2-DCA and DCM are not treated by ZVI.

17.2.4 Treatment of Chlorinated Aliphatic Hydrocarbons

17.2.4.1 Degradation Pathways. Considerable research during the initial stages of technology deployment focused on the degradation of CAHs such as TCE and PCE. The ZVI-mediated degradation process for CAHs is widely accepted as an abiotic reductive dechlorination, involving corrosion of ZVI, which serves as an electron donor in this redox reaction, coupled with a reduction of the CAH. ZVI metal is a powerful reductant with a reduction potential of −0.44 V (USEPA 1998). These reactions result in the degradation of a CAH (denoted as RCl) and formation of a dechlorinated (i.e., chemically reduced) hydrocarbon (denoted as RH), chloride, and ferrous iron (Fe^{2+}) products:

Oxidation Reaction:

$$Fe^0 \rightarrow Fe^{2+} + 2e^- \tag{17-1}$$

CAH Reduction Reaction:

$$RCl + 2e^- + H^+ \rightarrow RH + Cl^- \tag{17-2}$$

Net Reaction:

$$Fe^0 + RCl + H^+ \rightarrow Fe^{2+} + RH + Cl^- \tag{17-3}$$

The reaction pathways of CAH dechlorination by ZVI have been verified for the major classes of CAHs that occur at contaminated sites (Table 17-2 and Figure 17-2). This information is crucial for the optimal design of ZVI PRB, as incomplete dechlorination of a parent compound (e.g., PCE or TCE) may produce intermediate products (e.g., cDCE and VC) that are more toxic and typically react in a PRB at slower rates than the parent compounds. The ZVI PRB design should provide sufficient residence time to ensure that no undesirable by-products are created and released from the PRB during its lifetime.

Figure 17-2 details degradation pathways for major classes of CAHs: chlorinated ethenes, ethanes, and methanes. In general, the dechlorination of CAHs occurs via two general pathways: reductive elimination, which results in dechlorinated end products, and sequential hydrogenolysis. The latter pathway results in the formation of lower chlorinated products that typically degrade further, but at a slower rate than the parent product. For example, PCE and TCE degradation by ZVI may produce cDCE and VC at an amount equal to up to 10% of the parent concentration. Because of the slower degradation rates, these degradation products will require a longer residence time for complete

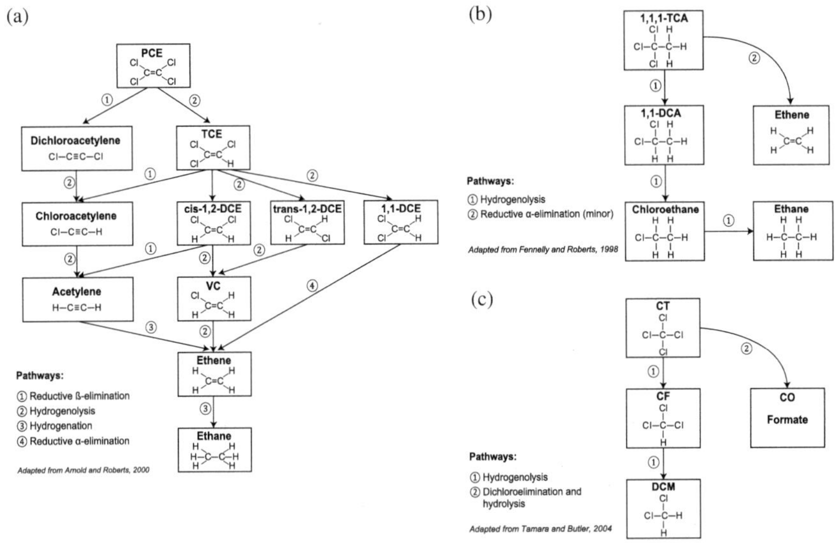

Figure 17-2. ZVI-mediated dechlorination pathways for major CAHs: (a) chlorinated ethenes, (b) chlorinated ethanes, (c) chlorinated methanes. Source: Arnold and Roberts (2000), Fennelly and Roberts (1998), Tamara and Butler (2004).

destruction compared to the parent materials. The degradation of 1,1,1-TCA by ZVI results in an intermediate formation of 1,1-DCA, typically at < 20% molar of the parent product, which then also degrades at a slower rate. Notably, ZVI does not treat 1,2-DCA that may be created by the degradation of 1,1,2-TCA or present in chlorinated ethane plumes (ITRC 2005). Carbon tetrachloride (CT) degrades in the presence of ZVI via sequential hydrogenolysis to TCM and DCM. The formation of TCM accounts for about 40% to 70% of the CT lost, depending on the ZVI type and water chemistry (Vidumsky et al. 2002). Dichloromethane formed in these reactions typically accounts for about 50% of the chloroform lost. No further degradation reactions of DCM are known to occur in documented ZVI systems (Matheson and Tratnyek 1994, ITRC 2005).

17.2.4.2 Degradation Rates. On the basis of previous research, the degradation of volatile organic compounds (VOCs) in contact with ZVI appears to be first-order with respect to the concentration of the contaminant (i.e., pseudo first-order, assuming an excess of ZVI present) (Gillham et al. 2010):

$$\frac{\partial C}{\partial t} = -kt \qquad (17\text{-}4)$$

where

C = Concentration in solution at a particular time (t),
C_o = Initial concentration, and
k = First-order rate constant.

The measured rate constants (which may alternatively be expressed as half-lives) are variable and can be influenced by (1) the ZVI source, (2) chemical composition of the treated groundwater, and (3) temperature (ITRC 2005, Gillham et al. 2010). A discussion of these reactivity factors is provided in the following.

17.2.4.3 Zero-Valent Iron Type. The ZVI-mediated CAH reduction reaction is a surface process, and the reaction rates depend on the reactive surface area of ZVI. In general, larger particle size ZVI materials (i.e., granular materials) have a lower reactivity compared to fine-grained ZVI, such as microscale materials (Table 17-1). Although the correlation between ZVI reactivity and grain size or specific surface area provides a useful basis for comparison, it is far from perfect. For example, a previous evaluation of anaerobic iron corrosion rates has shown a positive correlation between the particle size and the anaerobic corrosion rates for various types of commercial ZVIs used in PRBs, but substantial differences were observed for ZVI sources within specific size categories (Reardon 1995, 2005). From a practical perspective, the type of PRB application will determine the applicable ZVI source. That is, granular ZVI sources are applicable for trench-and-backfill PRB applications, whereas microscale ZVI sources are used for injected PRB applications. Screening tests for various commercial ZVI sources in the applicable particle size categories are recommended to select the most suitable material, because material cost and availability is likely to be equally significant in the selection process as the material's reactivity.

17.2.4.4 Groundwater Chemistry. Previous research has shown that the inorganic composition of the target groundwater can have a profound influence on the reactivity of commercial granular ZVI materials. Several common constituents of groundwater may influence the initial reactivity, including dissolved oxygen, nitrate, and bicarbonate. These constituents also affect long-term performance and are discussed further subsequently. However, groundwater from various sites using the same ZVI material have resulted in highly variable CAH degradation rates, as illustrated in Figure 17-3. Wide ranges in the degradation half-lives (i.e., the time

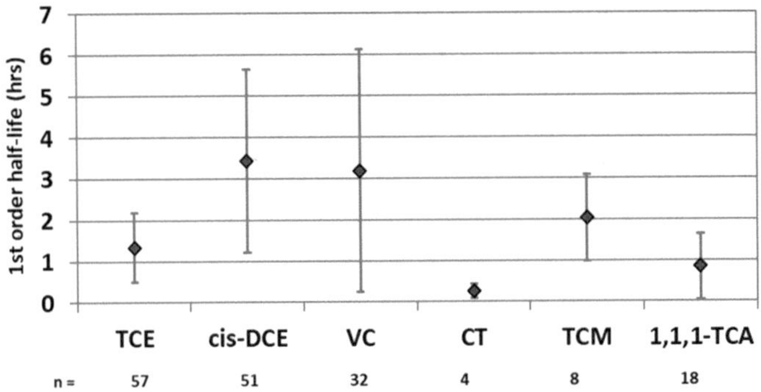

Figure 17-3. Ranges in CAH degradation half-lives obtained from column tests at room temperature using groundwater from multiple sites and commercial granular ZVI materials (Connelly GPM and Peerless Metal Powders).
Source: Gillham et al. (2010), presented as averages ± standard deviation; n denotes the number of tests performed.

required to reduce the initial concentration by 50%) are shown for common CAHs in column tests performed with granular ZVI materials (Connelly and Peerless materials), which demonstrates the complex, interwoven nature of remediation (Gillham et al. 2010).

PRBs are designed to provide adequate residence time for the degradation of the parent CAHs and all toxic intermediate products that may be generated. At sites where the groundwater contamination includes a mixture of CAHs, the design residence time in the PRB is usually determined by the least reactive constituent. As shown in Figure 17-3, higher-chlorinated CAHs (e.g., TCE and CT) tend to be reduced more rapidly, with less variability in degradation rates, than their less-chlorinated congeners. Also, the dechlorination reaction occurs more rapidly at saturated carbon centers (e.g., CT) than at unsaturated carbons (e.g., TCE). Therefore, the determination of site-specific degradation rates of all relevant CAHs and their potential chlorinated breakdown products is a crucial stage of the ZVI PRB design. For design purposes, laboratory treatability tests should be conducted using groundwater from the anticipated location of the PRB to obtain site-specific degradation parameters (i.e., rate constants and conversion factors from parents to daughter products) for the ZVI material selected for the PRB application.

17.2.4.5 Temperature. Empirical data have shown that CAH degradation rates by ZVI are sensitive to water temperature. O'Hannesin et al. (2004) reported that the degradation half-lives for chlorinated ethenes in contact with granular ZVI increased (i.e., the rate constants decreased) by 100%

per every 6 °C to 8 °C of temperature drop within the range of 5 °C to 25 °C. The relationship between the rate constant and the temperature were shown to follow an exponential trend described by the Arrhenius equation (Gillham et al. 2010). Similar trends have been observed in other laboratory tests and at field sites. Because of the temperature influence, corrections are required to adjust the degradation rates obtained in laboratory tests, which are typically performed at room temperature (Figure 17-3) to derive values appropriate for a field application where the groundwater temperatures may be lower.

17.2.5 Treatment of Dissolved Metals

Zero-Valent Iron has been shown effective for the remediation of various dissolved metals and metalloids that occur as anions or oxyanions under natural groundwater conditions, including arsenic, selenium, chromium, uranium, technetium, and antimony (ITRC 2005). Hexavalent chromium, arsenic, and uranium have been targeted in previous full-scale ZVI PRBs (Wilkin et al. 2009, 2014; Beak and Wilkin 2009; Morrison et al. 2001). Other metals, such as selenium, copper, cadmium, nickel, zinc, mercury, and manganese, and radionuclides have been shown to be treatable by ZVI in pilot-scale and laboratory tests (Wilkin and McNeil 2003, Morrison et al. 2002). In contrast to the dechlorination of CAHs that are destroyed by ZVI, dissolved metal oxyanions are sequestered within the ZVI zone via nondestructive immobilization processes involving reduction, precipitation, and/or adsorption.

Chromium occurs naturally in two oxidation states: hexavalent chromium (Cr[VI]) and trivalent chromium (Cr[III]). Dissolved chromium in groundwater is composed mainly of Cr[VI], while (Cr[III]) is insoluble in the pH range measured at most sites (USEPA 2007). The mechanism of Cr(VI) removal by ZVI involves the oxidation of Fe^0 to Fe(II) and Fe(III), coupled with the reduction of Cr(VI) to Cr(III), and the subsequent precipitation of a sparingly soluble solid (Blowes et al. 2000). Under moderate pH conditions (between pH 5 and 11), removal of Cr(III) can occur through the precipitation of $Cr(OH)_3$ or mixed Fe(III)–Cr(III) (oxy) hydroxide (Jeen et al. 2008).

Previous investigations have shown that removal of dissolved arsenic in a ZVI PRB occurs via surface adsorption, complexation, and precipitation. Reductive precipitation of arsenic in the presence of ZVI is associated with iron oxyhydroxide minerals such as goethite that form on the ZVI grain surfaces (Manning et al. 2002). Sorptive removal of arsenic on the secondary ZVI corrosion products (i.e., iron oxides and oxy-hydroxides) may also occur (Manning et al. 2002). When sulfate reduction occurs (often microbially mediated), arsenic can also be removed from solution by the formation of sparingly insoluble iron-arsenic sulfide solids and adsorption onto iron sulfides.

Removal kinetics of chromium and arsenic observed in flow-through column tests have been described using zero- and first-order models (Lien and Wilkin 2005, Melitas et al. 2001). Lien and Wilkin (2005) showed that the capacity of arsenic removal appeared to be relatively constant along the ZVI test column. This implies that arsenic removal capacity (i.e., a zero-order kinetics parameter), expressed as the mass of arsenic removed normalized for the mass of ZVI, is a better design parameter than the arsenic removal rate constant (i.e., a first-order rate constant) to estimate the amount of reactive material required for field-scale application of a ZVI PRB. The removal capacity (mg_{metal}/g_{ZVI}) is therefore the most commonly used PRB design parameter to quantify ZVI capacity for the remediation of both arsenic and chromium. A summary of values obtained in various laboratory column studies is provided in Table 17-3. The removal capacities vary, depending on the ZVI material, concentrations of the target metal, and the groundwater chemistry.

Similar to the degradation of CAHs, metal removal capacity generally increases with decreasing particle size and increasing specific surface area of ZVI (Gheju 2011). However, no dependence on water temperature has been observed in various treatability studies performed within the natural groundwater temperature range. Certain groundwater constituents are expected to compete with metal oxyanions for ZVI surface reduction sites (e.g., dissolved oxygen and nitrate), whereas others may create secondary precipitates that will gradually block the reactive surface sites on ZVI particles (e.g., calcium and iron carbonates, iron oxy-hydroxides), which, in turn, will affect the oxyanion removal capacities. The expected variability in ZVI performance in various water chemistries underscores the need for the design effort to include a site-specific laboratory evaluation of ZVI capacity for metal treatment once the applicable ZVI material has been selected.

Caution should be taken when implementing ZVI PRB at sites where solids in the targeted aquifer contain natural or anthropogenic arsenic. ZVI-treated water egressing from the PRB may cause the mobilization of arsenic and result in intermittent increases in dissolved arsenic concentrations downgradient, despite effective arsenic sequestration within the ZVI zone itself. Potential arsenic mobilization processes include (1) desorption from iron and manganese oxy-hydroxides and/or clays caused by a ZVI-mediated pH shift to alkaline and a subsequent charge reversal on those mineral surfaces, (2) reduction of oxy-hydroxides adsorbed arsenic(V) to arsenic(III) and the subsequent desorption of this reduced species, and (3) reductive dissolution of these iron and manganese mineral phases containing arsenic caused by contact with reducing water outflowing from the ZVI zone.

Table 17-3. Comparison of Chromium and Arsenic Removal Capacities Obtained for Commercial Granular ZVI Sources in Column Studies.

Reference	Influent concentration (mg/L)	Removal capacity (mg_{metal}/g_{ZVI})	ZVI type	Water type
Chromium				
USEPA (1999)	8	3.0	Peerless	GW
Lai and Lo (2008)	25	0.7–4.3	Connelly	Simulated GW
Jeen et al. (2008)	10	4.1	Connelly	Simulated GW
Unpublished*	90	1.8	Connelly	GW
Arsenic				
Lien and Wilkin (2005)	50	7.5	Peerless	Simulated GW
Köber et al. (2005)	5	1.3	ISPAT	Simulated GW
Nikolaidis et al. (2003)	0.3–0.5	0.77	Connelly	GW
Unpublished*	50, 95	5.3, 13.8	Connelly	GW
	49–88	1.9	Peerless	GW

*Internal testing by Geosyntec Consultants.

Notes: Peerless = Granular ZVI from Peerless Metal Powders (Detroit, MI); Connelly = Granular ZVI from Connelly GPM (Chicago); ISPAT = Hamburg, Germany; GW = groundwater; mg/L = milligrams per liter; mg_{metal}/g_{ZVI} = milligrams of metal (chromium or arsenic) per gram of ZVI.

17.3 ZERO-VALENT IRON PERMEABLE REACTIVE DESIGN

The design of a PRB comprises several consecutive steps that include the following: a preliminary technical and economic feasibility assessment, hydrogeologic and geochemical characterization of the site at the location of the PRB, selection of the ZVI type, treatability studies, engineering design, selection of the construction method, and formulation of a monitoring plan (ITRC 2011).

17.3.1 Predesign Considerations

A review of an existing conceptual site model (CSM) and a site visit are recommended to obtain sufficient information to evaluate the feasibility of a ZVI PRB at a given site. Existing site reports and observations should be scrutinized for the information summarized in Table 17-4, and these are described as follows:

- **Types of Contaminants**. Are the contaminants present in the targeted groundwater plume treatable by ZVI? The information provided previously can be referred. Other resources cited within this document and current research papers can be reviewed if the targeted chemicals are less common than those referenced herein.
- **Land Use**. Site access to a suitable location for the PRB is a major consideration for this technology to be feasible. Overlying buildings or other aboveground structures may limit the feasibility of a PRB. Underground utility lines can make the installation of a PRB difficult, especially if a trenched PRB is proposed.
- **Contaminant Distribution**. Although the plume dimensions alone will not probably disqualify the use of a PRB, very deep plumes or very wide plumes will result in high implementation costs. Where a trench-and-backfill PRB is infeasible because of a large plume depth, an injected ZVI PRB may be applicable. If the bottom of the target treatment zone does not coincide with a competent aquitard, a hanging PRB can be implemented (only considered for continuous ZVI installation), although caution would be necessary to evaluate the potential for the plume to migrate under the PRB.
- **Groundwater Velocity**. High groundwater flow rates may require a thicker barrier to ensure adequate residence contact time between the contaminants and ZVI, which, in turn, may result in prohibitively high material cost. Extremely low velocities may result in fouling caused by of the accumulation of hydrogen gas formed by aerobic corrosion of ZVI (Equation 17-1), or microbial fouling. Low velocities may also require unrealistically long time frames for remediation. On the basis of experience, most sites fall between these two extremes in hydrogeologic conditions.
- **Geotechnical Considerations**. The presence of cobbles, flowing sands, or highly consolidated sediments in the subsurface may impede the construction process. Trenched PRBs, for example, are infeasible in consolidated bedrock aquifers, although jetting applications of ZVI have been performed in fractured bedrock.
- **Geochemical Considerations**. Certain constituents if present at elevated concentrations (e.g., nitrate, carbonates, certain types dissolved organic carbon, and silica) may cause rapid ZVI passivation and are generally not amenable to PRB applications. If these

Table 17-4. Site Characteristics for Feasibility Evaluation of ZVI PRBs.

Site characteristics	Preferable conditions	Further evaluation required
Land use	No infrastructure or utilities along the line of installation with trenching or excavation	Some utilities (e.g., sewer or gas lines) or roadways may be rerouted or disconnected during trenched PRB construction. An injected PRB may be installed around those obstacles
PRB depth	< 11 m below ground surface (bgs) for conventional trenched applications	11–21 m bgs achievable by advanced trenching methods. Up to 61 m bgs and deeper achievable by direct injections
Maximum concentrations (CAHs)	< 10,000 μg/L (CAHs)	Solubility limits. Degradation rates decrease at higher concentrations and competition effects are possible. Dense nonaqueous phase liquid (DNAPL) may physically coat and passivate ZVI particles
Lithology	Unconsolidated sediments (silts and sands), weathered bedrock	Challenging conditions include clays; highly permeable gravel zones, flowing sands, and boulders. Consolidated bedrock not amenable to trenched PRB use, although an injected zone may be created
Stratigraphy	Confining layer present at the depth of the plume	Multiple water-bearing units targeted. No confining unit below the plume, vertical variability in permeability in the targeted saturated depth
Groundwater velocity	> 0.03 and < 0.3 m/day	< 0.03 m/day or > 0.3 m/day. At low values: permeability issues caused by hydrogen gas buildup and mineral/biological fouling, as well as impractical treatment timeframes. At high values: ZVI thickness may be cost-prohibitive and the mass flux of passivating constituents may be high; longevity issues

(*Continued*)

Table 17-4. (*Continued*)

Site characteristics	Preferable conditions	Further evaluation required
pH	Neutral or slightly acidic	pH > 8. High pH may inhibit ZVI corrosion and/or exacerbate formation of secondary precipitates on ZVI grains (e.g., Fe hydroxides and Ca/Fe carbonates)
Redox state	Anaerobic	Aerobic. Dissolved oxygen is one of the passivating factors for PRB design/longevity considerations
Nitrate concentration	Below detection limits	< 10 mg/L NO_3–N. Nitrate tends to passivate ZVI at a high rate, proportional to the mass flux. Not recommended at > 10 mg/L NO_3–N
Carbonate alkalinity	< 300 mg/L (as $CaCO_3$)	> 300 mg/L (as $CaCO_3$). Increased potential for carbonate precipitation and related loss in permeability and porosity
Sulfate concentration	< 500 mg/L	> 500 mg/L sulfate reduction to sulfide and the formation of Fe sulfides may occur once microbial activity establishes inside the PRB
Dissolved organic carbon	< 5 mg/L	> 5 mg/L. Certain types of DOC may cause the development of passivating films on ZVI grains with long-term exposure
Silica	< 10 mg/L	> 10 mg/L. Deposition of silica on a ZVI grain surface adversely affects ZVI performance

Source: Revised from ITRC (2011).

Notes: These factors can be accommodated with an adaptive design process and skillful implementation, even if they fall outside of the "preferable" range of the PRB parameters (Table 17-4). Nevertheless, a realistic assessment of the feasibility of a ZVI PRB for a given site requires a consideration of these elements.

conditions are noted, the influence of site chemistry needs to be carefully evaluated in a flow-through column test, followed by a hydrogeochemical analysis, as described in the following. A treated groundwater downgradient of a ZVI PRB may be more reducing and have a higher pH than ambient conditions in the aquifer. Adverse side effects from these geochemical conditions, as noted in the following, should be considered as well.

17.3.2 Site Characterization for Permeable Reactive Barrier Design

Once a PRB has been installed, relocation or modification would be difficult and expensive. Therefore, an in-depth understanding of the aquifer and plume characteristics is necessary to optimize PRB design. The plume location and extent, groundwater flow direction and velocity, contaminant concentrations, and detailed groundwater geochemistry must be accurately known to achieve the required performance. As all this characterization information is used to develop PRB design, location, emplacement methodology, and estimated life expectancy, insufficient or faulty information could jeopardize the remediation effort. Typically, PRB design requires some supplementary site characterization, in addition to a site-wide CSM, with an emphasis on the regions surrounding the prospective PRB location.

Specific information needed for PRB design includes advective velocity parameters such as hydraulic gradients, hydraulic conductivity, and porosity. Seasonal changes in the depth to water, flow direction, flow rate, and potential preferential flow paths are also important, because the PRB should be designed to accommodate these variables. An evaluation of the vertical variability in stratigraphy will inform the selection of the saturated zone(s) that the PRB will intersect. If the targeted contaminant plume is migrating through a highly permeable layer as well as less-permeable layers, the ZVI flow-through thickness may be varied vertically to accommodate the change in permeability.

Most natural aquifers are heterogeneous, with properties varying laterally and vertically, particularly hydraulic conductivity. Hydraulic conductivity within an aquifer can vary by orders of magnitude, causing similar spatial variability in groundwater flow rates. Elder et al. (2002) performed Monte-Carlo hydraulic and transport modeling of PRB performance within numerically simulated aquifers having varying degrees of heterogeneity representing (1) nearly homogeneous aquifers, (2) moderately heterogeneous aquifers, and (3) highly heterogeneous aquifers. To evaluate the impact of heterogeneity alone in aquifer hydraulic conductivity, all the simulated aquifers had the same mean hydraulic conductivity and porosity, the hydraulic gradient was the same for all simulations, and the contaminant source concentration and ZVI reaction rate were identical for all scenarios. The model results were then used to back-calculate the ZVI thickness required to achieve the same PRB performance. In this case, performance was defined as 90% of the effluent from the PRB reaching concentrations $\leq$ 10 mg/L for an upgradient source of 1,000 mg/L. As given in Table 17-5, the PRB thickness needed to be three to nine times larger for moderately and highly heterogeneous aquifers, compared to a nearly uniform aquifer, to achieve the same performance metric. In addition, the PRB width increased in the highly heterogeneous aquifer to capture the greater dispersion of the plume in that scenario.

Therefore, a careful evaluation of the lithologic variability at the location of the PRB will provide confidence in the design and installation to achieve the performance goals.

Particular attention should be given to (1) the variability of hydraulic conductivity across the face of the PRB, because this can vary by orders of magnitude and this correlates directly to groundwater velocity, and (2) concentration across the face of the PRB, because this can vary from the plume core to edges and can also be affected by groundwater velocity (i.e., dilution and dispersion). Based on experience, heterogeneity can be addressed in the design stage by collecting paired hydraulic conductivity and concentration data across the proposed PRB alignment, assessing the PRB requirements to accommodate the variability seen in these design inputs and then using engineering judgment when selecting PRB thickness.

An assessment of the horizontal and vertical distributions of contaminants and the variability in concentrations during seasonal

Table 17-5. Dimensions of Continuous PRBs for Various Degrees of Aquifer Heterogeneity.

PRB dimensions	Nearly uniform aquifer ($s_{\ln K} = 0.5$)	Moderately heterogeneous aquifer ($s_{\ln K} = 1.5$)	Highly heterogeneous aquifer ($s_{\ln K} = 3.0$)
Width (m)	20	20	45
Depth (m)	10	10	10
Thickness (m)	0.23	0.62	2.1

Notes: Modeling assumptions: The PRBs were "designed" to achieve a complete capture of influent plume and for 90% of effluent concentrations to be below 10 μg/L). The source zone for all simulations was 10 m wide by 5 m deep, had a uniform concentration of 1,000 μg/L, and was located 20 m upgradient of the PRB.

$s_{\ln K}$ = Standard deviation of lnK, where lnK is the natural logarithm of hydraulic conductivity, K.

changes is also important. The PRB must be designed to eliminate the possibility that portions of the plume could flow around the barrier in any direction. This requires a complete understanding of the extent of the plume, its width, depth, and length, contaminant concentrations within these spatial dimensions, and how these can be expected to change over time.

17.3.3 Determination of Permeable Reactive Barrier Thickness for Chlorinated Aliphatic Hydrocarbons

Site-specific reaction rates or half-lives for contaminants in contact with the ZVI medium are necessary to design a PRB thickness that will provide an adequate residence time for CAHs to degrade to their target treatment goals. Both batch and continuous column tests have been used in the past to determine CAH reaction rates. Because flow-through tests better simulate the dynamic groundwater flow environment, testing in columns is by far the most common testing method (ITRC 2005).

Although there is no standardized ZVI column test methodology, an example of preferred laboratory procedures involves the use of 0.5-m long, 0.04-m diameter Plexiglass columns with several sampling ports along the column length (ITRC 2005). These tests should be performed using the ZVI type selected for field installation and contaminated site groundwater collected near the proposed PRB. For applications of injected PRBs, site aquifer material mixed with ZVI at the anticipated amendment rate is packed into the column. The ZVI column test should be performed for a sufficiently long period (i.e., 1.5 to 2 months), including several sampling events performed over the test period, to obtain CAH concentration trends along the ZVI bed length and to evaluate the effects of continuous exposure to inorganic constituents on ZVI reactivity. The output of the tests is degradation parameters in the form of degradation rates or half-lives, for all CAHs present in groundwater, and molar conversion rates from the parent to the daughter products that may be created via the sequential reductive dechlorination pathways described previously.

As noted, pseudo-first-order kinetics are typically observed for CAH degradation in the presence of excess amounts of ZVI. The residence time of a first-order reaction can be calculated using (Gillham et al. 2010):

$$RT = -\ln(C_T/C_o)/k \tag{17-5}$$

where

C_T = Target concentration downgradient of the PRB (mg/L),
C_o = Concentration of the contaminant entering the PRB (mg/L), and
k = First-order rate constant of reaction (h^{-1}).

Because of the multiple pathways of the ZVI-induced dechlorination mechanisms for CAHs, the residence time calculations based on the laboratory-derived data (degradation rates and conversion rates) need to quantify the residence times for both the compounds that are present in the water initially and the degradation products, on the basis of the measured conversion rates. Table 17-6 gives typical results obtained from a ZVI column test. A more detailed description of multicomponent CAH degradation models is provided in Gillham et al. (2010).

The residence time required and the anticipated groundwater velocity through the PRB are then used to establish the width of the PRB needed to achieve the desired treatment targets. The required width of the ZVI in the direction of flow (w), assuming plug flow, is given by

$$w = v_{\mathrm{PRB}} \times \mathrm{RT} \times \mathrm{SF} \tag{17-6}$$

where

v_{PRB} = Groundwater velocity within the PRB (m/day),
RT = Residence time required to meet the discharge concentration targets (days), and
SF = Safety factor (unitless).

The safety factor used for a PRB width design should incorporate (1) uncertainties in the groundwater flow rate used in the PRB design; (2)

Table 17-6. Example Results of Column Treatability for 100% Granular ZVI at Room Temperature.

Compound	Influent concentration (μg/L)	Half-life (h)	Molar conversion
PCE	905	0.95	–
TCE	59	0.79	PCE to TCE = 25%
cDCE	20	1.5	TCE to cDCE = 10%
VC	5	1.8	cDCE to VC = 7%
1,1,1-TCA	75	0.20	–
1,1-DCA	nd	8.4	1,1,1-TCA to 1,1-DCA = 18% to 35%

Notes: nd = Compound not detected in influent water. Data from internal testing by Geosyntec Consultants for a site in California.

variability in influent CAH concentrations and composition over time; and (3) the passivation of ZVI with time, as described in the following. On the basis of experience, SFs have ranged from 2 to 5 depending on the results of site-specific column tests and site conditions, although some have suggested that values of up to 12 may be needed in highly heterogeneous aquifers (Eykholt et al. 1999, Elder et al. 2002).

17.3.4 Long-Term Performance of Zero-Valent Iron Permeable Reactive Barriers for Chlorinated Aliphatic Hydrocarbons

Zero-valent iron is not a contaminant-selective reactive material. That is, the ZVI corrosion reaction (Equation 17-1) that drives the desired contaminant treatment mechanisms, including reductive dechlorination of CAHs and reduction and immobilization of metal oxy-anions, also causes changes in the chemical equilibrium of common constituents of groundwater by creating geochemical gradients between the surrounding aquifer and the ZVI PRB. The major geochemical gradients that have been observed in ZVI PRBs involve (Table 17-7): (1) dissolved oxygen, which is completely removed within a short distance with the ZVI zone; (2) dissolved H_2, created by the ZVI corrosion reaction (Equation 17-1), which increases over the width of the ZVI-bearing zone up to near saturation, depending on the flow rate; (3) pH, which increases because of the iron corrosion reaction and remains steady in the ZVI zone; (4) dissolved Fe(II), which increases rapidly inside the ZVI PRB and then declines with limited dissolved iron concentrations egressing from the PRB; (5) dissolved CO_2 (i.e., carbonate alkalinity), which precipitates near the upgradient interface as calcium and iron carbonates; (6) NO_3^-, which is abiotically reduced to ammonia, causing ZVI oxidation and deposition of ZVI-passivating iron oxy-hydroxides; and (7) SO_4^{2-}, which is reduced by anaerobic bacteria to sulfide once microbial activity is established in mature PRBs, much of which then precipitates as iron sulfides.

Because of these geochemical gradients created in ZVI PRBs, secondary minerals are likely to form in all groundwater environments. Bicarbonate (HCO_3^-) is present in virtually all groundwaters at concentrations ranging from a few tens to several hundred mg/L. Furthermore, in PRBs that have been examined, carbonates commonly appear to be the predominant secondary minerals present as aragonite, calcite ($CaCO_3$), or siderite ($FeCO_3$) (Gillham et al. 2010, Wilkin et al. 2003). Carbonate precipitation in ZVI PRBs is controlled by the water pH which, in turn, is regulated by alkalinity created from the anaerobic ZVI corrosion reaction (Table 17-7). Although the pH is expected to increase to 9 to 10 in freshly installed 100% ZVI zones, more neutral pH values and less favorable conditions for carbonate precipitation will occur if a ZVI-sand mix is used in the PRB design.

Table 17-7. Geochemical Gradients Created by ZVI.

Constituents	Main reaction	Chemical gradient in ZVI
Dissolved oxygen	$4Fe^{\circ} + 3O_{2(aq)} + 12H^{+} \rightarrow 4Fe^{3+} + 6H_2O$	Decreases rapidly
Dissolved hydrogen	$Fe^{\circ} + 2H_2O \rightarrow Fe^{2+} + H_{2(aq)} + 2OH^{-}$	Increases in ZVI
pH		Increases to up to pH 9–10 in 100% ZVI
Ferrous iron Fe(II)*		Increases, but is rapidly incorporated into Fe oxyhydroxide solids
Carbonates ($CO_{2(aq)}$)	$Ca^{2+} + CO_3^{2-} \rightarrow CaCO_{3(s)}$ $Fe^{2+} + CO_3^{2-} \rightarrow FeCO_{3(s)}$	Carbonate precipitation as a result of the pH increase
Nitrate	$NO_3^{-} + 9H^{+} + 4Fe^{0} \rightarrow NH_3 + 3H_2O + 4Fe^{2+}$	Decreases rapidly
Sulfate	$SO_4^{2-} + 9H^{+} + 8e^{-} \rightarrow HS^{-} + 4H_2O$ $Fe^{2+} + HS^{-} \rightarrow FeS_{(s)} + H^{+}$	Reduction of sulfate has been shown in mature ZVI PRBs only, as a microbial process

Source: Tratnyek et al. (2003), Wilkin et al. (2003).

*Fe(II) is also formed by a dechlorination of CAHs and nitrate reduction by ZVI.

The effect of specific iron oxide mineral phases created by oxygen and nitrate reduction on the long-term performance of ZVI is variable. Ritter et al. (2002) found that in the presence of nitrate reduction by ZVI, a shift occurred in the thermodynamic equilibrium of iron oxide phases, including Fe_2O_3 (hematite and maghemite) and magnetite (Fe_3O_4). This caused a conversion of Fe^{2+} (produced from ZVI corrosion) into maghemite and/or goethite (α-FeOOH) deposited on ZVI grains. Goethite and Fe_2O_3 are known to inhibit further iron corrosion, resulting in a passivation of the ZVI grain surfaces. Thus, it is believed that as these minerals form, they reduce the surface area available for nitrate and CAH reduction.

The creation of biogenic iron sulfides in ZVI PRBs may be beneficial for long-term performance. For instance, Przepiora et al. (2022) showed that sections of a ZVI PRB that received influent groundwater containing dissolved sulfate performed significantly better in terms of long-term CAH degradation, compared to a PRB section that received a similar CAH loading in the absence of dissolved sulfate. This phenomenon has been attributed to the ability of biogenic iron sulfides to abiotically reduce CAHs and thus enhance the ZVI reactivity (He et al. 2015). In contrast, accelerated iron sulfide precipitation caused by the use of carbon-based PRB construction stabilizers (i.e., guar gum), in the presence of high sulfate, resulted in a clogging of pore space within the PRB and surrounding formation (Johnson et al. 2008).

Other common groundwater constituents that have been observed to be retained within ZVI and adversely affect its reactivity include certain types of dissolved organic carbon (DOC) and silica (Tratnyek et al. 2001, Klausen et al. 2003). Because of these changes in groundwater geochemistry in contact with ZVI, authigenic precipitates, films, and possibly microbial buildup develop on ZVI grains along the flow path of PRBs.

In the context of the long-term performance of ZVI PRBs, the effect of these processes is two-fold: (1) the reactivity of ZVI decreases with time as the grain surfaces are gradually blocked and electron transfer from the metal core is occluded, and (2) the porosity of the PRB decreases because of the buildup of secondary precipitates. Despite the relatively long application history of the PRB technology since its inception in the mid-1990s, few, if any, comprehensive studies document mature ZVI PRBs that have performed successfully in various groundwater chemistries for more than 10 years. Recent publications have documented that some of the early ZVI PRB applications, when properly designed and installed under amenable hydrogeochemical conditions, have provided effective treatment of CAHs and metals for more than 20 years (Wilkin et al. 2019, Przepiora et al. 2022).

Existing literature reflects a disagreement about which of the two factors listed previously controls PRB longevity, defined as the length of time that a PRB continues to treat groundwater to design concentration levels (Henderson and Demond 2007). However, controlled experimental data and field performance results indicate that secondary precipitation may have a greater effect on the reactivity of the ZVI than on the hydraulic performance of the PRB (Henderson and Demond 2007, Gillham et al. 2010). Field and laboratory data support this argument, showing that once the upgradient ZVI zone is passivated by a layer of precipitates such as carbonate minerals, the ZVI corrosion, which drives these processes, stops. Subsequently, there is no further precipitation in that region of the ZVI zone, and a precipitation front migrates downgradient to where a fresh ZVI zone remains. Therefore, it is expected that these processes will result in only a partial loss in PRB permeability rather than ZVI plugging. Given that granular ZVI possesses a relatively high permeability (higher than most typical unconsolidated aquifers), the effect on system hydraulics may be limited. Nevertheless, if applied at sites with highly permeable aquifers, even a partial loss of permeability may cause a change in flow patterns and potentially result in the impacted water bypassing the PRB. Construction methods used for PRB emplacement may also affect the PRB permeability, as described subsequently.

Numerical models of various complexity have been developed to predict ZVI PRB long-term performance, but they have generally not been able to fully account for the wide variability of geochemical conditions at various application sites (Gillham et al. 2010). Various secondary minerals will form at any ZVI PRB application site and will affect ZVI long-term performance, which will be predominantly expressed as a gradual loss in reactivity. The ZVI deactivation process is dependent on the type and mass flux of the passivating constituents.

If a treatability column test has been performed with site groundwater, the obtained results should be evaluated with respect to these metrics. For example, a typical column test may be performed over a period of 1.5 months with approximately 40 pore volumes of flow exchanged through a 0.5 m-long ZVI column. Depending on the groundwater flow rate in the field, this amount of volumetric flow may represent weeks, months, or years of field performance. Therefore, even if degradation rates have not changed with time during the column test period, the volumetric flow in the expected lifetime of the PRB may be much higher and the deactivation processes will probably be more pronounced. The ZVI width design needs to account for the loss of reactivity to ensure that fresh ZVI remains downstream of the deactivated zone. In other words, the residence time calculated on the basis of the degradation parameters obtained in a relatively short-term column test need to be corrected in the final PRB design to account for the expected long-term ZVI deactivation to ensure effective performance for 10 to 20 years. At many sites, the need for this correction may result in a ZVI thickness that is technically and/or economically infeasible.

Gavaskar et al. (2002) reported on an accelerated ZVI column test performed at a velocity of 3.8 m/day, using groundwater containing CAHs from two US sites, Moffett Field and Lowry Airforce Base (AFB). The tests were performed for approximately 1.5 years (1,300 pore volumes) to simulate 20 to 30 years of groundwater flow through PRBs at those sites. TCE half-lives increased (i.e., degradation rate constants decreased) continuously throughout the test period, and the observed loss in ZVI reactivity toward TCE was site-dependent (Figure 17-4). The half-life increased by a factor of approximately 5 for the Lowry AFB groundwater and approximately 3 for the Moffett Field groundwater. The more pronounced decrease in ZVI reactivity for the Lowry AFB groundwater was consistent with ZVI deactivation by higher concentrations of carbonate alkalinity, sulfate, and calcium, compared to the Moffett Field groundwater. Results from rapid column experiments may not be representative of field performance. For instance, the amount of secondary precipitate generated in those tests was probably lower and more dispersed than the lower flow

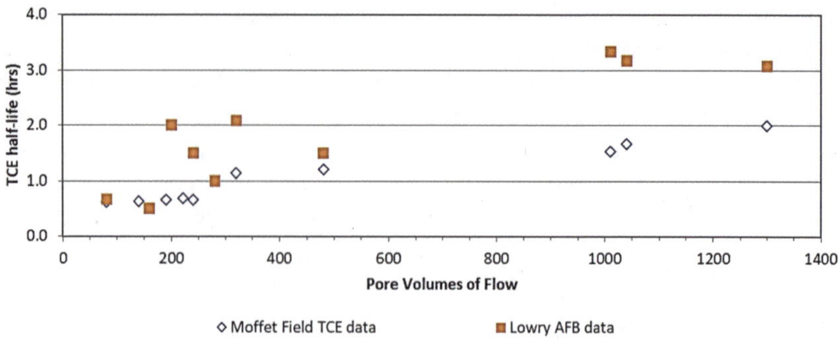

Figure 17-4. Change in TCE degradation half-lives in ZVI columns performed for groundwaters from Moffett Field and Lowry Airforce Base.
Source: Gavaskar et al. (2002).

rate expected under site conditions. Also, long-term effects on the conversion of TCE to breakdown products and degradation of those products were not reported. Nevertheless, these results illustrate that some applications for ZVI PRBs may require a relatively high degree of conservatism in the designs to ensure that degradation efficacy is sustained.

In summary, the influence of secondary minerals on PRB performance is strongly dependent on the groundwater velocity and the concentration of precipitating species in the groundwater. An evaluation of these effects is a crucial step in PRB design to ensure the desired lifetime. Laboratory column tests with site groundwater provide representative measures of ZVI reactivity, but the duration of these tests is typically not long enough to simulate the performance over the expected PRB lifetime in the field. For most sites, the long-term deactivation of ZVI can be accommodated in the design process through an application of design safety factors that will increase the width of the PRB, on the basis of a critical review of site hydrogeochemical conditions.

17.3.5 Determination of Zero-Valent Iron Width for Metals

As described, ZVI material deactivates over time with respect to chromium and arsenic removal. Therefore, any field-scale design needs to consider the influent dissolved mass fluxes, so that the PRB thickness can be designed to allow for a passivation of the ZVI material over the desired lifetime. Column testing is the most appropriate method for obtaining site-specific ZVI removal capacities for metals. The tests need to be performed long enough to observe a dissolved metal breakthrough in at least the first sampling port within the column, using the position of the 50% breakthrough as an indication of the front arrival (Karvonen 2004).

The removal capacity is given by way of Equation (17-7):

$$\text{Capacity}\left(\text{mg}_{\text{metal}}/\text{g}_{\text{ZVI}}\right) = (C_{\text{metal}} \times \text{Vol}_{\text{Sol}})/(\text{Vol}_{\text{ZVI}} \times \text{BD}_{\text{ZVI}}) \quad (17\text{-}7)$$

where

C_{metal} = Average metal concentration (mg/L),
Vol_{Sol} = Volume of water passed through the column before the front arrival (cm^3),
Vol_{ZVI} = Volume of ZVI in the column section where the front migrated to (cm^3), and
BD_{ZVI} = Bulk density of ZVI in the column (g/cm^3).

Figure 17-5 shows an example of column test results for dissolved chromium, showing migration of the breakthrough front with increasing volumes of flow. The test results were used to calculate a chromium removal capacity of 1.8 $\text{mg}_{\text{Cr}}/\text{g}_{\text{ZVI}}$. For the purposes of PRB design, the rate of the dissolved metal front migration [d_{metal} (m/year)] can then be calculated as

$$d_{\text{metal}} = (v_{\text{GW}} \times n \times C_{\text{metal}})/(\text{Capacity}_{\text{metal}} \times \text{BD}_{\text{ZVI}}) \quad (17\text{-}8)$$

where v_{GW} is the groundwater velocity, n is the aquifer porosity, C_{metal} is the influent metal concentration, $\text{Capacity}_{\text{metal}}$ as defined in Equation 17-7, and BD_{ZVI} is the bulk density of ZVI in the field PRB.

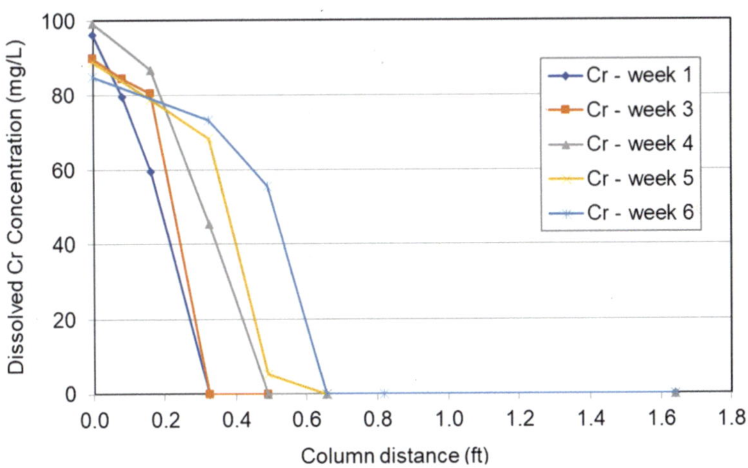

Figure 17-5. Example of dissolved chromium concentrations along a 100% ZVI column during the test period.
Source: Data from unpublished internal testing by Geosyntec Consultants.

The required width of the ZVI in the direction of flow (w) for the anticipated remedial period is then calculated as follows:

$$w = d_{\text{metal}} \times \text{LT} \times \text{SF} \tag{17-9}$$

where LT is the PRB lifetime, and SF is a safety factor.

Design safety factors that consider uncertainties in groundwater velocity and influent concentrations of the targeted metal should also be incorporated into the design PRB thickness. However, if a site-specific capacity for metal removal has been derived from the results of a column test, then the effect of other geochemical constituents on ZVI deactivation is already included in the capacity value, provided that the influent concentration in the field is equivalent to or greater than the influent concentration in the groundwater used in the test. Therefore, an additional correction for ZVI longevity, such as the one recommended for CAHs, would not be required.

17.4 ZERO-VALENT IRON CONSTRUCTION METHODS

Two common configurations of ZVI PRBs are a continuous reactive barrier and a funnel-and-gate system. A continuous PRB consists of a reactive ZVI zone. In a funnel-and-gate system, the funnel, which consists of low-permeability walls (discussed in Chapter 16), is designed to direct the captured groundwater to the treatment zones (gates), and the reactive gates contain the ZVI in which the treatment occurs. The majority of ZVI PRB field installations have been of the continuous reactive barrier configuration (ITRC 2011). The continuous barrier configuration is easy to construct, is less expensive than the funnel-and-gate system for many site-specific settings, and has little effect on the groundwater flow regime. Besides additional construction cost, funnel-and-gate systems require vertical key-in to an underlying confining unit and are prone to hydraulic capture failure because of changes in ZVI permeability with time. The construction methodologies described subsequently apply to both continuous ZVI PRBs and ZVI reactive gates.

The choice of the construction method depends on the total PRB depth, PRB dimensions (i.e., ZVI flow-through thickness), the type of lithology in the targeted depth, and the site access. Two general categories of construction methods have been applied for ZVI PRBs: (1) trench-and-backfill methods, and (2) direct injection methods. A detailed description of various PRB construction methods is provided in ITRC (2011) and is also summarized in the following.

17.4.1 Trenched Zero-Valent Iron Permeable Reactive Barriers

Trench-and-backfill PRB construction requires the creation of an open trench to the target PRB depth, followed by backfilling the trench with the ZVI material. The simplest and least expensive technique is unsupported excavation, but this method is limited to a depth of approximately 3 to 6 m bgs, depending on the site lithology and depth to groundwater. Physical trench wall supports are required for deeper installations, using interlocking sheet piles, trench boxes, or hydraulic shoring, which extend the maximum depth of trenched PRBs to approximately 9 m bgs. Continuous trenching machines allow simultaneous excavation and backfilling and have been used to install ZVI PRB to depths of approximately 12 m bgs. In this method, excavation is performed by a cutting chain immediately in front of a trench box (boot) that is loaded with ZVI material and extends the flow-through thickness and depth of the PRB.

Biodegradable polymer (biopolymer)–supported excavation has been used to install trenched ZVI PRBs to a depth of up to 21 m bgs (Gillham et al. 2010). As the excavation progresses, guar gum biopolymer slurry is added, which through density and viscosity effects, stabilizes the exposed trench walls. ZVI (or a ZVI/sand mixture) is then placed into the guar gum filled trench. Following installation, the pore space of the PRB is filled with guar gum and is relatively impermeable. To increase permeability, recirculation wells are emplaced in the PRB to flush out the guar gum by adding an enzyme to assist in breaking down the residual guar gum. Relatively little attention has been devoted to potential adverse effects of this construction method on ZVI reactivity. For instance, most construction contractors use guar gum slurry stabilizers that include biocides and high pH buffers, such as soda ash, to maintain the slurry viscosity. Moreover, the postconstruction recirculation process introduces dissolved oxygen and stimulates microbial activity through the formation of guar gum degradation products that are readily fermentable. The net effect may be an alteration of ZVI properties by deactivating the ZVI and/or reducing its permeability. Velimirovic et al. (2014) described a loss in ZVI reactivity in column tests performed in the presence of guar gum. Guar-gum supported excavation has been used for the installation of ZVI PRBs that performed successfully in the long term. For instance, a ZVI PRB installed in the northeast United States in 2000 has provided an effective barrier to the migration of a CAH-impacted groundwater for 21 years (Przepiora et al. 2022). If that construction method is used, column testing should be performed in the presence of guar gum, and construction procedures should ensure a complete breakdown of the guar contained within the backfilled PRB, without excessive aeration, immediately after emplacement.

The width of a trenched PRB is controlled by the construction equipment used, such as the excavator bucket width or the trenching cutting chain width, which typically ranges from 0.5 to 1 m. If the design flow-through thickness of ZVI is less than the construction width, ZVI can be mixed with construction sand of a similar particle size distribution as a bulking material to create a wall width equal to the width of the construction equipment, but with a ZVI mass equal to the mass associated with the design flow-through thickness. Previous research has shown that the reactivity of ZVI in a mix with sand is generally maintained, and it decreases proportionally to the ZVI content (Gillham et al. 2010). An empirical minimum–recommended ZVI content in a ZVI-sand PRB used in previous installations ranged from 10% to 30%. This empirical minimum aims at ensuring that sufficient direct contact between dissolved contaminants and ZVI particle surfaces is provided as groundwater migrates through the PRB.

Construction methods used for PRB emplacement may adversely affect PRB permeability, including smearing of clay from overlying layers across the PRB face area during trenching, deposition of suspended solids entrenched in biopolymer (guar gum) slurry, if this trenching method is used, or incomplete guar gum breakdown following construction resulting in excessive biomass or bioprecipitation. Johnson et al. (2008) reported substantial losses of permeability in a ZVI PRB installed using guar gum slurry, which was attributed to microbially mediated precipitation of iron sulfides in a sulfate-rich groundwater. These observations underscore the need for a careful review of selected contractor experience and procedures to minimize the potential side effects of the selected construction method.

17.4.2 Injected Zero-Valent Iron Permeable Reactive Barriers

Direct ZVI injection methods are used if the targeted PRB depth exceeds the capability of trench-and-backfill methods, the site is not accessible to trenching equipment because of surface/subsurface obstructions (e.g., buildings and utilities), and/or the ZVI amount requirements are relatively low. In contrast to the trenching methods where soil is removed prior to adding ZVI, the injection methods emplace ZVI directly into the aquifer matrix. As indicated previously, microscale ZVI (mZVI) is used in injected application methods to facilitate material distribution.

Direct injection methods rely on the horizontally radial distribution of micro-ZVI from the injection point. Hydraulic and pneumatic fracturing are proprietary methods of emplacing micro-ZVI in the subsurface under high pressures through vertical access points, resulting in a propagation of ZVI via fracturing. Direct push technology (DPT) uses DPT tooling equipped with an injection head to both access the targeted depths and inject mZVI. Multiple injection points are required to create a continuous mZVI PRB. Using fracturing and DPT injection, at least two rows of injection points with vertically staggered injection intervals are typically installed for a PRB. The injection point spacing is dependent on the injection radius of influence (ROI), which varies with the method used, the targeted depths, lithology, and ZVI loading. For example, a typical ROI for DPT injection varies from 1.5 to 3 m, whereas ROIs of up to tens of meters have been reported for the fracturing methods (ITRC 2011). Because soil material is not removed from the subsurface prior to injections, there is a physical limitation on how much mZVI can be emplaced to avoid ground surface heaving and/or daylighting. ITRC (2011) reported that because of injectability and cost concerns, typical field applications have involved an injection of 0.4% to 0.8% of mZVI by mass of soil. Higher mZVI application rates of up to $1.5\%_{wt}$ may have been used in certain conditions (Przepiora et al. 2021).

Apart from pneumatic fracturing, all direct mZVI injection methods use the material suspended in a guar gum slurry. As indicated for the guar-gum-supported trenching, guar gum may interfere with ZVI reactivity. The presence of a fermentable carbon is expected to intermittently stimulate anaerobic microbial activity. This may be beneficial by providing another mechanism of CAH degradation. However, if microbes capable of complete dechlorination (e.g., *Dehalococcoides* and *Dehalobacter*) are absent in the aquifer, partially dechlorinated breakdown products may be formed and then migrate untreated downgradient of the PRB. Commercial dechlorinating cultures may be used in those applications to accelerate the guar gum breakdown and use the created labile carbon source productively for intermittent CAH biodegradation.

17.5 PERMEABLE REACTIVE BARRIER MONITORING

Performance monitoring of a ZVI PRB involves collecting and analyzing groundwater samples from a network of groundwater monitoring wells. Site-specific PRB monitoring plans should consider the PRB dimensions and site conditions, as summarized subsequently for a continuous ZVI PRB. More detailed guidance is included in USEPA (1998) and ITRC (2005, 2011).

17.5.1 Monitoring Well Network

The PRB monitoring well network should target the depth intervals corresponding to the ZVI vertical distribution. If the PRB intercepts more than one hydrogeological unit, then the groundwater monitoring well network should include separate wells screened in the relevant hydrogeological units. An ideal monitoring well network additionally relies on the transects of monitoring wells along the groundwater flow path (perpendicular to the PRB) upgradient, within, and downgradient of the PRB. The transects are typically installed approximately every 30 m along the alignment or a minimum of two transects across the PRB. The upgradient well should be placed about 1.5 to 3 m upgradient and the downgradient well about 1.5 to 3 m downgradient. For hanging PRBs (not keyed into a confining unit at the bottom), wells may be placed below the PRB on the upgradient or downgradient side to monitor for plume bypass beneath the PRB. Monitoring wells are also typically placed at each end of the PRB to monitor for plume capture. If possible, these cross-gradient wells should be installed and sampled prior to PRB construction to confirm whether they are indeed clean prior to PRB installation.

It is important to consider that most PRBs are placed across an existing plume and not at the front edge of the plume. For this reason, the groundwater downgradient of the PRB may not show a significant change in contaminant concentrations for some time following PRB installation because of the time required to flush the downgradient aquifer with treated groundwater. In addition, the desorption of contaminants from the downgradient aquifer solids into treated groundwater may also delay the observation of decreasing contaminant concentrations in downgradient groundwater monitoring wells. This condition must be accounted for in the development of a monitoring plan for the PRB. Therefore, one or more monitoring wells should ideally be placed within the PRB as part of the monitoring transects to demonstrate that a degradation/sequestration of contaminants in the influent groundwater is occurring.

17.5.2 Groundwater Sample Collection and Analysis Plan

Sampling within and around a PRB requires special techniques to collect representative samples. For groundwater samples within the PRB, sampling methods that ensure consistent groundwater residence times and flow rates and that minimize disruption of the groundwater flow field during sampling and between sampling events are recommended. Low-flow sampling and purging techniques and passive sampling devices, such as passive diffusion bag samplers, are therefore recommended where appropriate to accomplish these goals. Standard sampling procedures can be used for wells installed in the aquifer upgradient and downgradient of the PRB, provided the purging and sampling will not influence groundwater flow through the PRB.

The potential change in the hydraulic performance of a PRB can be monitored by recording water levels in the vicinity of the PRB and by completing hydraulic testing on wells within the PRB, if needed. Water levels are used to compute hydraulic gradients, which, after accounting for seasonal variations, can be used to monitor potential changes in the hydraulic

Table 17-8. Guide for a PRB Monitoring Program.

Parameter	Monitoring frequency	Monitoring location
Field parameters • Water level • pH, temperature, redox potential, and dissolved oxygen	Quarterly for first 2 years. The frequency may be reduced on the basis of operational stability	All wells
Targeted contaminants • CAHs (including possible breakdown compounds) • Dissolved metals (chromium and arsenic)	Quarterly to semiannually for first two years. The frequency may be reduced on the basis of operational stability	All wells
Inorganic analytes • Calcium, iron, magnesium, manganese, potassium, silicon, and sodium • Alkalinity, chloride, nitrate (as N), and sulfate • Dissolved organic carbon • Total dissolved solids	Quarterly to semiannually for first 2 years. The frequency may be reduced on the basis of operational stability	One or two representative transects

Source: USEPA (1998), ITRC (2005).

conductivity of a PRB. Water levels can also be used to develop potentiometric maps to assess potential changes in flow through or adjacent to a PRB.

Table 17-8 lists the recommended parameters, monitoring frequency, and monitoring location for a basic groundwater monitoring program. This table is meant to be a guide and should be modified to suit site-specific needs and regulatory requirements.

17.6 SUMMARY

ZVI PRB is an established groundwater remediation technology with numerous documented full-scale applications installed in the last 20 years, predominantly for the treatment of CAHs, hexavalent chromium, and arsenic. The technology development included extensive research that initially focused on the contaminants treatable by ZVI and then transitioned to investigations of geochemical factors controlling the long-term performance of ZVI. The findings from the research and applied experience have been summarized in a variety of guidance documents and monographs that are readily available to practicing geoenvironmental engineers to guide and inform the design of ZVI PRB implementation at new sites.

Limitations of the technology should be considered during initial remedial feasibility evaluations. If deemed applicable, a ZVI PRB design involves three critical steps: (1) collecting hydrogeochemical site data from the location of the PRB; (2) obtaining site-specific degradation rates (CAHs) or removal capacities (metals) and an initial calculation of the ZVI flow-through width required to achieve target remedial concentrations; and (3) developing the design safety factors to account for the uncertainties related to the variability of site conditions (i.e., groundwater velocity and contaminant concentrations) and, most of all, the long-term deactivation of ZVI caused by interactions with groundwater constituents.

REFERENCES

Arnold, W. A., and A. L. Roberts. 2000. "Pathways and kinetics of chlorinated ethylene and chlorinated acetylene reaction with Fe(0) particles." *Environ. Sci. Technol.* 34 (9): 1794–1805.

Beak, D. G., and R. T. Wilkin. 2009. "Performance of a zerovalent iron reactive barrier for the treatment of arsenic in groundwater: Part 2. Geochemical modeling and solid phase studies." *J. Contam. Hydrol.* 106 (1-2): 15–28.

Blowes, D. W., C. J. Ptacek, S. G. Benner, C. W. T. McRae, et al. 2000. "Treatment of inorganic contaminants using permeable reactive barriers." *J. Contam. Hydrol.* 45 (1-2): 123–137.

Elder, C., C. Benson, and G. Eykholt. 2002. "Effects of heterogeneity on influent and effluent concentrations from horizontal permeable reactive barriers." *Water Resour. Res.* 38 (8): 27-1 to 27-7.

Eykholt, G. R., C. R. Elder, and C. H. Benson. 1999. "Effects of aquifer heterogeneity and reaction mechanism uncertainty on a reactive barrier." *J. Hazard. Mater.* 68 (1–2): 73–96.

Fennelly, J. P., and A. L. Roberts. 1998. "Reaction of 1,1,1-trichloroethane with zero-valent metals and bimetallic reductants." *Environ. Sci. Technol.* 32 (13): 1980–1988.

Fu, F., D. D. Dionysiou, and H. Liu. 2014. "The use of zero-valent iron for groundwater remediation and wastewater treatment: A review." *J. Hazard. Mater.* 267 (2014): 194–205.

Gavaskar, A., B. Sass, N. Gupta, E. Drescher, et al. 2002. "Evaluating the longevity and hydraulic performance of permeable reactive barriers at the Department of Defense sites." Columbus, OH: Battelle. https://apps.dtic.mil/sti/citations/ADA409969.

Gheju, M. 2011. "Hexavalent chromium reduction with zero-valent iron (ZVI) in aquatic systems." *Water Air Soil Pollut.* 222 (1-4): 103–148.

Gillham, R., J. Vogan, L. Gui, M. Duchene, et al. 2010. "Iron barrier walls for chlorinated solvent remediation." In *In situ remediation of chlorinated solvent plumes*, 805. Edited by H. F. Stroo and C. H. Ward. New York: Springer.

Guan, X., Y. Sun, H. Qin, J. Li, et al. 2015. "The limitations of applying zero-valent iron technology in contaminants sequestration and the corresponding countermeasures: The development in zero-valent iron technology in the last two decades (1994–2014)." *Water Res.* 75 (2015): 224–248.

He, Y. T., J. T. Wilson, C. Su, and R. T. Wilkin. 2015. "Review of abiotic degradation of chlorinated solvents by reactive iron minerals in aquifers." *Groundwater Monit. Rem.* 35 (3): 57–75.

Henderson, A. D., and A. H. Demond. 2007. "Long-term performance of zero-valent iron permeable reactive barriers: A critical review." *Environ. Eng. Sci.* 24 (4): 401–423.

ITRC (Interstate Technology and Regulatory Council). 2005. *Permeable reactive barriers: Lessons learned/new directions*. PRB-4. Washington, DC: Interstate Technology and Regulatory Council, Permeable Reactive Barriers Team.

ITRC. 2011. *Permeable reactive barrier: Technology update*. PRB-5. Washington, DC: Interstate Technology and Regulatory Council.

Jeen, S.-W., D. W. Blowes, and R. W. Gillham. 2008. "Performance evaluation of granular iron for removing hexavalent chromium under different geochemical conditions." *J. Contam. Hydrol.* 95 (1–2): 76–91.

Johnson, R. L., R. B. Thoms, R. O'Brien Johnson, and T. Krug. 2008. "Field evidence for flow reduction through a zero-valent iron permeable reactive barrier." *Groundwater Monit. Rem.* 28 (3): 47–55.

Karvonen, A. 2004. "Cation effects on chromium removal in permeable reactive walls." *J. Environ. Eng.* 130 (8): 863–866.

Klausen, J., P. J. Vikesland, T. Kohn, D. R. Burris, et al. 2003. "Longevity of granular iron in groundwater treatment processes: Solution composition effects on reduction of organohalides and nitroaromatic compounds." *Environ. Sci. Technol.* 37 (6): 1208–1218.

Köber, R., E. Welter, M. Ebert, and A. Dahmke. 2005. "Removal of arsenic from groundwater by zerovalent iron and the role of sulfide." *Environ. Sci. Technol.* 39 (20): 8038–8044.

Lai, K. C. K., and I. M. C. Lo. 2008. "Removal of chromium(VI) by acid-washed zero-valent iron under various groundwater geochemistry conditions." *Environ. Sci. Technol.* 42 (4): 1238–1244.

Lien, H., and R. T. Wilkin. 2005. "High-level arsenite removal from groundwater by zero-valent iron." *Chemosphere* 59 (3): 377–386.

Manning, B. A., M. L. Hunt, C. Amrhein, and J. A. Yarmoff. 2002. "Arsenic(III) and arsenic(V) reactions with zerovalent iron corrosion products." *Environ. Sci. Technol.* 36 (24): 5455–5461.

Matheson, L. J., and P. G. Tratnyek. 1994. "Reductive dehalogenation of chlorinated methanes by iron metal." *Environ. Sci. Technol.* 28 (12): 2045–2053.

Melitas, N., O. Chuffe-Moscoso, and J. Farrell. 2001. "Kinetics of soluble chromium removal from contaminated water by zerovalent iron media: Corrosion inhibition and passive oxide effects." *Environ. Sci. Technol.* 35 (19): 3948–3953.

Morrison, S. J., D. R. Metzler, and C. E. Carpenter. 2001. "Uranium precipitation in a permeable reactive barrier by progressive irreversible dissolution of zerovalent iron." *Environ. Sci. Technol.* 35 (2): 385–390.

Morrison, S. J., D. R. Metzler, and B. P. Dwyer. 2002. "Removal of As, Mn, Mo, Se, U, V and Zn from groundwater by zero-valent iron in a passive treatment cell: Reaction progress modeling." *J. Contam. Hydrol.* 56 (1–2): 99–116.

Nikolaidis, N. P., G. M. Dobbs, and J. A. Lackovic. 2003. "Arsenic removal by zero-valent iron: Field, laboratory and modeling studies." *Water Res.* 37 (6): 1417–1425.

O'Hannesin, S. F., A. Przepiora, and R. W. Gillham. 2004. "Effect of temperature and iron content on iron PRB design." In *Proc., 4th Int. Conf. on Remediation of Chlorinated and Recalcitrant Compounds*, 24–27. Columbus, OH: Battelle.

Przepiora, A., L. Nissen, T. H. Jørgensen, and N. Durant. 2021. "Implementation and performance of injected ZVI PRB to control chlorinated hydrocarbon plume migration in Denmark." In *AquaConSol Conf. Book of Abstracts*, 108. https://aquaconsoil.com/assets/Scientific-Programme/ACS2021_BookofAbstracts_incl-proceedings.pdf.

Przepiora, A., S. O'Hara, S. Wadley, and S. Huda. 2022. "Somersworth superfund ZVI PRB—20+ years of performance monitoring." In *Proc., 12th Int. Conf. on Remediation of Chlorinated and Recalcitrant Compounds*, Abstract 201. Columbus, OH: Battelle.

Reardon, E. J. 1995. "Anaerobic corrosion of granular iron: Measurement and interpretation of hydrogen evolution rates." *Environ. Sci. Technol.* 29 (12): 2936–2945.

Reardon, E. J. 2005. "Zerovalent irons: Styles of corrosion and inorganic control on hydrogen pressure buildup." *Environ. Sci. Technol.* 39 (18): 7311–7317.

Ritter, K., M. S. Odziemkowski, and R. W. Gillham. 2002. "An in situ study of the role of surface films on granular iron in the permeable iron wall technology." *J. Contam. Hydrol.* 55 (1-2): 87–111.

Tamara, M. L., and E. C. Butler. 2004. "Effects of iron purity and groundwater characteristics on rates and products in the degradation of carbon tetrachloride by iron metal." *Environ. Sci. Technol.* 38 (6): 1866–1876.

Tratnyek, P. G., M. M. Scherer, B. Deng, and S. Hu. 2001. "Effects of natural organic matter, anthropogenic surfactants, and model quinones on the reduction of contaminants by zero-valent iron." *Water Res.* 35 (18): 4435–4443.

Tratnyek, P. G., M. M. Scherer, T. L. Johnson, and L. J. Matheson. 2003. "Permeable reactive barriers of iron and other zero-valent metals." In *Chemical degradation methods for wastes and pollutants: Environmental and industrial application*, edited by M. A. Tarr. Environ. Sci. Pollut. Control Series 26, 371–421. New York: Marcel Dekker.

USEPA (US Environmental Protection Agency). 1998. *Permeable reactive barrier technologies for contaminant remediation*. EPA/600/R-98/125. Washington, DC: USEPA.

USEPA. 1999. *An in situ permeable reactive barrier for the treatment of hexavalent chromium and trichloroethylene in ground water*. Vol. 1: Design and installation. EPA/600/R-99/095a. Washington, DC: USEPA.

USEPA. 2007. *Monitored natural attenuation of inorganic contaminants in ground water*. Vol. 2: Assessment for non-radionuclides including arsenic, cadmium, chromium, copper, lead, nickel, nitrate, perchlorate, and selenium. EPA/600/R-07/140. Washington, DC: USEPA.

Velimirovic, M., Q. Simons, and L. Bastiaens. 2014. "Guar gum coupled microscale ZVI for in situ treatment of CAHs: Continuous-flow column study." *J. Hazard. Mater.* 265 (2014): 20–29.

Vidumsky, J. E., M. M. Thomson, R. C. Landis, and E. E. Mack. 2002. "Field performance of a permeable reactive barrier treating carbon tetrachloride and chlorofluorocarbons." In *Proc., 3rd Int. Conf. on Remediation of Chlorinated and Recalcitrant Compounds*, Paper 2A-18. Monterey, CA. Columbus, OH: Battelle.

Warner, S. D., B. L. Longino, M. Zhang, P. Bennett, et al. 2005. "The first commercial permeable reactive barrier composed of granular iron: Hydraulic and chemical performance at 10 years of operation." In *Proc., Int. Symp. on Permeable Reactive Barriers*, 32–42. Wallingford, Oxfordshire, UK: International Association of Hydrological Sciences.

Wilkin, R. T., S. D. Acree, R. R. Ross, D. G. Beak, et al. 2009. "Performance of a zerovalent iron reactive barrier for the treatment of arsenic in groundwater: Part 1. Hydrogeochemical studies." *J. Contam. Hydrol.* 106 (1): 1–14.

Wilkin, R. T., S. D. Acree, R. R. Ross, R. W. Puls, et al. 2014. "Fifteen-year assessment of a permeable reactive barrier for treatment of chromate and trichloroethylene in groundwater." *Sci. Total Environ.* 468–469 (2014): 186–194.

Wilkin, R. T., T. Lee, M. Sexton, S. D. Acree, et al. 2019. "Geochemical and isotope study of trichloroethene degradation in a zero-valent iron permeable reactive barrier: A twenty-two-year performance evaluation." *Environ. Sci. Technol.* 53 (1): 296–306.

Wilkin, R. T., and M. S. McNeil. 2003. "Laboratory evaluation of zero-valent iron to treat water impacted by acid mine drainage." *Chemosphere* 53 (7): 715–725.

Wilkin, R. T., R. W. Puls, and G. W. Sewell. 2003. "Long-term performance of permeable reactive barriers using zero-valent iron: Geochemistry and microbiological effects." *Groundwater* 41 (4): 493–503.

INDEX

1D advection–dispersion–reaction equation (ADRE), 13–14

ACAP. *See* Alternative Cover Assessment Program
AC/DC converter, 321–322
Acid/base reactions, 18
Acid digestion methods, 64–66
Acid rock drainage (ARD), 289, 292–293
Action leakage rate (ALR), 95–97
Advection process, 3–5
Aerial photography, 32
Air rotary casing hammer (ARCH) drilling, 37
ALR. *See* action leakage rate
Alternative Cover Assessment Program (ACAP), 189
Alternative land disposal restrictions for contaminated soils, 101–102
Ammonia as nitrogen, 79
Analytical methods for common metals and metalloids in water, 72–77
Applicable or Relevant and Appropriate Requirements (ARARs), 108
ARARs. *See* Applicable or Relevant and Appropriate Requirements
ARD. *See* acid rock drainage
ASTM D698, 190
ASTM D5030, 39
ASTM D5198-17, 65
ASTM D5321, 167, 242
ASTM D5504-12, 81
ASTM D5607, 242
ASTM D5885/D5885M, 165
ASTM D6243, 242
ASTM D6243/D6243M, 167
ASTM D6349-13, 65
ASTM D6357-11, 65
ASTM D7238, 165
ASTM D8117, 165
ASTM E1621-13, 66
ASTM E2941-14, 65
As[V], 153
Atomic absorption (AA) spectrometry, 69–70
Atomic absorption flame (AAFL) spectrometry, 70
Atomic absorption graphite furnace (AAGF) spectrometry, 70
Average rate: nonreactive solutes transportation, 3–4; travel time, 5

Barrier walls, 355. *See also* horizontal barrier walls; vertical barrier walls

BDAT. *See* best-demonstrated available technology
Becker penetration test (BPT), 44
Best-demonstrated available technology (BDAT), 100
Biochemical oxygen demand (BOD), 80
Biogas generation, 120
Biological transformations, 19
Biomass fly ash, 146–147; chemical composition and classification of, 149–150; in concrete, 151; as mine backfill, 152; organic content, 150–151; particle morphology, 148–149; particle size, 148–149; for soil improvement, 152; as sorptive agents, 153; specific gravity of, 149; for waste solidification/stabilization, 152–153
Bishop's method, modified, 228–229
BOD. *See* biochemical oxygen demand
Boiler slags, 129, 131
Borehole electrical and nuclear testing, 42–43
Borehole hydraulic conductivity testing, 39–41
Borehole pressuremeter testing, 43–44
Borehole sampling and testing, 35–49; BPT, 44; complications inherent to, 36–37; electrical and nuclear testing, 42–43; general, 35–36; health and safety considerations, 37; hydraulic conductivity testing, 39–41; IDW, 37–38; internal deformation measurements, 45–46; internal moisture content and temperature measurements, 46–49; pit sampling, 38; pit unit weight testing, 38–39; pressuremeter testing, 43–44; in situ direct shear tests, 44–45; SPT, 44; types of, 36; wave propagation velocity testing, 41–42
Borehole wave propagation velocity testing, 41–42
Bottom ash, 129
Boutwell permeameter. *See* two-stage borehole permeameter (TSB)
BPT. *See* becker penetration test
Bulk chemical analysis of solids, 64–66

CAHs. *See* chlorinated aliphatic hydrocarbons
Calcium carbonate ($CaCO_3$), 346
CCL. *See* compacted clay liner
CCRs. *See* coal combustion residuals
C&DD. *See* construction and demolition debris
Cement–bentonite (CB) slurry trench cutoff walls, 361–362
CERCLA. *See* Comprehensive Environmental Response, Compensation and Liability Act
Chemical incompatibility, 4
Chemical oxygen demand (COD), 80
Chloride, 78–79
Chlorinated aliphatic hydrocarbons (CAHs), 374; degradation pathways, 378–379; degradation rates, 379–380; groundwater chemistry, 380–381; temperature, 381–382; treatment of, 378–382; ZVI-mediated, 380
Chlorine, 79
Chlorosulfonated polyethylene (CSPE), 163–164
Chromium, 382

Clean Water Act (CWA), 64
CMCs. *See* controlled modulus columns
Coal combustion residuals (CCRs), 129–139, 241; beneficial use of, 135–139; encapsulated, 136; unencapsulated, 136–139; chemical composition, 131; compaction, 134–135; environmental concerns, 139; environmental regulation for, 129–130; fly ashes, classification of, 131–133; loss on ignition, 134; overview, 129; particle size distribution, 133; shear strength of, 135; size and shape, 130–131; specific gravity of, 133–134
COD. *See* chemical oxygen demand
Columns of improved material, 344–345
Compacted clay liner (CCL), 161–163, 167–171; construction issues, 170–171; hydraulic performance, 169; mechanical performance, 170; overview, 167–169; single resistive/barrier layers, 196–197; specifications, 170–171
Compaction grouting, 345–346
Complexation reactions, 18
Composite covers resistive/barrier layers, 199–200
Comprehensive Environmental Response, Compensation and Liability Act (CERCLA), 88; basics, 103–104; removal and remedial actions, 104–108; requirements, 108–109
COMSOL, 14
Concave isotherms, 16
Cone penetration tests (CPTs), 49–51, 346
Construction and demolition debris (C&DD), 241
Construction quality assurance (CQA) program, 160
Construction quality control (CQC), 164
Continuous surface wave system (CSWS), 342–343
Controlled modulus columns (CMCs), 344–345
Convex isotherms, 16
Coupled flow processes, mass transport, 11–12
CPTs. *See* cone penetration tests
Critical slip surface, 228
CSPE. *See* chlorosulfonated polyethylene
CSWS. *See* continuous surface wave system
Cyanide, 79

Dam/embankment, 296–299
Darcian velocity, 4
Darcy's law, 4
DDC. *See* deep dynamic compaction
Deep dynamic compaction (DDC), 343–344
DEMs. *See* discrete element methods
Dense nonaqueous phase liquids (DNAPLs), 15
Depth of improvement *(D)*, 343
Differential thermal analysis (DTA), 67–68
Diffusion coefficient, 6–7
Diffusive transport, 5–7
Digestion methods, 64–65
Dilatometer Marchetti test (DMT), 51
Direct ZVI injection methods, 400–401
Discrete element methods (DEMs), 231

Dispersion (mechanical), mass transport, 7–8
Dissolution/precipitation process, 15–17
Dissolved oxygen (DO), 69
DMT. *See* dilatometer Marchetti test
DNAPLs. *See* dense nonaqueous phase liquids
DO. *See* dissolved oxygen
Double-liner system for hazardous waste, 93–95
Drainage systems, 175–176; hydraulic performance, 175–176; overview, 175
Drop energy, 343
Dry-bulk density, 13
Dry density of soil, 13
DTA. *See* differential thermal analysis

Effective ionic mobility, 320
Effective porosity, 5
EGCs. *See* exposed geomembrane covers
EICP. *See* enzyme-induced calcite precipitation
EK. *See* electrokinetic (EK) remediation
EK-SALT, 320
EK-TAP, 320
Electrical resistance tomography (ERT), 31
Electrical resistivity and conductivity surveys, 31–32
Electrokinetic (EK) remediation, 317–330; application in field, 321–323; DC in, 319–320; electrode installation in field, 322; electrolysis in, 320–321; electroosmosis in, 320; electrophoresis in, 320; examples of, 326–329; ex situ physical techniques, 317; future prospect, 329–330; governing factors: contamination level, 325–326; electric voltage and current, 323–324; electrode placement, 324; soil type, 324–325; total costs for, 326; schematic of, 322; in situ chemical treatment techniques, 317; in situ physical removal of volatile organic compounds, 317; works/process, 319–321
Electrolysis, 320–321
Electromagnetic measurements (EM), 31
Electroosmosis, 320
Electrophoresis, 320
Environmental Protection Agency (EPA), 87–89
Enzyme-induced calcite precipitation (EICP), 346
EPA. *See* Environmental Protection Agency
EPA 2E, 81
EPA 3C, 81
EPA 18, 81
EPA 25C, 81
EPA 1668, 78
EPA 3050B, 65
EPA 3051A, 65
EPA 3052, 65
EPA 8082A, 78
EPA 8270D, 78
EPA SW-846, 78
EPDM. *See* ethylene propylene diene terpolymer
ERT. *See* electrical resistance tomography
Ethylene propylene diene terpolymer (EPDM), 164
Evapotranspiration (ET) covers. *See* water balance covers
Excavation and replacement, 305–315; advantages of, 305–306; construction vibrations

and, 310–313; groundwater control, 313; open water excavation (dredging), 314–315; overview, 305; protection of adjacent structures, 310–313; sheet pile installation, 309–310; subsurface investigation and design, 313; support, 306–309
Excavation shoring, 306
Exposed geomembrane covers (EGCs), 191
Exposed geomembrane covers resistive/barrier layers, 201–202

Factor of safety (FS), 225–227
Fate, 15–19; acid/base reactions, 18; biological transformations, 19; complexation reactions, 18; dissolution/precipitation, 17; hydrolysis/substitution reactions, 18–19; radioactive decay, 17–18; redox/oxidation/ reduction reactions, 19; sorption, 15–17
Favorable isotherms. *See* concave isotherms
FEA. *See* finite-element analysis
Federal Code of Regulations, 87
Federal Register (FR), 129
FGD. *See* flue gas desulfurization
Fick's first law, 5–6
Field capacity, 121
Final covers on landfills and waste piles, 183–214; control of fluid pressures below barrier layers, 204–205; definition of, 184; erosion control of, 202–203; fluid barriers, 184–185; functions of, 183; landfill cover thermal regime, 208–209; landfill gas control in, 209–211; liquid pressures above barrier layer, 205; overview, 183–184; physical isolation of waste, 184; postclosure maintenance, 211–214; buildings, roads, and security, 214; cover soil cracking, 212–213; cover soil perforations, 212–213; cover soil slumping, 212–213; deed restrictions, 214; documentation, 211–212; electrical/mechanical systems, 213–214; end use, 214; environmental monitoring, 214; inspections, 212; intrusion by humans/animals, 213; permits, 214; settlement, 213; stormwater control, 213; vandalism by humans/ animals, 213; vegetation, 212; regulatory requirements of, 185–186; resistive/barrier layers, 196–202; composite covers, 199–200; exposed geomembrane covers, 201–202; single CCLs, 196–197; single GCLs, 197–198; single-geomembrane, 198–199; resistive covers, 186, 191–196; example of, 191–192; foundation layer in, 192; gas collection layer in, 192–196; layers in, 192–196; settlement impact on performance of, 207–208; slope stability considerations for veneer cover systems, 205–207; soil covers, 186–187; surface water management of, 202–203; types of, 186–196; water balance covers, 186, 187–191; configuration, 189; *vs.* double-capillary barrier, 190; long-term effectiveness, 191; *vs.* monolithic covers, 189; numerical models of, 188–189; principle of, 188

Finite-element analysis (FEA), 230–231
F-listed wastes, 89
Flue gas desulfurization (FGD), 129, 131
Fluoride, 79
Fly ashes, 129, 131; beneficial use of, 135–139; encapsulated, 136; unencapsulated, 136–139; classification of, 131–133; compaction, 134–135; environmental concerns, 139; shear strength of, 135; specific gravity of, 133–134
Foundation layer in resistive covers, 192
FR. *See* Federal Register
Freundlich sorption isotherm equation, 16–17
FS. *See* factor of safety
Fundão failure, 299

Gas analysis, 80–83; questionnaire for, 81–83
Gas collection layer in resistive covers, 192–196; drainage layer, 194–195; hydraulic barrier layer, 193–194; protection layer, 195; surface layer, 196
GCL. *See* geosynthetic clay liners
Geomembrane liners, 163–167; hydraulic performance, 165–166; mechanical performance, 166–167; overview, 163–165
Geomembranes, 164
Geophones, 28–30, 41
Geosynthetic-based drainage systems, 163
Geosynthetic clay liners (GCL), 161–163, 171–172; construction issues, 172; mechanical performance, 172; overview, 171–172; single resistive/barrier layers, 197–198; specifications, 172
Geotechnical engineers, 223
GHGs. *See* greenhouse gases
Greenhouse gases (GHGs), 93
GRI GM13 (GRI 2021), 165
Ground improvement for geoenvironmental engineering practice, 339–348; columns of improved material for, 344–345; deep dynamic compaction for, 343–344; goal of, 339; grouting for, 345–346; preloading in, 340–342; roller compaction for, 342–343; in situ stabilization/solidification for, 347–348; soil mixing for, 346–347
Groundwater contamination, 317
Groundwater pump-and-treat, 317
Grouted barrier walls, 366–367
Grouting for ground improvement, 345–346

Hazardous and Solid Waste Amendments (HSWA) of 1984, 87
Hazardous chemicals, 91
Hazardous constituents, 91
Hazardous substance, 91
Hazardous waste, 87–112, 160, 241–242; action leakage rate, 95–97; alternative land disposal restrictions for contaminated soils, 101–102; basic requirements and treatment standards for, 100–101; CERCLA: basics, 103–104; removal and remedial actions, 104–108; requirements, 108–109; characteristics, 90–91; corrosivity, 90; ignitability, 90; reactivity, 90–91; toxicity, 91; closure and postclosure care, 97–100; constituents/chemicals/substances, 91–92;

definition, 88–89, 91; double-liner system for, 93–95; generation, 92; identification process, 88; landfill, configuration for, 161; landfill final cover system based on EPA, 99; listed, 89–90; F-list (40 CFR §261.31), 89; K-list (40 CFR §261.32), 89–90; P-list (40 CFR §261.33), 90; U-list (40 CFR §261.33), 90; mixed wastes, special requirements for, 111–112; polychlorinated biphenyl wastes, 110–111; recycling, 92–93; redevelopment of remediated comprehensive environmental response compensation and liability act sites, 109; response action plan, 95–97; transportation, 92; TSDFs, 87, 92–93, 102

HDPE. *See* high-density polyethylene

HELP model. *See* Hydraulic Evaluation of Landfill Performance

High-density polyethylene (HDPE), 163–164

Horizontal barrier walls, 355

Hybrid field measurement techniques, MSW, 49–53; cone penetration test soundings, 49–51; DMT, 51; general, 49; SDMT, 51; in situ measurement of shear modulus reduction, 51–53

Hydraulic conductivity, 121–122

Hydraulic Evaluation of Landfill Performance (HELP) model, 166

Hydrodynamic dispersion, 9–10

Hydrodynamic dispersion coefficient, 9

Hydrologic Evaluation of Landfill Performance (HELP) model, 189

Hydrolysis/substitution reactions, 18–19

HYDRUS, 14

ICP. *See* inductive coupled plasma

ICP-MS. *See* inductively coupled plasmamass spectrometry

IDW. *See* investigation-derived waste

Impounded tailings, 296–299

Inductive coupled plasma (ICP), 69–70

Inductively coupled plasmamass spectrometry (ICP-MS), 70

Infinite slope stability analysis, 226–227

InSAR, interpretation of, 32

In situ chemical oxidation (ISCO), 317, 320–321

In situ chemical reduction (ISCR), 317, 320–321

In situ direct shear tests, 44–45

In situ measurement of shear modulus reduction, 51–53

In situ testing of MSW, 27

Interface shear strength, 238–243

Interface shear strength parameters, 206, 242–244

Internal deformation measurements, 45–46

Internal moisture content and temperature measurements, 46–49

Intrusive field measurement techniques, 35–49; borehole and test pit sampling, 38; borehole and test pit unit weight testing, 38–39; borehole electrical and nuclear testing, 42–43; borehole hydraulic conductivity testing, 39–41; borehole pressuremeter testing, 43–44; borehole sampling and testing, 36–49;

Intrusive field measurement techniques (*Continued*)
borehole wave propagation velocity testing, 41–42; BPT, 44; complications inherent to, 36–37; general, 35–36; health and safety considerations, 37; IDW, 37–38; internal deformation measurements, 45–46; internal moisture content and temperature measurements, 46–49; in situ direct shear tests, 44–45; SPT, 44; types of, 36
Investigation-derived waste (IDW), disposition of, 37–38
Ionic migration, 319–320
Ionic mobility, 320
ISCO. *See* in situ chemical oxidation
ISCR. *See* in situ chemical reduction

Jet grouting, 346, 366–367

Kjeldahl nitrogen, 79
K-listed wastes, 89–90

Langmuir sorption isotherm equation, 16–17
LCRS. *See* leachate collection and removal system
LDS. *See* leak detection system
Leachate collection and removal system (LCRS), 94–95
Leak detection system (LDS), 96, 163
Life of waste containment system, 225
Light detection and ranging (LiDAR) imaging, 32
Light nonaqueous phase liquids (LNAPLs), 15
Limit analysis, 230
Limit equilibrium analyses, 225–230
Linear low-density polyethylene (LLDPE), 163
Liner systems for containment facilities, 159–176; CCL, 167–171; construction issues, 170–171; hydraulic performance, 169; mechanical performance, 170; overview, 167–169; specifications, 170–171; drainage systems, 175–176; hydraulic performance, 175–176; overview, 175; GCL, 171–172; construction issues, 172; mechanical performance, 172; overview, 171–172; specifications, 172; geomembrane liners, 163–167; hydraulic performance, 165–166; mechanical performance, 166–167; overview, 163–165; for hazardous solid waste landfills, 161–162; hydraulic performance of, 173–174; for nonhazardous solid waste landfills, 161–162; overview, 159–160; regulatory requirements for, 160–163; background, 160–161; hazardous solid waste landfills, 161–162; nonhazardous solid waste landfills, 161–162; prescriptive designs, 161–163; temperature and desiccation effects, 174
Liquid analysis methods, 67–80; general parameters for, 68–69; color, 68; dissolved oxygen, 69; electrical conductance, 69; hardness, 69; pH, 68–69; redox potential, 69; temperature, 68; turbidity, 68; metals and metalloids, analysis of, 69–71; organics, analysis of, 71–78; other compounds and parameters, analysis of, 78–80

LLDPE. *See* linear low-density polyethylene
LNAPLs. *See* light nonaqueous phase liquids
Load transfer platform (LTP), 344–345
Loam, 189
LOI. *See* loss on ignition
Loss on ignition (LOI), 134
LTP. *See* load transfer platform

MAM. *See* microtremor analysis method
Mass transport, 3–15; advection, 3–5; coupled flow processes in, 11–12; diffusion, 5–7; mechanical dispersion in, 7–8; nonaqueous phase liquids (NAPLs), 15; total mass flux in, 8–11; transient (time-dependent) advective–dispersive–reactive transport in, 12–14; unsaturated porous media, 14
MASW. *See* multichannel analysis of surface waves
Mechanical dispersion in mass transport, 7–8
Mechanically stabilized earth (MSE), 248
Methane oxidation, 211
Methanotrophs, 211
MICP. *See* microbially induced calcite precipitation
Microbially induced calcite precipitation (MICP), 346
Microscale ZVI (mZVI), 376
Microtremor analysis method (MAM), 28
Mineralogical analysis of solids, 66–68
Mine tailings, 290–292, 293–299
Mine waste geotechnics, 289–299; management, 296–299; tailing storage facilities, 296–299; waste rock piles, 296; mine tailings, 293–295; overview, 289–290; waste rock, 292–293
Mine waste rock, 290–293, 296
Mixed wastes, 111–112
Mix-in-place methods (continuous trench mixing), 364–365
Mix-in-place methods (panel mixing), 362–364
Mohr–Coulomb shear strength envelope, 238–241
Mohr–Coulomb strength intercept, 227
Mohr–Coulomb strength slope angle, 227
Moisture content, 120–121
Mount Polley and Cadia failures, 299
MSE. *See* mechanically stabilized earth
MSW. *See* municipal solid waste
MT3D, 14
Multichannel analysis of surface waves (MASW), 29
Municipal solid waste (MSW), 27–28, 117–125, 160, 241; biogas generation, 120; composition, 119; compressibility, 124–125; definition of, 117; dynamic properties of, 125; excavation in, 34; failure in, 35; field capacity, 121; generation intensity and comparison with US GDP, 118; hybrid field measurement techniques, 49–53; cone penetration test soundings, 49–51; DMT, 51; general, 49; SDMT, 51; in situ measurement of shear modulus reduction, 51–53; hydraulic conductivity, 121–122; intrusive field measurement techniques, 35–49;

Municipal solid waste (MSW) (*Continued*) borehole and test pit sampling, 38; borehole and test pit unit weight testing, 38–39; borehole electrical and nuclear testing, 42–43; borehole hydraulic conductivity testing, 39–41; borehole pressuremeter testing, 43–44; borehole sampling and testing, 36–49; borehole wave propagation velocity testing, 41–42; BPT, 44; complications inherent to, 36–37; general, 35–36; health and safety considerations, 37; IDW, 37–38; internal deformation measurements, 45–46; internal moisture content and temperature measurements, 46–49; in situ direct shear tests, 44–45; SPT, 44; types of, 36; moisture content of, 120–121; nonintrusive field measurement techniques, 28–35; electrical resistivity and conductivity surveys, 31–32; general, 28; shear strength estimation based on stability observations, 33–35; strong ground motion measurements, 33; surface deformation measurements, 32; wave propagation velocity measurements, 28–31; at Novy Svet, 35; recommendations, 53–56; settlement, 124–125; shear strength of, 122–124; in situ testing of, 27; structure, 119; in the United States, 117–118; unit weight of, 120

Nanoscale ZVI, 376–377
NAPLs. *See* nonaqueous phase liquids
Newmark-type displacement analysis, 275–279; chart-based approach to, 275–277; site-specific, 277–279
Newmark-type seismic displacement analysis, 275–277
Nitrates, 79
Nitrites, 79
Nonaqueous phase liquids (NAPLs), 15
Nonhazardous waste, 160
Nonintrusive field measurement techniques, 28–35; electrical resistivity and conductivity surveys, 31–32; general, 28; shear strength estimation based on stability observations, 33–35; strong ground motion measurements, 33; surface deformation measurements, 32; wave propagation velocity measurements, 28–31
Nonreactive chemical tracer, 10

OII landfill, 30, 39
Other industrial waste, 241–242
Oxidation, 211

PAHs. *See* polycyclic aromatic hydrocarbons
Pb[II], 153
PBSD. *See* performance-based seismic design
PCBs. *See* polychlorinated biphenyls
Peak ground acceleration (PGA), 260
Performance-based seismic design (PBSD), 280–281
Permeable reactive barrier (PRB), 373; definition, 373; material, 373–374; schematic of, 373; zero-valent iron, 374–404

PFAS. *See* poly- and per-fluoroalkyl substances
PGA. *See* peak ground acceleration
Phosphorus, 79–80
P-listed wastes, 90
Poisson's ratio, 30
Poly- and per-fluoroalkyl substances (PFAS), 71, 78
Polychlorinated biphenyls (PCBs), 71, 78
Polychlorinated biphenyl wastes, 110–111
Polycyclic aromatic hydrocarbons (PAHs), 71, 78
Polymeric materials, 163
Polypropylene (PP), 163
Polyvinyl chloride (PVC), 163
Pore pressure, 243–246
Porosity, 5
Porous media: macroscopic dispersion in, 8; microscopic dispersion in, 7; unsaturated, 14
PP. *See* polypropylene
PRB. *See* permeable reactive barrier
Preloading in ground improvement, 340–342
Probabilistic analyses, 232–233
Pseudo-static limit equilibrium, 270–275
PVC. *See* polyvinyl chloride

Questionnaire for gas analysis, 81–83

Radioactive decay, 17–18
Radioactive waste, 241–242
Rammed aggregate piers (RAPs), 344–345
Random field modeling, 232
RAP. *See* response action plan
RAPs. *See* rammed aggregate piers; recycled asphalt pavement
RCAs. *See* recycled concrete aggregates
RCRA. *See* Resource Conservation and Recovery Act
Reclaimed pavement materials (RPMs), 135
Recycled asphalt pavement (RAPs), 135
Recycled concrete aggregates (RCAs), 135
Redox/oxidation/reduction reactions, 19
Refraction microtremor (ReMi), 28
ReMi. *See* refraction microtremor
Representative elementary volume (REV), 12–13
Resistive/barrier layers, final covers, 196–202; composite covers, 199–200; exposed geomembrane covers, 201–202; single CCLs, 196–197; single GCLs, 197–198; single-geomembrane, 198–199
Resistive covers: foundation layer in, 192; gas collection layer in, 192–196
Resistivity piezocone (RCPTu), 31
Resource Conservation and Recovery Act (RCRA), 87–89, 91–94, 97, 129–130, 160; CERCLA and, 102–110; HSWA, 100–101; land disposal restrictions, 100–102; LDR program, 100–101; remediation of contamination at, 102; requirements for hazardous waste landfills, 93–100; in Subtitles C and D, 161; TSDFs, 93
Response action plan (RAP), 95–97
REV. *See* representative elementary volume
Richard's equation, 189
Rigid inclusion columns, 345
Roles, static slope stability of containment systems, 223–224
Roller compaction for ground improvement, 342–343

RPMs. *See* reclaimed pavement materials
RT3D, 14

Sand cone density test, 38
Sandia National Laboratories, 188
SASW. *See* spectral analysis of surface waves
Satellite imagery, 32
Scanning electron microscopy with energy-dispersive X-ray (SEM/EDX), 67–68
SDMT. *See* seismic dilatometer Marchetti test
Seismic coefficient, 260
Seismic dilatometer Marchetti test (SDMT), 51
Seismic performance standard, 261–263
Seismic refraction tests (SRT), 28, 30–31
Seismic stability of waste containment systems, 259–281; design criteria, 263–266; design ground motion, 266–270; hierarchy of seismic stability analyses, 259–261; Newmark-type displacement analysis, 275–279; chart-based approach to, 275–277; site-specific, 277–279; overview, 259; pseudo-static limit equilibrium, 270–275; seismic performance standard, 261–263; stress–deformation dynamic analyses, 279–281
Sensitivity study, 237
Se[VI], 153
Shear stiffness, 30
Shear strength estimation, 33–35
Single CCLs resistive/barrier layers, 196–197
Single compacted clay liners, 196–197
Single GCLs resistive/barrier layers, 197–198
Single-geomembrane resistive/barrier layers, 198–199
Situ stabilization/solidification for ground improvement, 347–348
Slag–CB slurry wall, 366
SLAMMER, freeware program, 277
Slices slope stability analysis, 228–229
Slickensides, 240
Soil–bentonite (SB) slurry trench cutoff walls, 357–360
Soil-cement mixture, 345
Soil covers, 186–187
Soil excavation-and-disposal/treatment, 317
Soil mixing for ground improvement, 346–347
Soil pH, 321
Soil reinforcement, 248–250
Soil shear strength, 238–243
Soil vapor extraction, 317
Soil washing, 317
Solid analysis methods: bulk chemical analysis, 64–66; mineralogical analysis, 66–67
Solid-phase concentration of contaminant, 13
Solute concentration, 5
Sorption process, 15–17
Sorptive agents, 153
Spectral analysis of surface waves (SASW), 28–29
SPT. *See* standard penetration test
SRT. *See* seismic refraction tests
Stable landfill waste slopes, back-analysis of, 35
Standard penetration test (SPT), 44
Static slope stability of containment systems, 221–254;

documentation of, 254; forensic evaluations, 246–248; input parameters, 236–246; pore pressure, 243–246; slope geometry, 237–238; soil shear strength/interface shear strength, 238–243; limitations, 223; methods of analysis, 225–233; DEMs, 231; FEA, 230–231; limit, 230; limit equilibrium, 225–230; probabilistic analyses, 232–233; three-dimensional analyses, 231–232; modes of failure in, 233–236; overview, 221–222; review guidelines, 250–254; roles, 223–224; soil reinforcement, 248–250; typical applications, 222–223; workflow, 224–225
Stress–deformation dynamic analyses, 279–281
Strong ground motion measurements, 33
Sulfate, 80
Sulfide, 80
Surface deformation measurements, 32
Surface wave testing, 30
SW-846 5000 series, 71
SW-846 manual, 64

Tailing storage facilities (TSFs), 289, 296–299
TCLP. *See* Toxicity Characteristic Leaching Procedure
TDS. *See* total dissolved solids
TGA. *See* thermogravimetric analysis
Thermogravimetric analysis (TGA), 67–68
Three-dimensional analyses, 231–232
TOC. *See* total organic carbon
TOD. *See* total oxygen demand
Tortuosity factor, 6
Total dissolved solids (TDS), 69
Total mass flux in mass transport, 8–11
Total mass of contaminant per unit total volume, 13
Total organic carbon (TOC), 71
Total oxygen demand (TOD), 80
Total petroleum hydrocarbons (TPH), 71, 77
Toxicity Characteristic Leaching Procedure (TCLP), 91
Toxic Substances Control Act (TSCA) of 1976, 110
TPH. *See* total petroleum hydrocarbons
Transient (time-dependent) advective–dispersive–reactive transport of solute, 12–13
TRD. *See* trench remixing and deep wall
Treatment, storage, and disposal facilities (TSDFs), 87
Trench remixing and deep wall (TRD), 364–365
TSB. *See* two-stage borehole permeameter
TSDFs. *See* treatment, storage, and disposal facilities
TSFs. *See* tailing storage facilities
Two-stage borehole permeameter (TSB), 41

UAV. *See* unmanned aerial vehicle
U-listed wastes, 90
Ultraviolet (UV) radiation, 164–165
Unfavorable isotherms. *See* convex isotherms
Unified Soil Classification Systems (USCS), 133
Universal treatment standards (UTS), 101
Unmanned aerial vehicle (UAV), 32

Unsaturated porous media, 14
Unstable landfill waste slopes, back-analysis of, 34–35
USCS. *See* Unified Soil Classification Systems
US Department of Transportation (USDOT), 92
USDOT. *See* US Department of Transportation
US Environmental Protection Agency (USEPA), 129
USEPA. *See* US Environmental Protection Agency
UTS. *See* universal treatment standards

Veneer, 194
Vertical barrier walls, 355–369; deployment options, 355–357; longevity considerations, 367–369; other methods, 366–367; types, 357–367; cement–bentonite (CB) slurry trench cutoff walls, 361–362; grouted barrier walls, 366–367; mix-in-place methods (continuous trench mixing), 364–365; mix-in-place methods (panel mixing), 362–364; soil–bentonite (SB) slurry trench cutoff walls, 357–360
Very low-density polyethylene (VLDPE), 163
Vibrating beam thin cutoff wall, 366
VLDPE. *See* very low-density polyethylene
VOCs. *See* volatile organic compounds
Volatile organic compounds (VOCs), 71, 77–78, 80–81
Volumetric moisture content, 121

Waste rock piles, 296
Water balance covers, 186, 187–191; configuration, 189; *vs.* double-capillary barrier, 190; long-term effectiveness, 191; *vs.* monolithic covers, 189; numerical models of, 188–189; principle of, 188
Water flux, 4
Wave propagation velocity measurements, 28–31
Wet gravimetric moisture content, 121
Woodward Clyde Consultants, 30
Woody biomass fly ash, 145–153; biomass ash, 146–147; chemical composition and classification of, 149–150; in concrete, 151; as mine backfill, 152; organic content, 150–151; particle morphology, 148–149; particle size, 148–149; for soil improvement, 152; as sorptive agents, 153; specific gravity of, 149; for waste solidification/stabilization, 152–153
Workflow, slope stability analysis, 224–225

X-ray diffraction (XRD), 67–68
X-ray fluorescence (XRF), 66
XRD. *See* X-ray diffraction
XRF. *See* X-ray fluorescence

Zero-valent iron (ZVI) PRB, 374–404; applicability, 374–384; chlorinated aliphatic hydrocarbons, treatment of, 378–382; degradation pathways, 378–379; degradation rates, 379–380; groundwater chemistry, 380–381; temperature, 381–382;

ZVI-mediated, 380; commercial, 376; construction methods, 398–401; direct ZVI injection methods, 400–401; trench-and-backfill, 399–400; contaminants treated, 377; design of, 384–398; dissolved metals, treatment of, 382–384; feasibility evaluation, 374–376; materials, 376–377; microscale, 376; monitoring, 401–403; groundwater sample collection and analysis plan, 402–403; well network, 401–402; nanoscale, 376–377